DATE DUE	
ILL 332 7816	10/21/99

SIGNAL PROCESSING
USING OPTICS

THE JOHNS HOPKINS UNIVERISTY
Applied Physics Laboratory Series in Science and Engineering

FOUNDING SERIES EDITOR: John R. Apel
SERIES EDITOR: Kishin Moorjani

Previous Volumes

William H. Avery and Chih Wu
Renewable Energy from the Ocean: A Guide to OTEC

Bruce I. Blum
Software Engineering: A Holistic View

R. M. Fristrom
Flame Structure and Processes

Richard A. Henle and Boris W. Kuvshinoff
Desktop Computers: In Perspective

Vincent L. Pisacane and Robert C. Moore (eds.)
Fundamentals of Space Systems

Bruce I. Blum
Beyond Programming: To a New Era of Design

SIGNAL PROCESSING USING OPTICS

Fundamentals, Devices, Architectures, and Applications

Bradley G. Boone
The Johns Hopkins University

New York   Oxford
OXFORD UNIVERSITY PRESS
1998

OXFORD UNIVERSITY PRESS

Oxford New York
Athens Auckland Bangkok Bogotá Bombay
Buenos Aires Calcutta Cape Town Da res Salaam
Delhi Florence Hong Kong Istanbul Karachi
Kuala Lumpur Madras Madrid Melbourne
Mexico City Nairobi Paris Singapore
Taipei Tokyo Toronto Warsaw

and associated companies in

Berlin Ibadan

Copyright © 1998 by Oxford University Press, Inc.

Published by Oxford University Press, Inc.,
198 Madison Avenue, New York, New York, 10016
http://www.oup-usa.org
1-800-334-4249

All rights reserved. No part of this publication may be
reproduced, stored in a retrieval system, or transmitted,
in any form or by any means, electronic, mechanical,
photocopying, recording, or otherwise, without the prior
permission of Oxford University Press.

Library of Congress Cataloging-in-Publication Data
Boone, Bradley G. (Bradley Gilbert), 1950–
Signal processing using optics : fundamentals, devices,
architectures, and applications / Bradley G. Boone.
p. cm.
Includes bibliographical references and index.
ISBN 0-19-508424-1 (cloth)
1. Optical data processing. 2. Signal processing. I. Title.
TA1632.B66 1997
621.36'7–dc21   97-6961
CIP

9 8 7 6 5 4 3 2 1

Printed in the United States of America
on acid-free paper

*To my wife Sandra Jean and my
children April Elizabeth and Jonathan Christopher;
all the patience, understanding, and time that I borrowed
belongs to you.*

Contents

PREFACE *xi*
ACKNOWLEDGMENTS *xv*
INTRODUCTION *xvii*
    Bibliography *xxi*

1. TWO-DIMENSIONAL LINEAR SYSTEMS

    1.1 Fundamental Properties *1*
    1.2 Linear Superposition *2*
    1.3 Convolution and Correlation *3*
    1.4 Two-Dimensional Fourier Transforms and Properties *7*
    1.5 Rectangular and Polar Forms *9*
    1.6 Linear Coordinate Transformation and Fourier Theorem *13*
    1.7 Examples of Magnification and Rotation *14*
    1.8 Two-Dimensional Impulse Functions: Properties and Fourier Transforms *17*
    1.9 Elementary Images and Their Fourier Properties *24*
    Problem Exercises *32*
    Bibliography *35*

2. STOCHASTIC PROCESSES AND NONLINEAR SYSTEMS

    2.1 Basic Concepts of Stochastic Processes *36*
    2.2 Fundamental Probability Density Functions *42*
    2.3 Matched Filter Derivation and Properties of Correlation *45*
    2.4 Nonlinear Transformations and Operations *49*
    2.5 Mixing and Modulation *55*
    Problem Exercises *61*
    Bibliography *63*

3. MATHEMATICAL TRANSFORMS USED IN OPTICAL SIGNAL PROCESSING

    3.1 Overview *64*
    3.2 Fresnel Transform *66*

3.3 Hilbert Transform   *69*
3.4 Radon Transform   *75*
3.5 Mellin Transform   *79*
3.6 Wavelet Transform   *82*
Problem Exercises   *87*
Bibliography   *92*

4. FUNDAMENTAL PROPERTIES OF LIGHT AND GEOMETRICAL OPTICS

4.1 Overview   *93*
4.2 Fundamental Scalar and Vector Properties of Light   *93*
4.3 Polarization   *96*
4.4 Rectilinear Glass Structures and Their Properties   *97*
4.5 Simple Lenses and Lens Combinations   *106*
Problem Exercises   *114*
Bibliography   *117*

5. SUMMARY OF PHYSICAL OPTICS

5.1 Overview   *118*
5.2 Coherence and Interference   *118*
5.3 Scalar Diffraction Theory   *122*
5.4 Fraunhofer Diffraction   *125*
5.5 Fresnel Diffraction   *131*
Problem Exercises   *134*
Bibliography   *136*

6. FOURIER TRANSFORM AND IMAGING PROPERTIES OF OPTICAL SYSTEMS

6.1 Overview   *137*
6.2 Effect of Lens on a Wavefront   *137*
6.3 Fourier Transform Property of a Single Lens   *139*
6.4 Imaging Property of Lenses   *142*
6.5 Linear System Properties of Imaging Systems   *143*
6.6 Point Spread Function   *144*
6.7 Optical Transfer Function   *146*
6.8 Signal Processing Analogies for Optics   *151*
Problem Exercises   *157*
Bibliography   *162*

7. LIGHT SOURCES AND DETECTORS

7.1 Overview   *163*
7.2 Laser Principles of Operation   *164*
7.3 Light Emitting Diodes and Laser Diodes   *168*
7.4 Laser Diode Arrays   *176*
7.5 Output Light Detectors   *178*
7.6 Single Detectors   *179*
7.7 Linear and Matrix Arrays   *184*
7.8 Optical Signal Processing Requirements   *190*
Problem Exercises   *192*
Bibliography   *196*

8. SPATIAL LIGHT MODULATORS
   8.1 Overview   *198*
   8.2 Acousto-Optic Bragg Cells   *200*
   8.3 Liquid Crystal Spatial Light Modulators   *211*
   8.4 Magneto-Optic Spatial Light Modulator   *216*
   8.5 Other Spatial Light Modulators   *218*
   Problem Exercises   *223*
   Bibliography   *228*

9. OPTICAL SPECTRUM ANALYSIS AND CORRELATION
   9.1 Overview   *230*
   9.2 Time- and Space-Integrating Architectures   *230*
   9.3 Coherent and Incoherent Architectures   *231*
   9.4 Spectrum Analysis   *232*
   9.5 Space-Integrating Spectrum Analyzer   *239*
   9.6 Time-Integrating Spectrum Analyzer   *242*
   9.7 Correlation   *244*
   9.8 Incoherent Optical Correlator Architectures   *244*
   9.9 Coherent Optical Correlator Architectures   *248*
   Problem Exercises   *255*
   Bibliography   *257*

10. IMAGE AND MATCHED SPATIAL FILTERING
   10.1 Overview   *258*
   10.2 VanderLugt Filter   *258*
   10.3 Image Spatial Filtering   *260*
   10.4 Matched Spatial Filter and Binary Phase-Only Correlators   *264*
   10.5 Techniques for Circumventing Geometric Distortions   *268*
   10.6 Spatial Multiplexing   *268*
   10.7 Distortion-Invariant Transformations   *270*
   10.8 Angular Correlation   *273*
   Problem Exercises   *276*
   Bibliography   *278*

11. RADAR SIGNAL PROCESSING APPLICATIONS
   11.1 Overview   *280*
   11.2 Radar Signal Processing   *280*
   11.3 Ambiguity Function Processing   *281*
   11.4 Synthetic Aperture Radar   *292*
   Problem Exercises   *305*
   Bibliography   *310*

12. PATTERN RECOGNITION APPLICATIONS
   12.1 Overview   *312*
   12.2 Feature Extraction   *313*
   12.3 Matrix–Vector Multiplication   *324*
   12.4 Optical Neural Networks   *329*
   Problem Exercises   *341*
   Bibliography   *346*

APPENDIX A: MATHEMATICAL TABLES *348*
APPENDIX B: ANNOTATED BIBLIOGRAPHY *353*
APPENDIX C: SOFTWARE FOR MODELING AND VISUALIZATION *356*
APPENDIX D: HINTS AND SOLUTIONS TO SELECTED PROBLEMS *376*
INDEX *390*

Preface

This textbook covers the basic aspects of optical signal processing at an introductory level, yet it should help the student bridge the gap to current technical literature. It is intended for senior-level undergraduate or first-year graduate students in electrical engineering or applied physics as well as practicing engineers and scientists. Although the student or professional should have some exposure to one-dimensional signals and systems, and optics beyond general physics, the text reviews the essentials of signal processing, optics, and imaging to make the book self-contained and the necessary background information readily accessible. The text is geared toward providing the student with insight on the underlying mathematical and physical principles, practical understanding of component technology and performance, a grasp of fundamental system design and analysis, and familiarity with architectures for selected but representative applications. Problem exercises and a list of references keyed to each chapter are intended to steer the student in the direction of the major sources and experts in the technical literature. Numerous architectural diagrams and MATLAB-based software tools are provided to assist in the understanding and visualization of important concepts and their implementation.

Optical signal processing technology continues to grow at a rapid pace. With any new technology, however, there are fundamentals that underpin the technology that are essential for understanding and applying it. These fundamentals, including the basic mathematical and physical concepts of signal processing and optics, are treated in the first half of the book. Since major technological impediments to the development of optical signal processing stem from electro-optical devices, such devices and their intrinsic limitations are key research and development topics, and hence should be described as well. Traditional applications of optical signal processing depend on the basic functions of spectrum analysis and correlation. Many more sophisticated algorithms and architectures depend on these two basic functions or elaborations of them, so they deserve special treatment. Although the choice of applications to address is a matter of individual interest, certain applications appear to be more popular and practical historically and have progressed further. These applications will be covered in some detail so as to provide case studies for the reader to see the connection between fundamentals, devices, and architectures.

There are eight major topics treated in this book which correspond to the 12 major chapters outlined in the table of contents. These topics are:

- Two-dimensional linear systems
- Stochastic processes and nonlinear systems
- Integral transforms used in optical signal processing
- Fundamentals of geometrical and physical optics
- Fourier and imaging properties of optical systems
- Electro-optical devices
- Fundamental architectures for correlation and spectrum analysis
- Selected applications in radar signal processing and pattern recognition

The first topic (Chapter 1) is a fundamental one, which emphasizes an analytical understanding of two-dimensional linear systems concepts, including convolution, correlation, and the Fourier transform. Linear transformations of coordinates are also considered, as well as basic impulse function descriptions of elementary images. This is the mathematical underpinning for the remainder of the text and is closely related to the chapters on optics and imaging.

The second chapter covers a collection of ideas important to understanding stochastic processes and nonlinear systems. These topics are important for understanding real electro-optical devices and systems, such as photodetectors, spatial light modulators, and electronic modulation of signals. The matched filter is treated from a statistical point of view. The theory of nonlinear devices and their effect on correlation and signal spectra are also described in detail.

Mathematical transforms frequently used in optics or related applications are described in detail in Chapter 3. These include: the Fresnel transform, which is used in near-field diffraction, chirp radar, and synthetic aperture radar (SAR); the Hilbert transform, which derives from consideration of causal and positive frequency signals and emerges in Schlieren imaging; the Radon transform, which appears in computer-aided tomography and SAR; the Mellin transform, which is used to make the recognition and correlation of signals and images scale-invariant; and the wavelet transform, which provides a more physically consistent description of real (nonstationary) signals.

The fourth and fifth chapters cover the essentials of geometrical and physical optics. Rectilinear glass structures, such as prisms, are described in terms of geometrical image transformations. Simple and compound lenses as well as the newer multiple lenslet arrays are described. Coherence, an important concept for understanding laser diodes, is introduced. Fraunhofer and Fresnel diffraction concepts, which serve as the physical basis of image systems analysis and Fourier optics, are also treated.

Fourier transform properties of lenses, in which the effect of a simple thin lens on the amplitude and phase of a wavefront is treated, are described in Chapter 6. Imaging systems are covered from the standpoint of linear system theory, and the optical transfer function is defined and characterized. The interpretation of lenses in terms of signal processing analogies is emphasized at the end of this chapter.

Chapters 7 and 8 describe sources, detectors, and spatial light modulators (SLMs) currently used in optical signal processing systems. It surveys the principles of operation, designs, and limitations of laser diodes, liquid crystal, magneto-optic, and acousto-optic spatial light modulators, as well as charge coupled device imagers and linear self-scanned arrays. Generic device characteristics and specifications such as spatial resolution, dynamic range, linearity and noise sensitivity are discussed to anticipate their use in evaluating optical processing devices and system performance in particular applications.

Chapter 9 is a key chapter because it attempts to combine many aspects of the previous topics pertinent to optical signal processing. Design and analysis of architectures for spectrum analysis and correlation are treated extensively. Generic signal processing concepts and categories developed in the technical literature for viewing optical signal processing architectures are presented in an attempt to create a formal approach to the subject.

Chapter 10 covers a very central topic and one of perennial interest in optical signal processing: the matched filter and its variations in the form of image spatial filtering techniques. The important role of scale, rotation, and shift invariance in matched filtering is treated by looking at various ways to implement it. Low-pass, high-pass, and Schlieren imaging are described, since they are the most common spatial filtering techniques. Binary and phase-only matched filtering are also treated.

In the final chapters (11 and 12), specific cases for applying optical processing are described. Applications to radar signal processing and pattern recognition are treated. In particular, ambiguity function and synthetic aperture radar processing, optical feature extraction, matrix–vector multiplication, and neural networks are covered.

BRADLEY G. BOONE
Johns Hopkins University
Applied Physics Laboratory
July 1, 1997

IEEE Code of Ethics

We, the members of the IEEE, in recognition of the importance of our technologies in affecting the quality of life throughout the world, and in accepting a personal obligation to our profession, its members and the communities we serve, do hereby commit ourselves to conduct of the highest ethical and professional manner and agree:

1. to accept responsiblity in making engineering decisions consistent with the safety, health, and welfare of the public, and to disclose promptly factors that might endanger the public or the environment;

2. to avoid real or perceived conflicts of interest whenever possible, and to disclose them to affected parties when they do exist;

3. to be honest and realistic in stating claims or estimates based on available data;

4. to reject bribery in all of its forms;

5. to improve understanding of technology; its appropriate application, and potential consequences;

6. to maintain and improve our technical competence and to undertake technological tasks for others only if qualified by training or experience, or after full disclosure of pertinent limitations;

7. to seek, accept, and offer honest criticism of technical work, to acknowledge an correct errors, and to credit properly the contributions of others;

8. to treat fairly all persons regardless of such factors as race, religion, gender, disability, age, or national origin;

9. to avoid injuring others, their property, reputation, or employment by false or malicious action;

10. to assist colleagues and co-workers in their professional development and to support them in following this code of ethics.

Approved by IEEE Board of Directors, August 1990

For further information please consult the IEEE Ethics Committee WWW page:
http://www.ieee.org/committee/ethics

Copyright © 1997 by IEEE

Acknowledgments

To Dr. Gary L. Smith, Director of the Johns Hopkins University Applied Physics Laboratory (JHU/APL), I wish to convey my heartiest thanks for his support for this work through internally funded research opportunities at JHU/APL, as well as my Parsons professorship at Homewood and my subsequent Janney Fellowship at JHU/APL. To Dr. Rodger Westgate, I am grateful for the opportunity to have a quiet, unfettered visit in the Electrical and Computer Engineering Department of JHU/APL while on sabbatical. I would also like to thank Dr. Kishin Moorjani and Dr. John Apel, current and past editors for the APL book series, for their encouragement. I am grateful to my immediate supervisors, Dr. Willam J. Tropf of the Missile Engineering Branch for his support. I also appreciate the secretarial help of Machele Grace and Jo-Anne Kierkowski in the Electro-Optical Systems Group. For Jeffrey Robbins, Bill Zobrist, Krysia Bebick, and Karen Shapiro of Oxford University Press, thanks for your continued interest, support, and patience. To my colleagues inside and outside the Laboratory, especially Dr. Donald D. Duncan, Dr. John N. Lee, Dr. Anthony "Bud" VanderLugt, and Dr. Terry Turpin—I profited greatly from your interaction. To Oodaye B. Shukla, Jim Connelly, Scott A. Gearhart, and David H. Terry—thanks for your contributions to the original experimental results described in the book. Finally, I am grateful for my many students in The Johns Hopkins University G.W.C. Whiting School of Engineering, who served as guinea pigs for my initial exposure of the book and who offered many constructive criticisms.

The author would also like to acknowledge the use of several software packages in preparing this book, including: MacWrite II (1.1v5), MacDraw II (1.1v2), ClarisDraw (1.0v1), Microsoft Word (5.1a), Expressionist (2.07p), MathView Pro (1.0), Mathcad (3.1), MATLAB (4.2a), IP Lab Spectrum (2.4.1), NIH Image (1.44), FFT Vision (1.4), DAFRAC (3.0.1), Thunderworks (1.3.2), and Computer Eyes Pro (1.3).

Introduction

HISTORICAL ROOTS

The historical roots of optical processing extend as far back as 1859 when Foucault first developed the knife-edge test for lenses whereby direct image light is removed and scattered or diffracted light is kept. Subsequently, Abbe recognized in 1873 the importance of diffraction in coherent image formation, particularly for the case of microscopy. It was not until 1906 that Porter demonstrated Abbe's theory experimentally. Following this, in 1935 Zernike developed the concept of phase contrast microscopy, for which he later received the Nobel prize. In 1946 Duffieux described the application of the Fourier integral to optical problems.

By the 1950s, Elias and O'Neill connected the fields of optics and communication theory. Marechal, Tsujiuchi, Lohmann, and others expanded the applications of optical processing by addressing the image deblurring problem and the detection of two-dimensional signals in noise. In the 1960s, optical processing was applied to synthetic aperture radar. The development of the holographic matched spatial filter by VanderLugt and the computer-generated spatial filter by Lohmann and Brown further expanded the applications into pattern recognition. Optical transforms were also being applied to diffraction pattern sampling. Important work has also been carried out by Casasent, Psaltis, and many others using acousto-optics for correlation, spectrum analysis and radar signal processing using both time and space integrating architectures. More recently, optical computing has been extensively addressed, including discrete optical processors such as matrix–vector and matrix–matrix multipliers and processors utilizing serial, parallel, and systolic array concepts. Numerical processing, as well as nonlinear (Boolean logic) processing using binary representations and residue arithmetic, are the focus of very recent research efforts.

MOTIVATION AND BENEFITS

There are two fundamental reasons to employ optics for signal processing: bandwidth (or speed) and massive parallelism (or connectivity). These two features allow for direct, real-time image processing. An entire pattern can be processed at once through an optical processing architecture, which is much quicker than sequential processing of each picture element, or pixel, characteristic of most digital methods. As the number of pixels to be processed increases, the number of interconnections becomes so large that input/output bottlenecks are created for planar electronic devices. Optics can alleviate this bottleneck without cross-talk because light beams can pass through one another without interference.

In comparing optical and electronic processors we can cite a number of key differences. The speed of information carried optically is set by the velocity of light, $c = 3 \times 10^{10}$ cm/sec *in vacuo*, versus that carried electronically, which is $\sim 10^6$–10^7 cm/sec. Bandwidth advantage, however, really accrues from the much higher spectral bandwidth of optical photons than radio-frequency (RF) waves typically found in high-speed electronic circuits. Photons, the carriers of optical information, also have low mutual interaction (since they have no charge), whereas electrons, the carriers of electronic information, have high mutual interaction. Photons are naturally and even globally parallel, whereas electrons are naturally and usually locally serial. For optics there is a bottleneck at the input/output device [or spatial light modulator (SLM)], which often requires optical-to-electronic signal conversion for electrically addressable types. For electronics the bottleneck is at the interconnect level, if not the computational (gate) level.

The ultimate bandwidth of optical processing is set by the uncertainty principle for an optically modulated gate operating in a thermal environment at temperature T. The quantum of useful information, represented by an energy change from one state to another, ΔE, must be greater than $k_B T$, where k_B is Boltzmann's constant; that is,

$$\Delta E > k_B T$$

Heisenberg's uncertainty principle states that

$$\Delta E \Delta \tau \geq h$$

where h is Planck's constant, and $\Delta \tau$ is the time to change state. Therefore any processor operating at minimum energy must execute a gate state change in a time no less than

$$\Delta \tau \sim h/\Delta E = h/k_B T$$

If this state change is induced by the absorption or emission of a photon of frequency ν, the corresponding quantum energy is

$$E = h\nu$$

which must exceed $k_B T$. Therefore

$$\Delta \tau \sim h/h\nu \sim 1/\nu$$

represents the minimum response time possible for an optical processor gate. If the frequency of visible light is $\sim 10^{14}$ Hz, then $\Delta \tau \sim 10^{-14}$ sec. This extraordinary speed cannot be achieved in practice because real optical processor speed is limited by less fundamental optical-to-electronic signal conversion limitations, which must occur somewhere in the system to read-in or read-out information.

In practice, however, remarkable computational advantages can be achieved, even at modest modulation bandwidths (B). For instance, pixel speeds at television (TV) frame rates correspond to $B \sim 5$ MHz; and time–bandwidth products (BT) are $\sim 10^5$. When two-dimensional (2-D) optical correlators are implemented, for instance, with 2×10^5 points processed every 1/30 sec, the equivalent of a digital fast Fourier transform (FFT) (with 2×10^5 points) occurs every 1/30 sec. This is equivalent to 6×10^8 multiplications/sec. Typical digital processors compute 10^6–10^7 (64-bit) floating point operations/sec.

The elegance and simplicity of Fourier optics, as a primary exemplar of *analog* optical processing (as opposed to *digitally* oriented optical computing), is compromised in practice by speed and accuracy limitations of current input/output peripheral electro-optical devices, such as input light sources, spatial light modulators, and readout (detector) arrays. Thus, the technological aspects of implementing optical signal processors (particularly the construction of SLMs) become the focus of important research and development.

TRENDS AND APPLICATIONS

The direction of optical processing in the near future is likely toward hybrid (optical/electronic) systems. Smaller size and cost as well as more than just marginal gains in performance over electronics are necessary to justify further development. Not all application areas will benefit from the use of optical processing, but certain niche areas will continue to find uses, particularly those already using optics naturally for sensing or other functions.

Several general categories of applications can be identified that have been and will likely continue to be important: signal processing, image processing, pattern recognition, and computing. Traditional signal processing applications include spectrum analysis (and frequency excision), correlation (and convolution), and ambiguity function processing. Image processing includes frame-to-frame processing, bandwidth compression, and synthetic aperture radar image formation. Pattern recognition applications include preprocessing or feature extraction and optical neural network-based classification, among others. Optical computing applications include matrix–vector and matrix–matrix multiplications, and binary optical logic.

The utility of optical processing for the above applications will be advanced if developments occur in several categories. These categories include: architectures and algorithms, materials, spatial light modulators (SLMs), and sources and detectors. Algorithms are not independent of architectures if optimal design is to be carried out. For instance, the availability of excellent one-dimensional SLMs (acousto-optic Bragg cells) in recent years has stimulated the development of certain classes of separable 2-D processing algorithms, such as for synthetic aperture radar image formation.

Devices depend on materials. Materials with improved electro-optical efficiencies and spatial resolution are needed to make better SLMs (with higher dynamic range and numbers of pixels). Compact laser diode arrays with greater linearity and dynamic range in the current versus light intensity characteristics, enhanced modal stability, and greater output power are needed. Focal plane arrays with lower detection threshold sensitivity, higher dynamic range, greater pixel uniformity, reduced cross-talk, and parallel readout capability (specifically designed for optical processing) are also desirable. A goal of 1000×1000 pixels and 60-dB dynamic range for all electro-optical interface devices are reasonable goals. Improved postdetection and data acquisition electronics will also be needed to handle the high throughput rates.

GENERAL APPROACH OF THIS TEXTBOOK

The most fundamental way to characterize the subject of this book is to look at the essential components and properties of the simplest optical processing architecture as depicted in Fig. 0.1. This diagram is useful in understanding the basic functions of optical processor operation, especially for more complex systems for which it is a building block, and the requirements derived from considering particular applications. This basic architecture has three electro-optical devices for input (or display), intermediate spatial light modulation, and output detection. Although essential components enable the technology to work, it is just as important to understand optical propagation between input, intermediate, and output planes. This is typically expressed by the "free-space propagator," which can take on various forms (e.g., Fourier transform or imaging, etc.) depending on the character of the light distribution and the details of the architecture. In addition, we should allow for the possibility of feedback (as in any electronic system).

It is clear that although technology is an important aspect of the subject of optical signal processing, fundamentals remain the cornerstone of learning. Hence, we will address the fundamental mathematical and physical properties of devices and processes as well as the detailed analysis and design of architectures for specific applications. Therefore, as the student goes through this book, he or she should learn about the fundamentals and go on to more practical examples of architectures drawn from the technical journal literature that are intended to address "real-world" applications. Perhaps this is captured most simply by Fig. 0.2.

Perhaps a more inspiring message comes from an old friend and teacher of many students of electrical engineering and physics (who was around in the days when those subjects were less distinguishable): James Clerk Maxwell, who said the following in his introductory lecture at the Cambridge Cavendish Laboratory in October 1871:

> *It is not until we attempt to bring the theoretical part of our training into contact with the practical that we begin to experience the full effect of what Faraday has called "mental inertia"—not only the difficulty of recognizing among the objects before us, the abstract relations which we have learned from books, but the distracting pain of wrenching the mind away from the symbols to the objects, and from the objects back to the symbols. This, however, is the price we have to pay for new ideas.*

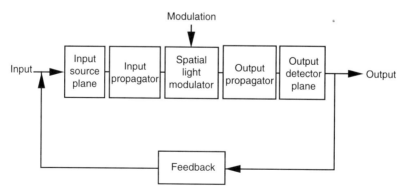

Figure 0.1

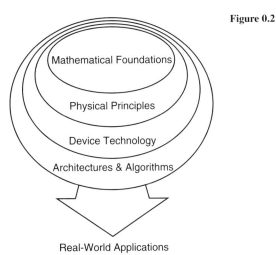

Figure 0.2

BIBLIOGRAPHY

E. Abbe, *Arch. Mikros. Anat.* **9**, 413 (1873).

R. Adler, "Interaction Between Light and Sound," *IEEE Spectrum*, pp. 42–54 (May 1967).

T. E. Bell, ed., "Optical Computing: A Field in Flux," *IEEE Spectrum*, pp. 34–57 (August 1986).

N. J. Berg and J. N. Lee, eds., *Acousto-Optic Signal Processing (Theory and Implementation)*, Marcel Dekker, New York (1983).

B. R. Brown and A. W. Lohmann, "Complex Spatial Filtering with Binary Masks," *Appl. Opt.* **5**, 967 (1966).

H. J. Caulfield, J. A. Neff, and W. T. Rhodes, "Optical Computing: The Coming Revolution in Optical Signal Processing," *Laser Focus/Electro-Optics*, pp. 100–110 (November 1983).

L. J. Cutrona, E. N. Leith, L. J. Porcello, and W. E. Vivian, "On the Application of Coherent Optical Processing Techniques to Synthetic-Aperture Radar," *Proc. IEEE* **54**, 1026 (1966).

P. M. Duffieux, *L'intégrale de Fourier et ses Applications à l'optique*, Faculte des Sciences, Université de Besancon (1946), reprinted by Masson, Editeur, Paris (1970) and as *The Fourier Transform and Its Applications to Optics*, John Wiley & Sons, New York (1983).

P. Elias, "Optics and Communication Theory," *JOSA* **43**, 229 (1953).

P. Elias, D. S. Grey, and D. Z. Robinson, "Fourier Treatment of Optical Processes," *JOSA* **42**, 127 (1952).

D. G. Feitelson, *Optical Computing*, MIT Press, Cambridge, MA (1988).

L. Foucault, *Ann. Obs. Imp. (Paris)* **5**, 197 (1859).

A. P. Goutzoulis and I. J. Abramovitz, "Digital Electronics Meets its Match," *IEEE Spectrum*, pp. 21–25 (August 1988).

J. L. Horner, ed., *Optical Signal Processing*, Academic Press, Orlando, FL (1987), pp. 165–241.

P. Kellman, "Time Integrating Optical Signal Processing," *Opt. Eng.* **19**, 370 (1980).

E. N. Leith, "Quasi-Holographic Techniques in the Microwave Region," *Proc. IEEE* **59**, 1305 (1971).

H. S. Lipson, ed., *Optical Transforms*, Academic Press, Orlando, FL (1972).

A. Marechal and P. Croce, *Compt. Rend.* **237**, 706 (1953).

E. L. O'Neill, "Spatial Filtering in Optics," *IRE Trans.* **IT-2**, 56 (1956).

A. B. Porter, *Philos. Mag.* **11**, 154 (1906).

A. VanderLugt, "Signal Detection by Complex Spatial Filtering," *IEEE Trans.* **IT-10**, 139 (1964).

F. Zernike, *Z. Tech. Mag.* **16**, 454 (1935).

Chapter 1
TWO-DIMENSIONAL LINEAR SYSTEMS

1.1 FUNDAMENTAL PROPERTIES

In order to develop the capability of processing signals using optics, especially two-dimensional signals (i.e., images), a fundamental mathematical understanding of two-dimensional (2-D) linear systems theory is required. Two-dimensional linear systems, which are described most succinctly and generally by Equation (1.1), obey essentially the same rules and properties that one-dimensional systems do. The important properties of linear systems that we should recall are linearity, shift invariance, stability, and causality, which will hold with restrictions noted below.

A system can be defined as linear if its output is determined purely by addition as described in Equation (1.2). Linearity is usually assumed to hold except when nonlinear electro-optical interfaces are used. Nonlinear effects on the signal output may result from raising the input signal level into the saturation region of an otherwise linear spatial light modulator, or a component may be deliberately nonlinear as in the case of square-law photodetectors or logarithmic amplifiers (designed to accommodate large changes in input signal power). These topics will be covered in Chapters 2, 7, and 8. Shift invariance will hold over limited regions in the (x, y) plane (known as *isoplanatic patches* in optics). This is an important consideration in real optical systems that are designed with large apertures or wide fields of view. In image processing the term describing nonisoplanatic systems is "space variant." Stability will be required for the same reasons as in one-dimensional systems: A bounded input must yield a bounded output. Instances where this may be a problem are optical systems with feedback or iterative computations that may saturate a limited dynamic range device embedded in the system. Causality does not have its usual meaning for two-dimensional spatial coordinate systems characteristic of optics since we are not working with time as a variable (except for sequential processing of images). In other words, negative space coordinates are acceptable.

As expected, within the context of optics, the input and output will be functions of two spatial coordinates, x and y, and these functions will be either intensity (or brightness) or just amplitude at a point (x, y). Intensity will always be non-negative, but amplitude can be negative as well as complex.

1.2 LINEAR SUPERPOSITION

A 2-D linear system is illustrated as a block diagram in Fig. 1.1, and symbolically it is represented by the equation

$$g(x, y) = T[f(x, y)] \tag{1.1}$$

where $T[\ldots]$ denotes the operation of the 2-D system on the input function $f(x, y)$ that yields the output function $g(x, y)$. Linearity is defined in the usual way by

$$T[af_1(x, y) + bf_2(x, y)] = ag_1(x, y) + bg_2(x, y) \tag{1.2}$$

where a and b are arbitrary constants. Shift invariance is defined by

$$g(x - x', y - y') = T[f(x - x', y - y')] \tag{1.3}$$

where x' and y' are the arbitrary shifts (or offsets) in x and y, respectively.

A 2-D system is completely characterized by the $T[\ldots]$ operation, which is usually denoted by $h(x, y)$ for linear systems [when acting under the integral representing convolution as described below in Equation (1.10)]. The function $h(x, y)$ is the so-called impulse response, and in general is written $h(x, y; x', y')$ in order to denote that it is the output at (x, y) due to a unit impulse input at (x', y'). The unit impulse in two dimensions is often denoted by the Dirac delta function symbol $\delta(x, y)$ such that

$$\delta(x, y) = \begin{cases} \infty, & x = y = 0 \\ 0, & x \neq 0, y \neq 0 \end{cases} \tag{1.4}$$

as shown in Fig. 1.2. Strictly speaking, the Dirac delta "function" is not a function since its value at $x = y = 0$ is infinite (or undefined). We will also use the term "unit impulse" to describe the Dirac delta "function," which is plotted with unity height because its height actually corresponds to the "area" (for 1-D) or "volume" (for 2-D) under the delta function. Thus the peak value corresponds to the volume under the integral of $\delta(x, y)$ as defined in Equation (1.5).

$\delta(x, y)$ has the following general properties.

$$\int_{-\varepsilon}^{\varepsilon} \int_{-\varepsilon}^{\varepsilon} \delta(x, y) \, dx dy = 1 \quad \text{(for any } \varepsilon > 0\text{)} \tag{1.5}$$

$$f(x, y) = \int_{-\infty}^{\infty} \int_{-\infty}^{\infty} f(x', y') \delta(x - x', y - y') \, dx' dy' \tag{1.6}$$

(at each point of continuity of f). Equation (1.6) merely states that the delta function sifts out of the integral on the right-hand side the value of the function $f(x', y')$ at $x' = x$ and $y' = y$. Other properties of the Dirac delta function are given in Table 1.1 in Appendix A.

Figure 1.1 Block diagram of a general linear system.

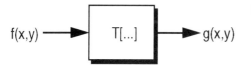

Figure 1.2 Two-dimensional Dirac delta function (or "thumbtack").

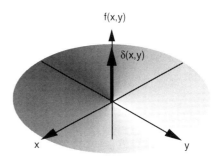

An important definition of the Dirac delta function useful in signal processing is

$$\delta(x, y) = \int_{-\infty}^{\infty} \exp(-j2\pi kx)\, dk \tag{1.7}$$

where k is spatial frequency, having units of inverse length. There are many other models for the Dirac delta function (see Table 1.2 in Appendix A).

For a 2-D linear system the output $g(x, y)$ is given in general terms by the superposition integral:

$$g(x, y) = \int_{-\infty}^{\infty}\int_{-\infty}^{\infty} f(x', y')\, h(x, y; x', y')\, dx'dy' \tag{1.8}$$

where $f(x, y)$ is the input and $h(x, y; x', y')$ is the kernel of the superposition integral, which acts as the "propagator" of the system. (It "propagates" the input through the system that it defines and yields the output.) Shift invariance [originally defined by Equation (1.3)] implies that $h(x, y; x', y')$ can be written as

$$h(x, y; x', y') = h(x - x', y - y') \tag{1.9}$$

Then the superposition integral becomes

$$g(x, y) = \int_{-\infty}^{\infty}\int_{-\infty}^{\infty} f(x', y')\, h(x - x', y - y')\, dx'dy' \tag{1.10}$$

which is the 2-D convolution integral. Thus $h(x - x', y - y')$ acts like a filter which can "smooth" or "sharpen" input information in the form of an image or irradiance distribution $f(x', y')$ to yield the resultant output $g(x, y)$.

1.3 CONVOLUTION AND CORRELATION

Convolution in electrical engineering is usually associated with electrical filtering of time-domain signals, in which the form of the kernel $h(t)$ (equivalent to $h(x, y)$) is entirely arbitrary (within the limits of causality) but selectable for a specific bandshaping purpose. The same could be said for optics and imaging systems in two dimensions (except for the causality restriction for electrical signals). Correlation, on the other hand, although very similar to convolution, differs because the kernel associated with correlation, denoted by $k(x, y)$, has shift variables (x', y') *not* reversed in sign as is the case with convolution. Thus correlation is given by

$$g(x, y) = \int_{-\infty}^{\infty} \int_{-\infty}^{\infty} f(x', y') k^*(x + x', y + y') \, dx' dy' \quad (1.11)$$

In general, the kernal of the correlation integral, $k^*(x, y)$, like the impulse response, can also be complex as indicated by the fact that $k^*(x, y)$ is the complex conjugate of $k(x, y)$. If we choose to constrain $k(x, y)$ to be the same function as the input image $f(x, y)$, then Equation (1.11) becomes the autocorrelation. Otherwise, $g'(x, y)$ is the cross-correlation. When Equation (1.11) is the autocorrelation, then the filter defined by $k(x, y)$ is the so-called matched filter, because it matches the input image $f(x, y)$. Such a filter is guaranteed to be optimal in detecting the presence and location of $f(x, y)$ embedded in an otherwise undetermined image in most situations. The undetermined image has often undesired components (noise) which obscure $f(x, y)$. Details on its derivation are given later in Chapter 2. The filter is so important, however, that it deserves mention here, and we will have much more to say about it later.

There are several variations on the matched filter (or correlator) that stem from the fact that signals are, in general, complex. These variations make use of the real, imaginary, or phase-only portions of the original complex signal in order to reduce computational (or dynamic range) requirements for correlation without significant loss of (and in some cases much improved) output signal-to-noise ratio. (This may be desirable when dealing with electro-optical interfaces of limited capability.) The fact that only a limited dynamic range phase-only signal can be used to correlate well with its full dynamic range complex counterpart is due, in part, to the fact that correlation is so sensitive to the spatial distribution of structural information in an image. This aspect of correlation is also intimately related to the fact that limited dynamic range phase-only (or real or imaginary only) information (in the Fourier domain) can be Fourier-transformed to reconstruct a desired image in the (x, y) plane (space domain) with reasonable fidelity to that obtainable from the full-complex image. We will return to this topic in Chapter 10.

General properties of correlation are summarized in Table 1.3 in Appendix A, where the symbol $\otimes$ denotes correlation. The asterisk ($*$) will be used to denote convolution (except as a superscript, which denotes complex conjugate).

Two examples of 2-D convolution are given below, including the convolution of a rectangular pillbox and circular pillbox. These two examples are the most common since they represent rectangular and circular apertures. They also represent the correlation in each case because the input functions are symmetrical.

EXAMPLE 1.1 Convolution of a Rectangular Pillbox Given the rectangular pillbox illustrated in Fig. 1.3 and defined by

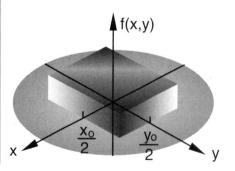

Figure 1.3 Two-dimensional rectangular pillbox function.

1.3 CONVOLUTION AND CORRELATION

$$g(x, y) = \text{rect}(x/x_0)\,\text{rect}(y/y_0) = \begin{cases} 1, & |x| \leq x_0/2, |y| \leq y_0/2 \\ 0, & |x| > x_0/2, |y| > y_0/2 \end{cases}$$

the convolution of the pillbox with itself is given by

$$f(x, y) * f(x, y) = [\text{rect}(x/x_0)\text{rect}(y/y_0)] * [\text{rect}(x/x_0)\text{rect}(y/y_0)]$$

$$= \int_{-\infty}^{\infty} \text{rect}\left(\frac{x'}{x_0}\right) \text{rect}\left(\frac{x'-x}{x_0}\right) dx' \int_{-\infty}^{\infty} \text{rect}\left(\frac{y'}{y_0}\right) \text{rect}\left(\frac{y'-y}{y_0}\right) dy'$$

where the notation rect(...), among others, was first introduced by Woodward. The area of overlap is illustrated in Fig. 1.4. Considering only the first integral (the second is identical), it can be solved with the aid of the graphical construction shown in Fig. 1.5(a). Only x-axis shifts are illustrated in the figure. Therefore,

$$\text{rect}(x/x_0) * \text{rect}(x/x_0) = \begin{cases} 0, & x < -x_0 \\ x + x_0, & -x_0 \leq x \leq 0 \\ -x + x_0, & 0 \leq x \leq x_0 \\ 0, & x_0 < x \end{cases}$$

or, written more compactly: $\text{rect}(x/x_0) * \text{rect}(x/x_0) = \text{tri}(x/x_0)$, which is plotted in Fig. 1.5(b). Therefore, $f(x, y) * f(x, y) = \text{tri}(x/x_0)\,\text{tri}(y/y_0)$, which is plotted in Fig. 1.6.

Figure 1.4 Area of overlap of rectangular pillbox with itself.

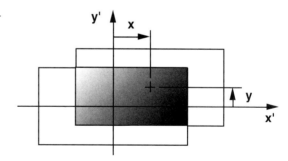

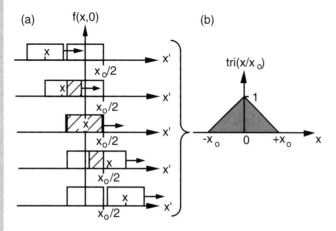

Figure 1.5 (a) Graphical construction of convolution process for rectangular pillbox for shifts along one axis, and (b) resulting function $\text{tri}(x/x_0)$

6   TWO-DIMENSIONAL LINEAR SYSTEMS

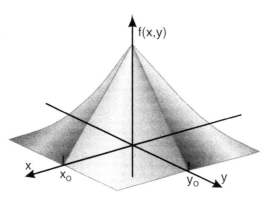

Figure 1.6 Two-dimensional convolution for rectangular pillbox function.

EXAMPLE 1.2 Convolution of a Circular Pillbox   The convolution of a circular pillbox of radius a is a little more involved than the previous example because the functional expression is not separable in x and y coordinates. The circular pillbox is defined as

$$\text{circ}\left(\frac{r}{a}\right) = \begin{cases} 1, & r \leq a \\ 0, & r > a \end{cases}$$

We can solve this problem using the geometric construction shown in Fig. 1.7. By looking at the plan view of the circular aperture and its displaced copy along one axis (the x axis in this case), we see that the area of overlap corresponding to the convolution is not as simple as the rectangular aperture.

The area of overlap is four times the portion of the circle area that is double cross-hatched (area B). The area of B is the area of $(A + B)-$ area of A. The area of $(A + B)$ is actually a sector of one circle. Therefore, given the expression area $(A + B) = (\theta/2\pi)(\pi a^2) = \theta a^2/2$, where $\theta = \cos^{-1}(x'/2a)$, and

$$\text{area}(A) = \frac{1}{2}\left(\frac{x'}{2}\right)\left[a^2 - \left(\frac{x'}{2}\right)^2\right]^{1/2}$$

we obtain

Figure 1.7 Geometric construction used as basis for calculating convolution of circular pillbox.

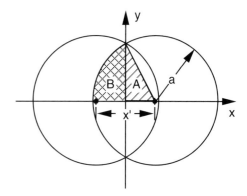

Figure 1.8 Result of circular pillbox convolution.

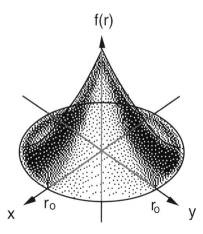

$$\text{Area of overlap} = 4 \times \text{area}(B) = 2a^2 \cos^{-1}(x'/2a) - x'a[1 - (x'/2a)^2]^{1/2}$$

Normalizing the area of overlap by the maximum area πa^2 yields the convolution result:

$$\text{circ}\left(\frac{r}{r_0}\right) * \text{circ}\left(\frac{r}{r_0}\right) = \begin{cases} \frac{2}{\pi} \left\{ \cos^{-1}\left(\frac{r'}{2a}\right) - \left(\frac{r'}{2a}\right)\left[1 - \left(\frac{r'}{2a}\right)^2\right]^{1/2} \right\}, & r' \leq 2a \\ 0, & r' > 2a \end{cases}$$

where x' is replaced by r'. This result is plotted in Fig. 1.8. This and the previous example will turn out to be useful results for calculating the spatial frequency transfer function of diffraction-limited apertures in Chapter 6.

1.4 TWO-DIMENSIONAL FOURIER TRANSFORMS AND PROPERTIES

In signal processing, two ways that convolution (or correlation) can be performed are in the time (or space) domain and in the corresponding Fourier domain. For similar reasons in optics we need to introduce the Fourier transform in two dimensions. The Fourier transform pair can be written

$$F(u, v) = \int_{-\infty}^{\infty} \int_{-\infty}^{\infty} f(x, y) \exp[-j2\pi(xu + yv)] \, dx \, dy \qquad (1.12)$$

and

$$f(x, y) = \int_{-\infty}^{\infty} \int_{-\infty}^{\infty} F(u, v) \exp[j2\pi(xu + yv)] \, du \, dv \qquad (1.13)$$

where (u, v) are spatial frequencies measured in units of cycles/unit length and (x, y) are the corresponding space variables. Thus, given a distance or scale x_0, the corresponding spatial frequency is $u_0 = 1/x_0$, and so on. Equations (1.12) and (1.13) are the Fourier transform pair in Cartesian coordinates.

TWO-DIMENSIONAL LINEAR SYSTEMS

Of course, the Fourier transform yields the spectrum of a signal $f(x, y)$ in amplitude $F(u, v)$ or power $|F(u, v)|^2$. Calculation of Equation (1.13) to yield $f(x, y)$ can be considered the process of "reconstruction" from $F(u, v)$ alluded to earlier. In this sense, calculating $F(u, v)$ is considered "analysis"—hence its association with "spectrum analyzer"—and calculating $f(x, y)$ is considered "synthesis." With this in mind it will be easier to understand the matched filter concept, which we will consider shortly.

In space coordinates the correlation (or convolution) is calculated just as we stated in Equation (1.11) [or (1.10)], without resorting to the Fourier domain. If, however, we Fourier transform Equation (1.11) we obtain $G'(u, v) = K^*(u, v) F(u, v)$ for the Fourier transform of the correlation of $k(x, y)$ with $f(x, y)$. Likewise, for convolution: $G(u, v) = H(u, v) F(u, v)$. This last expression is the so-called convolution theorem: Given the convolution of two functions, their Fourier transform is simply the product of the Fourier transforms of the respective functions. Clearly then, by applying the *inverse* Fourier transform we obtain the result

$$g(x, y) = \int_{-\infty}^{\infty} \int_{-\infty}^{\infty} H(u, v) F(u, v) \exp[j2\pi(xu + yv)] \, du \, dv \qquad (1.14)$$

which can be written formally as

$$g(x, y) = \text{IFT} \{\text{FT}[h(x, y)] \cdot \text{FT}[f(x, y)]\} \qquad (1.15)$$

where FT denotes Fourier transform and IFT denotes inverse Fourier transform. This relationship is shown as a block diagram in Fig. 1.9. Equation (1.15) shows that the convolution of $h(x, y)$ and $f(x, y)$ is given by the inverse Fourier transform of the product of the Fourier transforms of $h(x, y)$ and $f(x, y)$. This is the basis of the familiar fast convolution algorithm in digital signal processing when using the fast Fourier transform (FFT) algorithm.

By analogy, "fast" correlation is given by

$$g'(x, y) = \text{IFT} \{\text{FT}[k^*(x, y)] \cdot \text{FT}[f(x, y)]\} \qquad (1.16)$$

Now it is clear that correlation can proceed in three stages: Fourier transform the input image $f(x, y)$, multiply the result by the appropriate Fourier-domain image function FT $[k^*(x, y)]$, and inverse Fourier transform the result. FT $[k^*(x, y)]$ is said to be the matched filter for $f(x, y)$ when $k(x, y) = f(x, y)$. It is also understandable now why correlation (matched filtering) is intimately related to Fourier reconstruction. By picking a suitable matched filter form FT $[k^*(x, y)]$, that is,

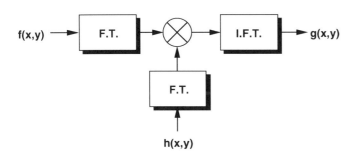

Figure 1.9 Block diagram of "fast" convolution algorithm in 2-D (where the Fourier transform is usually implemented by the FFT algorithm.)

$$\text{FT}\,[k*(x,y)] = 1 \tag{1.17}$$

the inverse Fourier transform operation in Equation (1.16) now yields

$$g'(x,y) = \text{IFT}\,\{\text{FT}\,[f(x,y)]\} = f(x,y) \tag{1.18}$$

which is the exact reconstruction of $f(x, y)$. In practice, $g'(x, y)$ is only approximately $f(x, y)$. The extent to which $f(x, y)$ is approximated depends upon the fact that Equation (1.17) holds true only over a finite region of support (or aperture size in optics), rather than over the entire (x, y) plane in the ideal case.

There are a number of important properties of the Fourier transform $F(u, v)$, most of which have to do with what happens to it when $f(x, y)$ is modified in some way. These modifications include: change in sign, scale, and location (or shift) of coordinates x or y, as well as conjugation, frequency shift (heterodyning), and space or frequency differentiation of the input function (or image) $f(x, y)$. All of these properties are summarized in Table 1.4 in Appendix A. Finally, Parseval's theorem holds in two dimensions. This theorem states that the energy of a signal in the space domain equals the energy of the corresponding signal in the Fourier domain.

1.5 RECTANGULAR AND POLAR FORMS

A useful (but not fundamental) property of 2-D Fourier transforms is the property of separability in Cartesian coordinates. If $f(x, y)$ is separable, then $f(x, y) = f_a(x) f_b(y)$, and its corresponding Fourier transform is separable: $F(u, v) = F_a(u) F_b(v)$. Some frequently used separable Fourier transform pairs are provided in Table 1.5 in Appendix A for reference. Corresponding pictorial examples and definitions of the one-dimensional functional notation (e.g., rect, sinc, tri, comb, and sgn) are provided in Fig. A.1 of Appendix A to aid in the visualization of shape and properties. An example of a separable two-dimensional Fourier transform is given in Example 1.3. It is the classic case of a rectangular pillbox.

EXAMPLE 1.3 Fourier Transform of a Rectangular Pillbox The rectangular pillbox illustrated in Fig. 1.3 is defined as $f(x, y) = \text{rect}(x/x_0)\,\text{rect}(y/y_0)$. Since this function is separable in x and y coordinates, and the separated parts are equivalent, we will carry out only one of the Fourier transform integrations explicitly. Therefore,

$$\int_{-\infty}^{\infty} \text{rect}\left(\frac{x}{x_0}\right) \exp(-j2\pi u x)\,dx$$

$$= \frac{1}{-j2\pi u} \int_{-x_0/2}^{x_0/2} \exp(-j2\pi u x)\,d(-j2\pi u x)$$

$$= \frac{1}{\pi u} \left[\frac{\exp(j\pi u x_0) - \exp(-j2\pi u x_0)}{2j}\right]$$

$$= x_0 \frac{\sin(\pi u x_0)}{\pi u x_0}$$

$$= x_0\,\text{sinc}(\pi u x_0)$$

Consequently,

$$F(u, v) = \text{FT}_{2D}\{f(x, y)\} = x_0 y_0 \, \text{sinc}(\pi u x_0) \, \text{sinc}(\pi v y_0)$$

which is illustrated in Fig. 1.10. Note from Fig. A.1 the relationship between the first zeroes of $F(u, v)$ (mainlobe width) and the width of $f(x, y)$: The sinc mainlobe width equals $2/x_0$, whereas the rectangle function has a width of x_0. Thus the wider the pillbox, the narrower the sinc-function mainlobe width.

Because many optical systems possess circular symmetry about their optical axis (i.e., lenses often have circular apertures), it is simpler and more useful to express the Fourier transform in polar coordinates. To do this we merely take the Fourier transform in Cartesian coordinates and transform the independent variables, namely, $x = r \cos \theta$ and $y = r \sin \theta$. The corresponding coordinate transforms in the Fourier domain are $u = \rho \cos \phi$ and $v = \rho \sin \phi$, where (r, θ) and (ρ, ϕ) are defined in Fig. 1.11. These coordinate transforms can be inverted to give the polar coordinates in terms of the Cartesian: $r = (x^2 + y^2)^{1/2}$ and $\theta = \tan^{-1}(y/x)$, as well as $\rho = (u^2 + v^2)^{1/2}$ and $\phi = \tan^{-1}(v/u)$.

Thus the Fourier transform pair can be written

$$F(\rho, \phi) = \int_0^\infty \int_0^{2\pi} f(r, \theta) \exp[-j2\pi r \rho \cos(\theta - \phi)] r \, dr d\theta \quad (1.19)$$

$$f(r, \theta) = \int_0^\infty \int_0^{2\pi} F(\rho, \phi) \exp[j2\pi r \rho \cos(\theta - \phi)] \rho \, d\rho d\phi \quad (1.20)$$

Equations (1.19) and (1.20) constitute the so-called Fourier–Bessel transform pair. For circularly symmetric functions there is no θ dependence and $f(r, \theta)$ is defined as $g(r)$. Then Equation (1.19) becomes

$$F(\rho, \phi) = \int_0^\infty g(r) r \left\{ \int_0^{2\pi} \exp[-j2\pi r \rho \cos(\theta - \phi)] \, d\theta \right\} dr \quad (1.21)$$

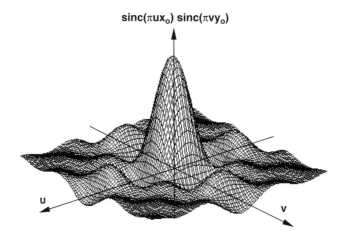

Figure 1.10 Sin x/x function in two dimensions representing the Fourier transform of a rectangular pillbox.

1.5 RECTANGULAR AND POLAR FORMS

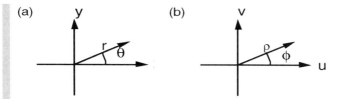

Figure 1.11 Polar coordinate systems in the space and Fourier domains.

where the exponential function in the integrand is periodic in θ. It turns out that the integral in braces is the definition of the zero-order Bessel function of the first kind, denoted by $J_0(x)$ and given by the identity

$$J_0(x) = (2\pi)^{-1} \int_0^{2\pi} \exp[-j2\pi r\rho \cos(\theta - \phi)] \, d\theta \tag{1.22}$$

where $x = 2\pi r\rho$ here. (See Table 1.6 in Appendix A for some of the properties of Bessel functions.) Thus $F(\rho, \phi)$ becomes $G(\rho)$ and is therefore circularly symmetric. Thus Equation (1.21) becomes

$$G(\rho) = 2\pi \int_0^\infty g(r) r J_0(2\pi r\rho) \, dr \tag{1.23}$$

The inverse transform follows from an analogous argument. Therefore,

$$g(r) = 2\pi \int_0^\infty G(\rho) \rho J_0(2\pi r\rho) \, d\rho \tag{1.24}$$

Equations (1.23) and (1.24) constitute the so-called Hankel transform pair. Several examples of Hankel transform pairs are shown in Table 1.7 in Appendix A. Similar properties hold for the Hankel transform as for the Fourier transform, and these are summarized in Table 1.8 in Appendix A.

An example of the Hankel transform of a circular pillbox, which represents a circular aperture, is given in Example 1.4.

EXAMPLE 1.4 Hankel Transform of a Circular Pillbox The circular pillbox illustrated in Fig. 1.12 is defined as

$$\text{circ}(r/r_0) = \begin{cases} 1, & r \leq r_0 \\ 0, & r > r_0 \end{cases}$$

Using the Hankel transform [Equation (1.23)] we obtain

$$H\{\text{circ}(r)\} = 2\pi \int_0^1 J_0(2\pi r\rho) r \, dr$$

A change of variables—that is, $r' = 2\pi r\rho$—puts this result into the form

$$H\{\text{circ}(r)\} = \frac{1}{2\pi\rho^2} \int_0^{2\pi\rho} J_0(r')r'\,dr'$$

Using one of the properties of the Bessel function cited in Table 1.6 of Appendix A, we obtain

$$H\{\text{circ}(r)\} = \frac{J_1(2\pi\rho)}{\rho}$$

which is plotted in Fig. 1.13(a). We will return to this result when we discuss optical diffraction from a circular aperture. Note the location of the zeroes of $J_1(2\pi\rho)/\rho$ as shown in the cross section in Fig. 1.13(b). They are located at 3.8317, 7.0156, 10.1735, 13.3237, 16.4706, and so on. As $\rho \to \infty$ the zeroes become more evenly spaced (like the zeroes of the sinc function).

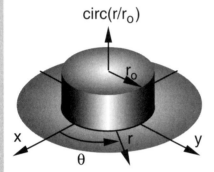

Figure 1.12 Circular pillbox function in polar coordinates.

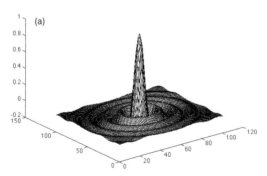

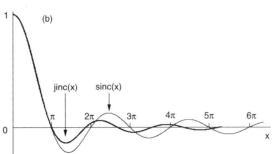

Figure 1.13 (a) Fourier transform of circular pillbox, also known as the "Sombrero" function, and (b) cross-section showing location of zeroes. Note the comparison to the sinc function.

1.6 LINEAR COORDINATE TRANSFORMATION AND FOURIER THEOREM

In addition to a linear transformation of amplitude there can also be a linear transformation of coordinates. [Here "linear" means operating on the vector of two-space coordinate variables (x, y) via an orthogonal matrix operation.] This concept is a 2-D extension of the notion of scaling or shifting in one dimension. Unlike a shift in coordinates, which is qualitatively the same in one and two dimensions, scaling in two dimensions yields a unique result because the scale factors along the two axes can differ. This leads to the notion of anamorphic magnification (different magnifications along orthogonal axes, analogous to astigmatism). If the scale factors are the same in both axes, then we have simple magnification. If they are equal and of opposite sign, then we have image inversion. In addition, in two dimensions we can also perform rotation, which has no corresponding equivalent in one dimension.

Consider a linear change of variables, which is expressed as a matrix operation on coordinates (x', y') yielding a resultant set of coordinates (x, y):

$$\begin{bmatrix} x \\ y \end{bmatrix} = \begin{bmatrix} a_{11} & a_{12} \\ a_{21} & a_{22} \end{bmatrix} \begin{bmatrix} x' \\ y' \end{bmatrix} \tag{1.25}$$

or $\mathbf{x} = \mathbf{A}\mathbf{x}'$, where $\mathbf{A}$ is the linear transformation matrix, and $\mathbf{x}' = (x', y')$ and $\mathbf{x} = (x, y)$ are vectors (which represent 2-D coordinates). If $\mathbf{A}$ corresponds to simple anamorphic magnification, then

$$\mathbf{A} = \begin{bmatrix} a & 0 \\ 0 & b \end{bmatrix} \tag{1.26}$$

where, in general, $a \neq b$. If $a = b > 1$, we have simple magnification, and if $a = b = -1$ we have inversion. If $\mathbf{A}$ corresponds to rotation, then

$$\mathbf{A} = \mathbf{R} = \begin{bmatrix} \cos\phi & \sin\phi \\ -\sin\phi & \cos\phi \end{bmatrix} \tag{1.27}$$

where ϕ is the angle of rotation of the matrix operator $\mathbf{A}$ or the equivalent angle between the (x', y') coordinate system and the (x, y) coordinate system as illustrated in Fig. 1.14. Note that a rotation by ϕ is, by definition, counterclockwise with respect to the initial coordinate system (x', y'). (Also note that if $\phi = \pi$ we have inversion again.)

An important consideration is the effect of a linear transformation of coordinates on a function $f(x, y)$ representing an image. This is expressed as follows. The function in the new

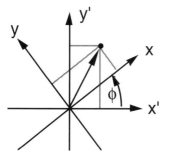

Figure 1.14 Coordinate system before (x', y') and after (x, y) rotation by angle ϕ.

coordinate system is given by $f(\mathbf{x}) = f(\mathbf{A}\mathbf{x}') = f_1(x', y')$. In Fourier optics we will often be interested in what happens to the Fourier transform of $f(\mathbf{x}')$ under a linear transformation $\mathbf{A}$. We can show that

$$F_1(\mathbf{u}') = |\det \mathbf{A}|^{-1} F(\mathbf{A}_T^{-1}\mathbf{u}') \qquad (1.28)$$

where $\mathbf{u}' = (u', v')$ is the spatial frequency vector corresponding to $\mathbf{x}'$; $F_1(\mathbf{u}')$ is the Fourier transform of $f_1(\mathbf{x}')$; $F(\mathbf{A}_T^{-1}\mathbf{u}')$ is the Fourier transform corresponding to $f(\mathbf{x})$; and $\mathbf{A}_T^{-1}$ is the inverse of the transpose of $\mathbf{A}$. This result is useful in deriving the rotational properties of the two-dimensional Fourier transform and in obtaining the Fourier transform of certain impulse arrays under rotation or anamorphic magnification.

Equation (1.28) can be proved by the following argument. By definition,

$$F_1(u', v') = \int_{-\infty}^{\infty} \int_{-\infty}^{\infty} f_1(x', y') \exp\left[-j2\pi(x'u' + y'v')\right] dx'dy' \qquad (1.29)$$

$$= \int_{-\infty}^{\infty} \int_{-\infty}^{\infty} f(\mathbf{A}\mathbf{x}') \exp\left[-j2\pi \mathbf{x}' \cdot \mathbf{u}'\right] d\mathbf{x}' \qquad (1.30)$$

Changing variables—that is, applying $\mathbf{x} = \mathbf{A}\mathbf{x}'$—yields

$$F_1(u', v') = \int_{-\infty}^{\infty} \int_{-\infty}^{\infty} f(\mathbf{x}) \exp(-j2\pi \mathbf{x}^T \cdot \mathbf{A}_T^{-1}\mathbf{u}') \left|\frac{\partial(x', y')}{\partial(x, y)}\right| dxdy \qquad (1.31)$$

where

$$\left|\frac{\partial(x', y')}{\partial(x, y)}\right| = \begin{vmatrix} \frac{\partial x'}{\partial x} & \frac{\partial x'}{\partial y} \\ \frac{\partial y'}{\partial x} & \frac{\partial y'}{\partial y} \end{vmatrix} = \det(\mathbf{A}^{-1}) = |\det(\mathbf{A})|^{-1} \qquad (1.32)$$

is the Jacobian of the coordinate transformation and $\mathbf{x}^T \cdot \mathbf{A}_T^{-1}\mathbf{u}' = \mathbf{A}^{-1}\mathbf{x} \cdot \mathbf{u}'$. Therefore,

$$F_1(\mathbf{u}') = |\det \mathbf{A}|^{-1} F(\mathbf{A}_T^{-1}\mathbf{u}') \qquad (1.33)$$

1.7 EXAMPLES OF MAGNIFICATION AND ROTATION

The following examples are provided to illustrate the validity and significance of the theorem stated in Equation (1.28).

EXAMPLE 1.5 Validation of Equation (1.28) for Rectangular Aperture Subject to Anamorphic Magnification Given the equation defining a rectangular aperture,

$$f(x') = \begin{cases} 1, & |x'| \leq X_0/2, |y'| \leq Y_0/2 \\ 0, & |x'| > X_0/2, |y'| > Y_0/2 \end{cases}$$

as shown in Fig. 1.15(a), where $X_0 = 2$ and $Y_0 = 4$, consider first what happens when the transformation representing anamorphic magnification is applied:

1.7 EXAMPLES OF MAGNIFICATION AND ROTATION

$$A = \begin{bmatrix} a & 0 \\ 0 & b \end{bmatrix}$$

We seek the Fourier transform of the rectangular aperture after magnification. Before magnification the Fourier transform is

$$F(u, v) = X_0 Y_0 \operatorname{sinc}(\pi X_0 u) \operatorname{sinc}(\pi Y_0 v) = X_0 Y_0 \frac{\sin(\pi X_0 u)}{(\pi X_0 u)} \frac{\sin(\pi Y_0 v)}{(\pi Y_0 v)}$$

$F(u, v)$ has zeroes at $u = n/X_0$ and $v = n/Y_0$, $n = \pm 1, \pm 2, \pm 3, \ldots$ and is plotted along the u and v axes in Fig. 1.16(a).

To apply the theorem to yield $F_1(u', v')$ using (1.28), note that we must use the transformed function $f(\mathbf{A}\mathbf{x}')$ or $f(\mathbf{x})$ defined as

$$f(\mathbf{x}) = \begin{cases} 1, & |x| \leq aX_0/2, |y| \leq bY_0/2 \\ 0, & |x| > aX_0/2, |y| > bY_0/2 \end{cases}$$

and illustrated in Fig. 1.15(b) for the case where the transformation matrix diagonal elements are $a = 3$ and $b = 0.5$.

In constructing the integrand of Equation (1.28) we must define the expression: $\mathbf{x}^T \cdot \mathbf{A}_T^{-1} \mathbf{u}'$ (or $\mathbf{A}^{-1}\mathbf{x} \cdot \mathbf{u}'$), which is

$$\mathbf{x}^T \cdot \mathbf{A}_T^{-1} \mathbf{u}' = \begin{bmatrix} x & y \end{bmatrix} \begin{bmatrix} \frac{1}{a} & 0 \\ 0 & \frac{1}{b} \end{bmatrix} \begin{bmatrix} u' \\ v' \end{bmatrix}$$

or

$$\mathbf{x}^T \cdot \mathbf{A}_T^{-1} \mathbf{u}' = xu'/a + yv'/b$$

Therefore (1.28) becomes

$$F_1(u', v') = \int_{-Y_0}^{Y_0} \left[\int_{-X_0}^{X_0} \exp[-j2\pi(u'\cos\theta + v'\sin\theta)] \, dx \right]$$
$$\exp[-2j\pi(u'\sin\theta - v'\cos\theta)] \, dy$$

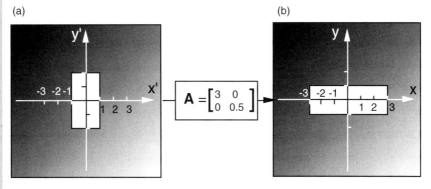

Figure 1.15 (a) Rectangular aperture function, and (b) rectangular aperture function after anamorphic scaling by amount $a = 3$ in x and $b = 0.5$ in y.

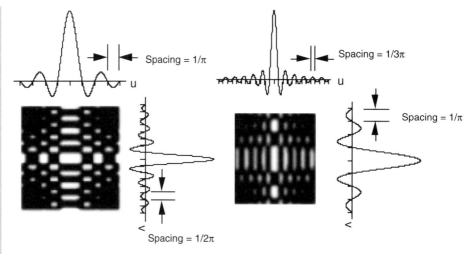

Figure 1.16 (a) Fourier transform of rectangular aperture function shown in Fig. 1.15(a) in the (u, v) plane along the u and v axes, and (b) corresponding Fourier transform for rectangular aperture scaled as shown in Fig. 1.15(b).

or

$$F_1(u', v') = X_0 Y_0 \frac{\sin(\pi a X_0 u')}{(\pi a X_0 u')} \frac{\sin(\pi b Y_0 v')}{(\pi b Y_0 v')}$$

a result which is consistent with the theorem since $\det(\mathbf{A}^{-1}) = 1/ab$, as well as being consistent with the similarity property of the Fourier transform. Plotting this equation along the u and v axes for the case of $a = 3$ and $b = 0.5$ in Fig. 1.16(b) clearly shows the zeroes to be at $u = n/3X_0, n = \pm 1, \pm 2, \ldots$, and $v = m/0.5X_0, m = \pm 1, \pm 2, \ldots$. Thus the sinc function has been compressed along the u axis and expanded along the v axis, as indicated.

EXAMPLE 1.6 Fourier Transform for Rotation of a Rectangular Aperture For the case of a rectangular aperture defined in the previous example [Fig. 1.15(a)] which is rotated clockwise by an angle θ about its center, the transformation matrix is

$$A = \begin{bmatrix} \cos\theta & \sin\theta \\ -\sin\theta & \cos\theta \end{bmatrix}$$

and

$$\mathbf{x}^T \cdot \mathbf{A}_T^{-1} \mathbf{u}' = [x\ y] \begin{bmatrix} \cos\theta & \sin\theta \\ -\sin\theta & \cos\theta \end{bmatrix} \begin{bmatrix} u' \\ v' \end{bmatrix}$$
$$= (x\cos\theta - y\sin\theta)u' + (x\sin\theta + y\cos\theta)v'$$

Then Equation (1.31) becomes

$$F_1(u', v') = \int_{-\infty}^{\infty} \int_{-\infty}^{\infty} f(x, y)\ \exp\{-2j\pi[(x\cos\theta - y\sin\theta)u' + (x\sin\theta + y\cos\theta)v']\}\,dx\,dy$$

where the Jacobian of the transformation is

$$\left|\frac{\partial(x', y')}{\partial(xy)}\right| = |\det(\mathbf{A})|^{-1} = 1$$

which can be easily verified for the rotation matrix $\mathbf{A} = \mathbf{R}$. Now that we have this result, we can proceed by using the definition of $f(x, y)$ in this example to set limits on the integration to yield

$$F_1(u', v') = \int_{-Y_0/2}^{Y_0/2} \left[\int_{-X_0/2}^{X_0/2} \exp\{-j2\pi[(u'\cos\theta + v'\sin\theta)]\} dx \right] \exp\{-j2\pi[(u'\sin\theta - v'\cos\theta)]\} dy$$

After performing the indicated integrations in x and y, we obtain

$$F_1(u', v') = X_0 Y_0 \left\{ \frac{\sin[\pi(u'\cos\theta + v'\sin\theta)]}{\pi(u'\cos\theta + v'\sin\theta)} \right\} \left\{ \frac{\sin[\pi(u'\sin\theta - v'\cos\theta)]}{\pi(u'\sin\theta - v'\cos\theta)} \right\}$$

Finally, noting that the corresponding rotation transformation for spatial frequency coordinates $\mathbf{u} = (u, v)$ (i.e., $\mathbf{u} = \mathbf{A}\mathbf{u}'$), is the same as for the spatial coordinates (i.e., $\mathbf{x} = \mathbf{A}\mathbf{x}'$) yields

$$F_1(u', y) = X_0 Y_0 \left\{ \frac{\sin(\pi X_0 u)}{(\pi X_0 u)} \right\} \left\{ \frac{\sin(\pi Y_0 v)}{(\pi Y_0 v)} \right\}$$

Thus, instead of the Fourier transform given by Fig. 1.17(a) we now have the rotated Fourier transform shown in Fig. 1.17(b), for the case where $\theta = \pi/4$.

1.8 TWO-DIMENSIONAL IMPULSE FUNCTIONS: PROPERTIES AND FOURIER TRANSFORMS

In order to treat optical imaging and processing systems we need to adopt a mathematical notation or language that describes images or two-dimensional patterns. Taking our cue from one-dimensional linear systems theory, we can adopt the notion of a two-dimensional unit impulse $\delta(x, y)$ introduced earlier in Section 1.2. This function was defined in Equation (1.4) and illustrated in Fig. 1.2. In optics one can think of it as representing a tiny but bright spot of light in the (x, y) plane. The corresponding two-dimensional Fourier transform of $\delta(x, y)$ is $FT[\delta(x, y] = 1$ for all (u, v). Conversely, in the Fourier domain the unit impulse is given by

$$\delta(u, v) = \begin{cases} \infty, & u = v = 0 \\ 0, & u \neq 0, v \neq 0 \end{cases} \quad (1.34)$$

and its Fourier transform is $FT[\delta(u, v)] = 1$ for all (x, y).

If the impulse function is shifted from the origin by amount (a, b), then it is written $\delta(x - a, y - b)$. If we now convolve this impulse function with the arbitrary function $f(x, y)$, we obtain $f(x, y) * \delta(x - a, y - b) = f(x - a, y - b)$, where we make use of Equation (1.6). If $f(x, y)$ represents the impulse response of a linear system, then the convolution operation represents the excitation of such a system with a unit impulse whose resulting output is the desired impulse response (commonly termed "system identification" because it represents a

18 TWO-DIMENSIONAL LINEAR SYSTEMS

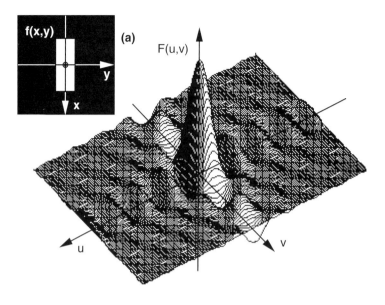

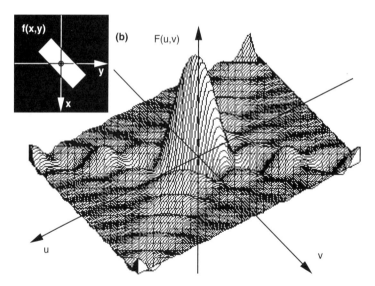

Figure 1.17 (a) Fourier transform of rectangular pillbox and (b) corresponding Fourier transform for a rotated pillbox, for the case of rotation angle $\theta = \pi/4$.

way of obtaining the impulse response of an unknown linear system or "black box"). Also note that the operation given by $f(x, y) * \delta(x - a, y - b) = f(x - a, y - b)$ can be interpreted as "replication," since a replicate (or copy) of $f(x, y)$ is created at location $(x = a, y = b)$.

1.8 TWO-DIMENSIONAL IMPULSE FUNCTIONS: PROPERTIES AND FOURIER TRANSFORMS

The notion of unit impulse can be extended in the following sense. A two-dimensional impulse "line" can be defined as

$$\delta(ax + by + c) = \begin{cases} \infty, & ax + by + x = 0 \\ 0, & \text{elsewhere} \end{cases} \quad (1.35)$$

which is illustrated in Fig. 1.18.

Thus graphically the unity height impulse "line" is oriented along a direction whose equation is given by $ax + by + c = 0$, which has a slope given by $-a/b$ and the y intercept of $-c/b$. This can represent a bright line of light of arbitrary position and orientation in the (x, y) plane. An impulse "line" located on a line defined by $y - ax = 0$ (where $a = \tan\theta$ is the slope) goes through the origin and can be written as $\delta(y - ax)$ or as $\delta(-x\sin\theta + y\cos\theta)$, although these expressions differ by a multiplicative constant. Such an impulse "line" runs through the origin. If we look at this impulse "line" oriented along the x axis, then $y = 0$, and the impulse "line" is described by $\delta(y)$ [see Fig. 1.19(a)]. The corresponding Fourier transform is given by

$$\text{FT}[\delta(y)] = \int_{-\infty}^{\infty}\int_{-\infty}^{\infty} \delta(y)\exp[-j2\pi(xu + yv)]\,dx\,dy \quad (1.36)$$

$$\text{FT}[\delta(y)] = \int_{-\infty}^{\infty} \exp(-j2\pi xu)\,dx \int_{-\infty}^{\infty} \delta(y)\exp(-j2\pi yv)\,dy \quad (1.37)$$

The first integral on the right-hand side is a general definition of the delta function $\delta(u)$ [recall Equation (1.7)], and the second integral is unity [using Equation (1.6)]. Therefore, $\text{FT}[\delta(y)] = \delta(u)$.

The resulting orientation of the Fourier transform impulse "line" $\delta(u)$ is along the v axis and perpendicular to $\delta(y)$, assuming (as is the convention) that the (x, y) and (u, v) axes would overlay each other, as suggested in Figs. 1.19(a) and 1.19(b). This result is consistent with an intuitive understanding of what the Fourier transform does: The frequency content along the y direction is high because we encounter a rapidly changing impulse "line," and that is reflected in the fact that along the v axis the Fourier-domain line height is constant (representing all frequencies). Of course, along the x axis there is no frequency content

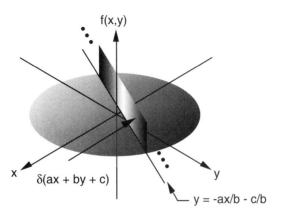

Figure 1.18 Impulse "line" defined by (1.35).

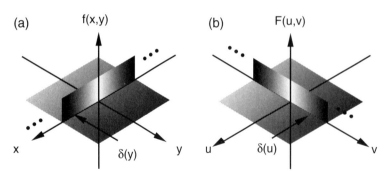

Figure 1.19 (a) Impulse "line" defined by $\delta(y)$, and (b) corresponding Fourier transform $\delta(u)$.

(except at zero frequency) since the impulse "line" is constant along x. The corresponding transform is $\delta(v)$, a "spike" along the u axis, reflecting the fact that there is energy only at zero frequency.

As previously mentioned, the impulse "line" can take on an arbitrary orientation θ with respect to the x axis. Therefore it can be expressed as $\delta(-x\sin\theta + y\cos\theta)$ and is illustrated in Fig. 1.20(a). By the rotational property of the Fourier transform, which derives from the general linear transformation property expressed in Equation (1.28), we can write

$$\text{FT}[\delta(-x\sin\theta + y\cos\theta)] = \delta(u\cos\theta + v\sin\theta) \tag{1.38}$$

This can be derived directly using the Fourier transform, namely,

$$\int_{-\infty}^{\infty}\int_{-\infty}^{\infty} \delta(-x\sin\theta + y\cos\theta) \exp[-j2\pi(xu+yv)]\,dx\,dy \tag{1.39}$$

$$= \int_{-\infty}^{\infty} \exp(-j2\pi yv)\,dy \int_{-\infty}^{\infty} (-1/\sin\theta)\delta(x - y\cot\theta) \exp(-j2\pi xu)\,dx$$

$$= (-1/\sin\theta) \int_{-\infty}^{\infty} \exp[-j2\pi(v - u\cot\theta)y]\,dy$$

$$= (-1/\sin\theta)\,\delta(v - u\cot\theta)$$

$$= \delta(u\cos\theta + v\sin\theta)$$

The impulse line $\delta(u\cos\theta + v\sin\theta)$ is perpendicular to the original impulse line $\delta(-x\sin\theta + y\cos\theta)$, a fact that is obvious by inspection of the arguments of both delta functions [see Fig. 1.20(b)]. If the original impulse line does not pass through the origin, then the Fourier transform of it equals $\delta(u\cos\theta + v\sin\theta)$ multiplied by a linear phase factor, a result derivable from the shift theorem.

Since impulse lines are not true functions, they can exist only under integration. When they do appear inside an integral by themselves, they yield a scalar multiplier representing their "strength" when the integration is performed, namely,

$$\int_{-\infty}^{\infty} A\delta(x)\,dx = A \tag{1.40}$$

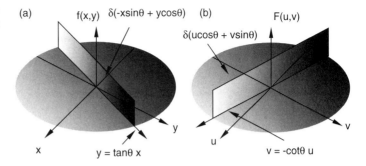

Figure 1.20 (a) Impulse line defined by $\delta(-x\sin\theta + y\cos\theta)$ and (b) corresponding Fourier transform impulse line defined by $\delta(u\cos\theta + v\sin\theta)$.

If the argument of the delta function is scaled by a, then $\delta(x) = |a|\delta(ax)$, and thus their strength is modifiable by a linear coordinate transform. Likewise, in two dimensions we obtain $\delta(\mathbf{x}) = |\det \mathbf{A}|\delta(\mathbf{Ax})$. Therefore,

$$\int_{-\infty}^{\infty}\int_{-\infty}^{\infty} \delta(\mathbf{Ax})\,dx\,dy = |\det \mathbf{A}|^{-1} \qquad (1.41)$$

where $|\det \mathbf{A}|$ is the "strength." When $|\det \mathbf{A}| = 1$ we have the "unit" impulse.

If two impulse lines—one on the x axis, $\delta(y)$, and one on the y axis, $\delta(x)$—are multiplied together, we get the unit impulse $\delta(x, y) = \delta(x)\delta(y)$.

We can form arrays of impulse lines in order to represent scanned images as in conventional television images. Denoting an array of unit impulse lines by $i(x, y)$, then

$$i(x, y) = \sum_{n=-\infty}^{\infty} \delta(y - nb) \qquad (1.42)$$

for a uniform height array with spacing b distributed along the y axis as shown in Fig. 1.21(a). The corresponding Fourier transform is given by

$$I(u, v) = \int_{-\infty}^{\infty}\int_{-\infty}^{\infty} \sum_{n=-\infty}^{\infty} \delta(y - nb)\,\exp[-j2\pi(ux + vy)]\,dx\,dy \qquad (1.43)$$

$$= \frac{1}{b}\sum_{n=-\infty}^{\infty} \delta(u, v - n/b) \qquad (1.44)$$

and is illustrated in Fig. 1.21(b), showing a series of delta functions ("spikes" not impulse "lines") with spacing $1/b$ along the v axis. [The integral in Equation (1.43) can be simplified by exchanging the order of the summation and integration by virtue of the fact that the integral is "well-behaved"; that is, it is not undefined (infinite).] By virtue of the rotational property of the Fourier transform of an impulse line, as discussed earlier, and by virtue of the result given in Equation (1.44), the Fourier transform of an impulse line array of arbitrary orientation θ is a set of impulses equally spaced on a line orthogonal to the direction of the sheets and passing through the origin.

Equipped with the above results we can now show how raster-scanned images can be described mathematically. An image scanned in a raster format can be represented by

22 TWO-DIMENSIONAL LINEAR SYSTEMS

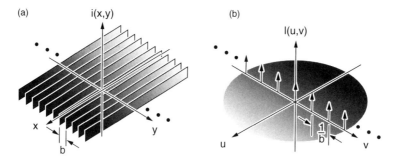

Figure 1.21 (a) Array of unit impulse lines defined by (1.42) and (b) corresponding Fourier transform defined by (1.44).

$$f_s(x, y) = i(x, y) f(x, y) \qquad (1.45)$$

where $f(x, y)$ is a continuous function representing the original image distribution. The Fourier transform of this scanned image is given by the convolution theorem, such that

$$F_s(u, v) = F(u, v) * I(u, v) \qquad (1.46)$$

which reduces to

$$F_s(u, v) = \frac{1}{b} \sum_{n=-\infty}^{\infty} F(u, v - n/b) \qquad (1.47)$$

where b must be $\geq v_{\max}/2$, where $v_{\max}$ is the maximum spatial frequency within $F(u, v)$ along the v axis. This requirement is just the Nyquist sampling theorem [since the function $f(x, y)$ is sampled along the y axis]. As a result of sampling along the y axis, the spectral content of $f(x, y)$, represented as a shaded region in the (u, v) plane shown in Fig. 1.22(a), is replicated as shown in Fig. 1.22(b). The Nyquist sampling criterion cited above is necessary and sufficient if $F(u, v)$ is circularly symmetric.

By extending Equation (1.42) to include impulse line arrays in both directions, we can introduce impulse arrays (or "bed of nails") defined by

$$s(x, y) = \sum_{m=-\infty}^{\infty} \sum_{n=-\infty}^{\infty} \delta(x - ma, y - nb) \qquad (1.48)$$

with corresponding Fourier transform

$$S(u, v) = \frac{1}{ab} \sum_{m=-\infty}^{\infty} \sum_{n=-\infty}^{\infty} \delta(u - m/a, v - n/b) \qquad (1.49)$$

which are illustrated in Figs. 1.23(a) and 1.23(b), respectively. Note that the spacings in the Fourier domain $(1/a, 1/b)$ are the reciprocal of the spacings in the original signal space (a, b). Note also that the expressions for $s(x, y)$ and $S(u, v)$ are made from the comb functions defined in Table 1.5 and illustrated in Fig. A.1 of Appendix A.

These results are useful in understanding the sampling of images characteristic of staring focal plane arrays. The sampling of a function $f(x, y)$ in x and y yields $f_{s2}(x, y) = f(x, y) s(x, y)$. The resulting Fourier transform $F_{s2}(u, v)$ is given by $F_{s2}(u, v) = F(u, v) * S(u, v)$. If this last equation is written out explicitly, we have

1.8 TWO-DIMENSIONAL IMPULSE FUNCTIONS: PROPERTIES AND FOURIER TRANSFORMS   23

Figure 1.22 (a) Original spectral region of support for $F(u, v)$ and (b) replicated versions of it $F_s(u, v)$.

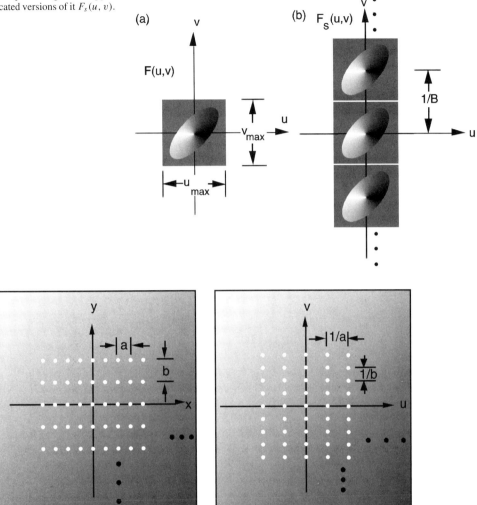

Figure 1.23 Plan views of (a) impulse array defined by (1.48) and (b) corresponding Fourier transform impulse array defined by (1.49).

$$F_{s2}(u, v) = \frac{1}{ab} \sum_{m=-\infty}^{\infty} \sum_{n=-\infty}^{\infty} F(m/a, n/b)\delta(u - m/a, v - n/b) \quad (1.50)$$

Equation (1.50) can also be written as

$$F_{s2}(u, v) = \frac{1}{ab} \sum_{m=-\infty}^{\infty} \sum_{n=-\infty}^{\infty} F(u - m/a, v - n/b) \quad (1.51)$$

indicating that $F(u, v)$ shown in Fig. 1.24(a) is replicated in both dimensions, as shown in Fig. 1.24(b). Thus the two-dimensional replicates are the "aliases" (or copies) of the original

24   TWO-DIMENSIONAL LINEAR SYSTEMS

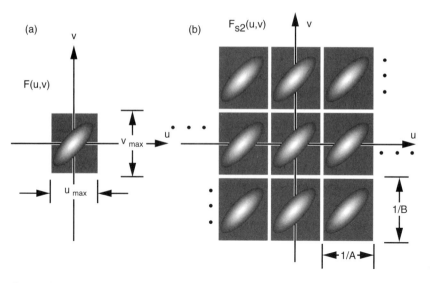

Figure 1.24 (a) Fourier transform of a continuous space-domain function $f(x, y)$ and (b) the Fourier transform of the corresponding sampled function $f_{s2}(x, y)$.

Nyquist-domain version. Again, the Nyquist criterion holds—namely, $1/a \geq u_{\max}/2$ and $1/b \geq v_{\max}/2$—and is a sufficient (but not necessary) condition for the recovery of $f(x, y)$ from $f_{s2}(x, y)$. If $F(u, v)$ is circularly symmetric, then the above criterion is both necessary and sufficient.

1.9 ELEMENTARY IMAGES AND THEIR FOURIER PROPERTIES

Now that we have described the basic principles of 2-D linear systems, it would be useful to consider some examples that illustrate key features of images and their Fourier properties. First consider a couple of simple synthetic images that illustrate the most basic Fourier properties of images. The first of these examples begins in the Fourier domain and shows how a simple "dipole" spectrum yields a "washboard" image for arbitrary orientation of the "dipole." The second example starts in the space domain and shows the subtle difference betweeen two "quadrupole" unit impulse arrays rotated with respect to each other. These examples should help to develop a sense of what to expect in interpreting more complex images developed by the optical Fourier transform, and how they might be filtered in the Fourier domain.

EXAMPLE 1.7 Inverse Fourier Transform of a Pair of Impulses   The illustration is as shown in Fig. 1.25 for a pair of impulses in the Fourier domain equally distant about the origin along a line of arbitrary orientation θ with respect to the u axis. To obtain the mathematical expression for this function, start with the simpler Fourier domain function $F(u, v)$ defined by

$$F(u, v) = \delta(u + a, v) + \delta(u - a, v)$$

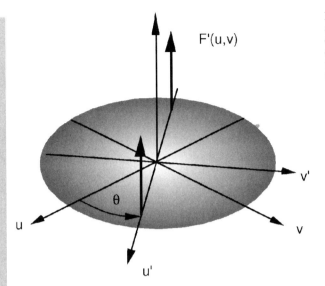

Figure 1.25 Pair of impulses oriented along a line at an arbitrary angle θ and equidistant from the origin, and the required rotation of coordinate system relative to the impulse pair to place them along the u' axis.

which is simply a pair of impulses along the u axis) and rotate the coordinate system by θ as shown. We obtain the desired impulse function pair: $F'(u, v) = \delta(u' + a, v') + \delta(u' - a, v')$ where

$$\begin{bmatrix} u' \\ v' \end{bmatrix} = \begin{bmatrix} \cos\theta & \sin\theta \\ -\sin\theta & \cos\theta \end{bmatrix} \begin{bmatrix} u \\ v \end{bmatrix} = \begin{bmatrix} u\cos\theta + v\sin\theta \\ -u\sin\theta + v\cos\theta \end{bmatrix}$$

Then the expression for the desired $F'(u, v)$ is

$$F'(u, v) = \delta(u\cos\theta + v\sin\theta + a, -u\sin\theta + v\cos\theta)$$
$$+ \delta(u\cos\theta + v\sin\theta - a, -u\sin\theta + v\cos\theta)$$

Thus the inverse Fourier transform is given by

$$f(x, y) = \int_{-\infty}^{\infty} \int_{-\infty}^{\infty} F(u, v) \exp[j2\pi(ux + vy)] \, du \, dv$$

To solve this integral substitute $F'(u, v)$ for $F(u, v)$ and note that $F'(u, v)$ is nonzero if and only if: $u\cos\theta + v\sin\theta \pm a = 0$ and $-u\sin\theta + v\cos\theta = 0$, which, when solved simultaneously yield: $u = \pm a\cos\theta$ and $v = \pm\sin\theta$. Therefore $F'(u, v)$ has impulse functions that "sift-out" the exponentials from the integral when $u = \pm a\cos\theta$ and $v = \pm a\sin\theta$. Thus,

$$f(x, y) = \exp[j2\pi a(x\cos\theta + y\sin\theta)] + \exp[-j2\pi a(x\cos\theta + y\sin\theta)]$$
$$= 2\cos[2\pi a(x\cos\theta + y\sin\theta)]$$

which is illustrated in Fig. 1.26. The value of $f(x, y)$ is a constant along any line defined by the relation $y = -x\cot\theta$.

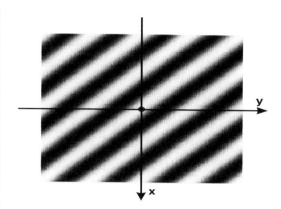

Figure 1.26 Inverse Fourier transform of a pair of impulses at angle $\theta = \pi/4$.

EXAMPLE 1.8 Fourier Transform of Delta Function Arrays Given the function

$$f(x, y) = \delta(x + x_0, y) + \delta(x - x_0, y) + \delta(x, y + y_0) + \delta(x, y - y_0)$$

illustrated in Fig. 1.27(a), the Fourier transform is given by

$$F(u, v) = \exp(j2\pi u x_0) + \exp(-j2\pi u x_0) + \exp(j2\pi v y_0) + \exp(-j2\pi v y_0)$$

Thus,

$$F(u, v) = 2\cos(2\pi u x_0) + 2\cos(2\pi v y_0)$$

which is a linear superposition of cosine waves, as illustrated in Fig. 1.27(b).

A similar array given by $g(x, y) = \delta(x + x_0, y + y_0) + \delta(x - x_0, y - y_0) + \delta(x - x_0, y + y_0) + \delta(x + x_0, y - y_0)$ is illustrated in Fig. 1.28(a). Its Fourier transform is given by $G(u, v) = 4\cos(2\pi u x_0)\cos(2\pi v y_0)$ and is illustrated in Fig. 1.28(b). The functional form is now a multiplication (or mixing) of two cosine waves. To distinguish these two results more clearly, note that the second array is equivalent to the first array rotated by 45° (apart from a scaling by $\sqrt{2}$). Similarly, by rotating the first result, $F(u, v)$, by 45° such that

$$\begin{bmatrix} u' \\ v' \end{bmatrix} = \frac{1}{2}\sqrt{2}\begin{bmatrix} 1 & 1 \\ -1 & 1 \end{bmatrix}\begin{bmatrix} u \\ v \end{bmatrix}$$

$F(u, v)$ can be transformed into $G(u, v)$. Furthermore, $F(u, v)$ is constant along a locus of points defined by $u = -v$, and $G(u, v)$ is constant along $u = $ constant (zero) or $v = $ constant (zero). Thus the difference between linear superposition and multiplication of these two spatial functions is expressed as a rotation between their corresponding Fourier transforms.

Next, consider some interesting but simple synthetic images and their corresponding Fourier transforms. We will consider both the real and imaginary parts, as well as the magnitude, of the Fourier transform.

1.9 ELEMENTARY IMAGES AND THEIR FOURIER PROPERTIES

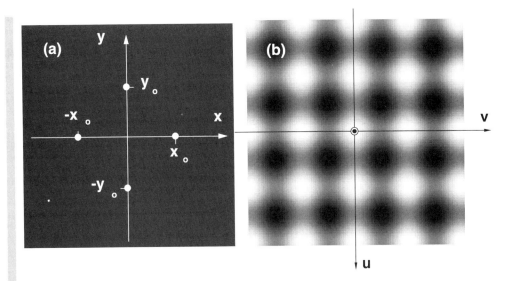

Figure 1.27 (a) Array of four impulses (forming a "plus" sign in plan view) and (b) corresponding Fourier transform.

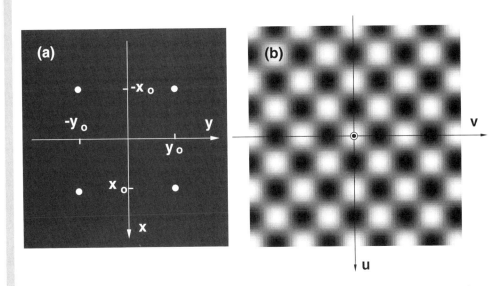

Figure 1.28 (a) Array of four impulses (forming an "X" in plan view) and (b) corresponding Fourier transform.

EXAMPLE 1.9 Two-Dimensional Fourier Transform: Examples of Simple Synthetic Images In this example we will consider several different object distributions and how each of them illustrates an important property of two-dimensional Fourier transforms. In the first case consider the rectangular box object distribution centered at the origin and offset from the origin as shown in Figs. 1.29(a) and 1.29(b). The magnitudes of their Fourier transforms are the same [Figs. 1.29(c) and 1.29(d)], but the real (or imaginary) parts [Figs. 1.29(e) and 1.29(f)] are not. A three-dimensional plot of the real part of the offset rectangle transform indicates a cosine modulation due to the separation of the center of the rectangle from the origin. The corresponding imaginary part has the sine modulation. This is clear from the equation

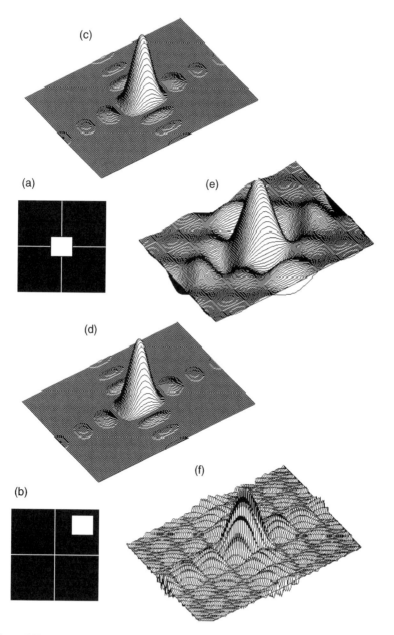

Figure 1.29 A rectangular pillbox located at the origin (a) and located offset from the origin (b), and corresponding Fourier transform magnitudes and real parts ((c), (d), (e) and (f)). This illustrates effect of shift on the Fourier transform.

$$\text{FT}\left\{\text{rect}\left(\frac{x-x'}{x_0}\right)\right\} = \exp(-j2\pi ux')\,\text{sinc}(\pi ux_0)$$
$$= [\cos(2\pi ux') - j\sin(2\pi ux')]\,\text{sinc}(\pi ux_0)$$

This is an important consequence of the shift theorem.

In the second example, two entirely different shapes, as indicated in Figs. 1.30(a) and 1.30(b), yield similar Fourier transform magnitudes [Figs. 1.30(c) and 1.30(d)]. In addition, by virtue of their symmetry about the origin, they have no imaginary parts.

As another interesting property of Fourier transform, consider the Fourier transform of the rectangular boundary shown in Fig. 1.30(b). Its Fourier transform magnitude is similar to that of the rectangle function except that the sidelobes are enhanced. This is approximately the result of differentiation, because the rectangle boundary is approximately the derivative

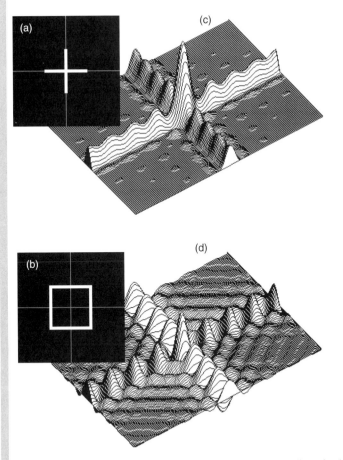

Figure 1.30 Two different shapes: (a) "plus" sign and (b) outline of a box and their corresponding Fourier transforms ((c) and (d)). Although their Fourier transforms are actually different, they might be confused under some circumstances.

(magnitude) of the rectangle function. As a result the Fourier transform magnitude is multiplied by a term linear in spatial frequency according to the differentiation property (given in Table 1.4 of Appendix A), namely,

$$|\text{FT}\{\partial f(x, y)/\partial x\}| = |2\pi u F(u, v)|$$

Thus, even though the sinc lobes go down, the product with spatial frequency u tends to enhance them for increasing values of u.

As a third example, consider the product of the Fourier transforms of two rectangle functions (i.e., two sinc functions) as shown in Figs. 1.31(a) and 1.31(b). The result is shown in Fig. 1.31(c), which has as its corresponding inverse Fourier transform the triangle function $\text{tri}(x/x_0)\,\text{tri}(y/y_0)$. This illustrates the convolution theorem.

An interesting property of some Fourier transforms is their approximate self-imaging property for some functions. The Gaussian is an example of this type of function where this relation holds exactly. Consider the function shown in Fig. 1.32(a), where this is only approximately true. The Fourier transform is shown in Fig. 1.32(b). Note also that the Fourier transform of this image is very close to the sum of the Fourier transforms of a plus sign and a multiplication sign, which demonstrates linear superposition.

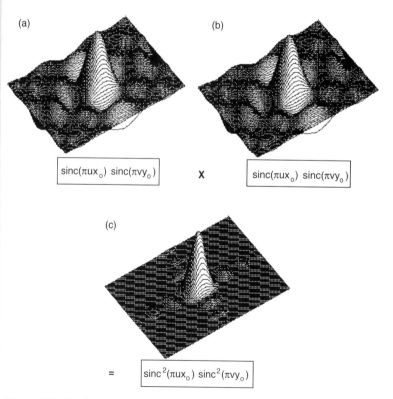

Figure 1.31 Fourier transform of a rectangular pillbox ((a) and (b)) multiplied to yield the Fourier transform corresponding to a triangle function in x and y axes, a result known from the convolution theorem.

1.9 ELEMENTARY IMAGES AND THEIR FOURIER PROPERTIES

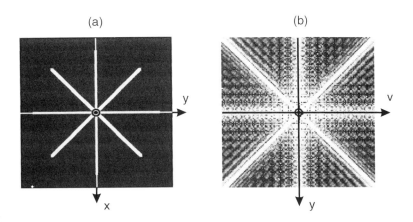

Figure 1.32 (a) Symmetric function consisting of a plus and "x" sign and (b) the corresponding Fourier transform, illustrating approximate self-imaging and linear superposition.

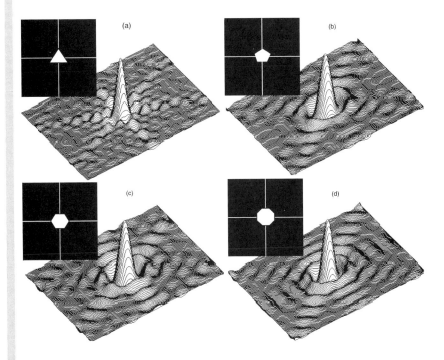

Figure 1.33 Fourier transforms of (a) a triangular prism, (b) pentagonal prism, (c) hexagonal prism, and (d) octagonal prism. In the limit as the number sides approaches infinity we obtain the jinc function for the Fourier transform.

Finally, consider the triangular prism pillbox shown in Fig. 1.33(a). For each of its three edges a sharp transition occurs versus radial distance from the center, yielding six radial spokes. As the number of sides increases to five [Fig. 1.33(b)], six [Fig. 1.33(c)], eight [Fig. 1.33(d)] (and so on), we see the Fourier transform begin to approximate the $J_1(x)/x$ function corresponding to the Fourier transfom of a circle.

In summary, a few major features of the Fourier transform of typical images will be described. If there is a large zero-frequency component in the Fourier transform $F(u, v)$, it is because the original image has a positive-definite value throughout the image domain (i.e., a "pedestal"). In fact, usually the magnitude of the Fourier transform $|F(u, v)|$ has a large amount of low-frequency content since most images have a background that varies gradually in intensity. The phase angle $\phi(u, v)$ of the Fourier transform often will vary much more rapidly than the magnitude, $|F(u, v)|$. If we write $F(u, v)$ in phasor notation,

$$F(u, v) = |F(u, v)| \exp[j\phi(u, v)] \tag{1.52}$$

then the inverse Fourier transform of just the $\exp[j\phi(u, v)]$ term usually contains a likeness of the original image $f(x, y)$. The inverse Fourier transform of $|F(u, v)|$, however, is often a low detail image. Fourier transform magnitudes are always symmetrical with respect to the frequency origin.

Each sharp edge in an image has a line in the Fourier domain orthogonal to the original edge. Periodic line patterns give rise to point pairs in the transform domain where the direction of the line joining the point pair is orthogonal to the direction of the periodic pattern, as illustrated in the previous examples.

PROBLEM EXERCISES

1. Prove the following relationships:

   (a) $$\text{comb}(x) = \lim_{N \to \infty} \left| \frac{\sin(N\pi x)}{\sin(\pi x)} \right|$$

   (b) $$\pi \delta(\sin \pi x) = \text{comb}(x)$$

   (c) $$\frac{\delta(\rho)}{\pi \rho} = 2\pi \int_0^\infty J_0(2\pi \rho r) r \, dr$$

2. Given the functions

$$f(x) = \text{rect}\left(\frac{x}{x_0}\right)$$

and

$$h(x) = \exp(-\alpha x) u(x)$$

calculate their (a) convolution and (b) cross-correlation. $u(x)$ is the unit step function. Plot the results for each case.

3. Determine the 2-D Fourier transform of (a) $\text{rect}(x/x_0)\text{rect}(y/y_0)$ and (b) $\text{tri}(x/x_0)\text{tri}(y/y_0)$ by direct computation. Use the convolution theorem to verify your result for part (b).

4. Given the functions

$$f_1(x) = \cos(k_0 x)$$

and

$$f_2(x) = \sin(k_1 x)/(\pi x)$$

find the Fourier transform of $f_1(x) f_2(x)$ and $f_1(x) * f_2(x)$.

5. Given the function

$$f(x, y) = \text{rect}\left(\frac{2x}{x_0}\right) \text{rect}\left(\frac{2y}{y_0}\right)$$

calculate and plot its Fourier transform $F(u, v)$. Then sketch the periodic function

$$g(x, y) = f(x, y) * \text{comb}\left(\frac{x}{x_0}\right) \text{comb}\left(\frac{y}{y_0}\right)$$

and calculate and plot its Fourier transform $G(u, v)$. Express $G(u, v)$ in terms of $F(u, v)$ for specific values of (u, v).

6. Consider a function $f(x, y)$ defined in polar coordinates such that

$$f(x, y) \to f(r\cos\theta, r\sin\theta) = u(r, \theta) = s(\mathbf{r})$$

Write out the convolution operation $s(\mathbf{r}) * \delta(\mathbf{r})$ explicitly and show that

$$s(\mathbf{r}) * \delta(\mathbf{r}) = s(\mathbf{r})$$

where bold $\mathbf{r}$ denotes a vector.

7. Calculate the 2-D Fourier transform of (a) the function $[\pi(x^2 + y^2)]^{-1/2}$ and (b) the function $\cos(2\pi xy/ab)$. What general property do these functions possess?

8. Show that the 2-D Fourier transform of a comb function is a comb function. What is the relationship between the spacing of the unit impulses of each comb?

9. Given the plane wave $\exp(jky)$ propagating in the positive y direction, express it in cylindrical coordinates and expand it into a Fourier series in θ.

10. Calculate the Fourier transform $G(\rho, \phi)$ of the function $g(r, \theta) = \text{circ}(r)\cos\theta$. Plot both $g(r, \theta)$ and $G(\rho, \phi)$.

11. Calculate the Hankel transform of the $1 - \text{circ}(r/r_0)$ and interpret the result. Sketch the function and its Hankel transform.

12. Determine the effect of a shift of θ' on the Fourier–Bessel transform of a function $f(r, \theta)$.

13. Calculate the combined effect of a scaling (by a) and rotation (by θ) on the Fourier transform of the function $f(x, y)$, where $f(x, y)$ is a 2-D rect function. Use the software tools in Appendix C. Sketch the result in plan view for each effect separately and combined.

14. For the anamorphic transformation of a point object, verify Equation (1.28).

15. Show explicitly that $\mathbf{A}^{-1}\mathbf{x} \cdot \mathbf{u} = \mathbf{x}^T \cdot \mathbf{A}_T^{-1}\mathbf{u}$ is true for

$$A = \begin{bmatrix} a_{11} & a_{12} \\ a_{21} & a_{22} \end{bmatrix}$$

16. Use the frequency shift, linearity, and rotational properties of the Fourier transform to get the result of Example 1.7.

17. Calculate the Hankel transforms of (a) $\cos(\pi r^2)$, (b) $\exp(-\pi r^2)$, and (c) $\exp(-ar)$. Sketch the function and its Hankel transform.

18. Evaluate the 2-D convolution:

$$g(r) = \delta(r-a) * \delta(r-b)$$

for the case of $a = b$. Plot the result versus r using MATLAB tools in Appendix C.

19. Show that the relation

$$\text{FT}\{\delta(-x'\sin\theta + y'\cos\theta)\} = \delta\{u'\cos\theta + v\sin\theta\}$$

is consistent with the linear transformation theorem

$$F_1(u', v') = |\det \mathbf{A}|^{-1} F[\mathbf{A}_T^{-1}(u', v')]$$

20. Use the moment generating function for $J_n(a)$, namely,

$$\exp\left[\frac{a}{2}\left(t - \frac{1}{t}\right)\right] = \sum_{n=-\infty}^{\infty} J_n(a) t^n$$

with the proper substitution for t, to demonstrate that

$$J_0(a) = \frac{1}{2\pi} \int_0^{2\pi} \exp[-ja\cos(\theta - \phi)] \, d\theta$$

21. Calculate the convolution and correlation of a rectangular pillbox of width x_0 and length y_0 with a finite-width, finite-length ramp function in the x axis defined by

$$f(x, y) = (x/x_0) \text{rect}[(x - x_0/2)/x_0] \text{rect}[(y - y_0/2)/y_0]$$

where the components of $f(x, y)$ are single-sided (not symmetrical about the origin). Plot this function using MATLAB. Calculate the convolution and correlation results in the space domain and in the spatial frequency domain using the convolution and correlation theorems, respectively. You may use the MATLAB tools in Appendix C. What happens when the rectangular pillbox is offset from the origin? What happens when the rectangular function becomes small or large with respect to the ramp function; that is, x_0 and y_0 of the rectangular function differ from x_0 and y_0 of the ramp function?

22. Using MATLAB calculate the 2-D convolution of the edge-enhanced verision of a rectangular pillbox with itself as shown in Fig. 1.30(b).

23. Using MATLAB calculate the 2-D convolution and correlation of the triangular pillbox shown in Fig. 1.33(b).

24. Using MATLAB calculate the 2-D Fourier transform of the triangular pillbox of Fig. 1.33(b).

BIBLIOGRAPHY

R. N. Bracewell, *The Fourier Transform and Its Applications*, McGraw-Hill, New York (1978).
K. R. Castleman, *Digital Image Processing*, Prentice-Hall, Englewood Cliffs, NJ (1979).
J. D. Gaskill, *Linear Systems Fourier Transforms and Optics*, Wiley-Interscience, New York (1978).
C. D. McGillem and G. R. Cooper, *Continuous and Discrete Signal and System Analysis*, Holt, Rinehart and Winston, New York (1984).
P. M. Woodward, *Probability and Information Theory, with Applications to Radar*, McGraw-Hill, New York (1955).

Chapter 2
STOCHASTIC PROCESSES AND NONLINEAR SYSTEMS

2.1 BASIC CONCEPTS OF STOCHASTIC PROCESSES

Why are random processes an important consideration in optical signal processing? Because there's always noise present in real systems, whether in the form of fluctuations in the photon arrival time (photon shot noise), fluctuations of the current in the load resistor following the detector (Johnson noise), or additional noise in the preamplifiers that contributes to an increased noise figure for the overall system.

From a purely theoretical standpoint, noise also presents a fundamental problem. Noise signals are not periodic; hence we would need infinite time to give them a complete description. But infinite signals don't have finite energy, hence the Fourier transform will not converge. What can we do about this? We can truncate all realizations of $X(t)$ to a finite interval T, which is equivalent to what happens spatially when we use finite apertures in optical architectures. Then we can take time averages whenever appropriate, which is what we do after detection in an optical processor. (In time-domain signal processing we then may take the limit as T approaches infinity.)

In addition to characterizing noise in terms of its magnitude relative to the signal [expressed as the signal-to-noise ratio (SNR)], higher-order statistics are needed to characterize the spatial and or temporal structure of noise-like (or stochastic) processes. This is provided by the correlation function introduced earlier. To arrive at a complete understanding of correlation, we need to understand the basics of statistics. For our purposes it will be sufficient to grasp first- and second-order statistics. Random processes are described by random variables in terms of their associated probability distributions and probability density functions, which are essentially the basis of first- and second-order statistics.

If X is a random variable, several "realizations" (or examples) of X denoted by $\{x_k(t)\}$ can be depicted as shown in Fig. 2.1. This picture represents an ensemble of a random process, which consists of specific realizations of that process. $X(t)$ represents many possible outcomes of an experiment conducted repeatedly. The set of all functions that make up $X(t)$ is an ensemble; a particular realization from a given $X(t)$—that is, $x_k(t)$—is unique. The cumulative distribution function (CDF) is the probability that a given value of some random variable $X \leq \theta$ will occur and is denoted by

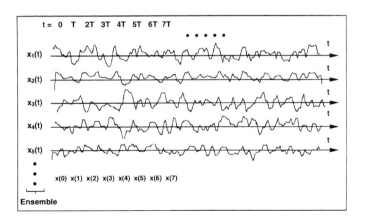

Figure 2.1 Several realizations of a random process (or time series) presented in a format suggesting that finite segments form an ensemble.

$$F_X(x_k) = P(X \leq \theta) \tag{2.1}$$

for a particular realization of X denoted by θ. The two basic asymptotic properties of $F_X(\theta)$ are

$$\lim_{\theta \to -\infty} F_X(\theta) = 0 \tag{2.2}$$

and

$$\lim_{\theta \to \infty} F_X(\theta) = 1 \tag{2.3}$$

In addition, $F_X(\theta)$ is monotonically increasing with θ. The probability density function (PDF) is then defined as

$$f_X(\theta) = \frac{dF_X(\theta)}{d\theta} \tag{2.4}$$

where $f_X(\theta)$ is non-negative and

$$\int_{-\infty}^{\infty} f_X(\theta)\, d\theta = 1 \tag{2.5}$$

The statistical moments (e.g., mean and mean-square) correspond to expectation values of the random variable x for weighting given by the PDF. Thus,

$$E\{x\} = \int_{-\infty}^{\infty} x f_X(\theta)\, d\theta = \bar{x} \tag{2.6}$$

is the mean value and

$$E\{x^2\} = \int_{-\infty}^{\infty} x^2 f_X(\theta)\, d\theta = \overline{x^2} \tag{2.7}$$

is the mean-square value. The variance denoted by var(x) or σ_x^2 is given by

$$\sigma_x^2 = E\left\{[x - E(x)]^2\right\} = E\{x^2\} - (E\{x\})^2 \tag{2.8}$$

where the so-called standard deviation σ_x is the square root of the variance.

The aforementioned properties correspond to the so-called first-order statistics. The second-order statistics, which include correlation, depend on the so-called joint probability distribution function defined as

$$F_{XY}(x, y) = P(X \leq x \text{ and } Y \leq y) \tag{2.9}$$

such that $F_{XY}(-\infty, -\infty) = 0$, $F_{XY}(\infty, \infty) = 1$, and F_{XY} is monotonically increasing with x and y. The joint probability distribution function $F_{XY}(x, y)$ is appropriate when comparing two random variables X and Y, which share a statistical relationship. This is in direct analogy with first-order statistics. Furthermore, the joint density function is defined as

$$f_{XY}(x, y) = \frac{\partial^2 F_{XY}(x, y)}{\partial x \partial y} \tag{2.10}$$

such that the probability that X and Y lie within the space R is given by

$$P(X, Y \leq R) = \int\int_R f_{XY}(x, y) \, dx dy \tag{2.11}$$

We can get an intuitive feel for what $f_{XY}(x, y)$ represents by the following illustrations and discussion. Consider a random process represented by a time series illustrated in Fig. 2.2(a), and break it up into N (e.g., 256) segments each of length T, and treat each as a distinct realization of an ensemble as shown in Fig. 2.2(b). Sample each individual waveform (or realization) into M (e.g., 256) points and digitize each point to a prescribed number of bits up to a maximum of, say, 8 bits. Display this multichannel array of points as an image as shown in Fig. 2.2(c). Each pixel in this digital image representing the original continuous-time and continuous-amplitude signal has a specific intensity from 0 to 255. (Bipolar signal values have been implicitly assumed to be shifted up in amplitude by a fixed amount to create a non-negative image.) Imagine this original gray-level image as an optical transparency. Now copy it and overlay it on top of the original. For perfect registration, count the co-occurrence of pixel pairs. For example, pixel pairs at coordinates (1, 2), (5, 9), (16, 23) might all have an intensity value of 50. The number of co-occurrences of 50 will thus be counted, regardless of their location, as 3. This co-occurrence number will be plotted versus the intensity value for each image, where for one image it is 50 (x coordinate) and for the other image it is also 50 (y coordinate) because they are the same image. Plotting all such cases yields the isometric plot of Fig. 2.3, which is termed the 2-D (co-occurrence) histogram. If we normalize this plot by the total volume under the surface we obtain the joint PDF. In plan-view it would look like a straight line of varying intensity. Because it *is* a straight line, we have perfect correlation between the two data sets (or images). Of course, they are the same data! If for any reason one set is not identical to the other (due to a spatial offset or just being different data), then the 2-D co-occurrence histogram would be smeared-out as shown in Fig. 2.4(b). If points in Fig. 2.4(b) occurred only once, then the data could represent the correlogram of data taken on one set of measurements. These points could be fit by a straight line of slope m and intercept b to yield $y = mx + b$, with a correlation coefficient ρ. For perfectly correlated images (or data), $\rho = 1$. For totally uncorrelated images, $\rho = 0$. When both signal ensembles are Gaussian—that is, they have first-order Gaussian PDFs—their normalized frequency of co-occurrence is bivariate Gaussian; that is,

$$f_{XY}(x, y) = \frac{1}{2\pi \sigma_x \sigma_y \sqrt{1 - \rho^2}}$$
$$\times \exp\left\{\frac{-1}{2[1 - \rho^2]} \left[\frac{(x - \mu_x)^2}{\sigma_x^2} - \frac{2\rho(x - \mu_x)(y - \mu_y)}{\sigma_x \sigma_y} + \frac{(y - \mu_y)^2}{\sigma_y^2}\right]\right\} \quad (2.12)$$

If x and y are statistically independent, this reduces to the product of two Gaussians, namely,

$$f_{XY}(x, y) = \frac{1}{\sqrt{2\pi} \sigma_x} \exp\left[\frac{-(x - \mu_x)^2}{2\sigma_x^2}\right] \frac{1}{\sqrt{2\pi} \sigma_y} \exp\left[\frac{-(y - \mu_y)^2}{2\sigma_y^2}\right] \quad (2.13)$$

Now, it is clear that if $f_{XY}(x, y)$ is integrated over x we get $f_Y(y)$ or vice versa—that is, the marginal PDFs (or first-order statistics), that are normalized histograms shown as shadows or projections in Fig. 2.3.

For the above discussion it is easy to see that correlation emerges naturally as a statistical concept when two data sets (or images) are compared. As it turns out, there is more than one way to determine the correlation of data: by ensemble average or time average. To understand these two types of averages we have to define two types of statistical processes: stationary and ergodic. Stationary processes are processes that yield the same distribution functions $f_X, f_{XY}, f_{XYZ}, \ldots$ regardless of the time (or space) origin where the set

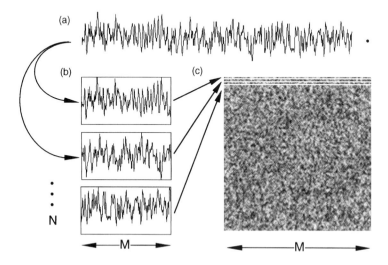

Figure 2.2 (a) Time serial signal, (b) its segmented form suggesting an ensemble of signals, and (b) a gray-level equivalent in an image format.

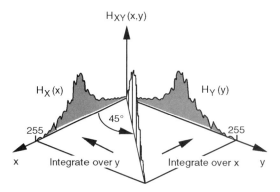

Figure 2.3 Two-dimensional frequency of co-occurrence (histogram) for two random variables which are exactly the same sample but derived from a randomly distributed set of values. Note marginal distributions (1-D histograms) on the projected planes.

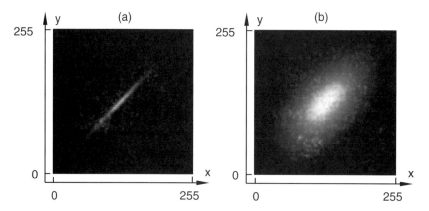

Figure 2.4 Plan views of (a) the 2-D histogram for similar (correlated) signals (or images) with gray scales ranging over 8-bits (256 gray levels), and (b) the 2-D histogram of two different (largely uncorrelated) signals (or images) each of which are randomly distributed in space.

of random variables $X, Y, Z, \ldots$ are evaluated—that is, regardless of whether $x(t_1), x(t_2), \ldots$ or $x(t_1 + \tau), x(t_2 + \tau), \ldots$ are used. If this is true for all orders of f (i.e., $f_X, f_{XY}, f_{XYZ}, \ldots$), then we have "strict stationarity." If it's true to second order in f (i.e., f_{XY}), then we have so-called "wide sense stationarity," which is what usually suffices in most cases. The relevant statistical measures (mean, mean square, correlation, etc.) are therefore independent of time (or space) origin. An ergodic process is simply a random process where the time (or space) average (over a given experiment) is equal to the ensemble average. An ergodic process is necessarily stationary, but a stationary process is not necessarily ergodic.

We will usually deal with "wide-sense" stationary processes. They are completely described by the following three parameters:

1. mean $= \mu_x = \langle x(t) \rangle =$ constant,
2. mean-square value $= x^2 = \langle x^2(t) \rangle =$ constant, and
3. autocorrelation function $= R_{xx}(t_1, t_2) = R_{xx}(\tau)$, where $\tau = t_1 - t_2$ so that $R_{xx}(\tau)$ is independent of the time origin.

Two processes are jointly stationary if

$$E\{x(t+\tau)y(t)\} = R_{xy}(\tau) \tag{2.14}$$

where $R_{xy}(\tau)$ is the cross-correlation. In most engineering applications, the above three quantities are sufficient in lieu of the second-order (joint) PDF.

In general, if measurements x and y are taken and the quantities $x - \mu_x$ and $x - \mu_y$ are computed, the correlation is a useful quantity to use and is equal to the covariance (because the mean has been removed). For most trials, is greater than $(y - \mu_y)$ or less than 0 when $(x - \mu_x)$ is greater than or less than 0 such that the product $(x - \mu_x)(y - \mu_y) > 0$, always. Then the average of $(x - \mu_x)(y - \mu_y)$ is the covariance (or correlation), and x and y are said to be correlated when it is greater than 0, uncorrelated when it is equal to 0, and anticorrelated when it is less than 0. Denoting the covariance by C_{xy}, we obtain

$$C_{xy} = \overline{xy} - \overline{x\overline{y}} - \overline{\overline{x}y} + \overline{\overline{x}\,\overline{y}} = \overline{xy} - \overline{x}\,\overline{y} \tag{2.15}$$

Thus x and y are not correlated if $\overline{xy} = \overline{x}\,\overline{y}$, and independent random variables are not correlated. However, the converse is not true; that is, uncorrelated random variables are not necessarily independent. Thus, independence is a stronger condition than uncorrelatedness; that is,

$$\overline{f_1(x)f_2(y)} = \overline{f_1(x)}\,\overline{f_2(y)} \tag{2.16}$$

as a requirement for independence is stronger than: $\overline{xy} = \overline{x}\,\overline{y}$ as a requirement for uncorrelatedness.

The correlation coefficient mentioned earlier in conjunction with the bivariate Gaussian joint PDF is related to the covariance above by the relation

$$\rho_{xy} = \frac{C_{xy}}{\sigma_x \sigma_y} \tag{2.17}$$

The ensemble average of the quantity $(x - \mu_x)(x - \mu_y)$ is expressed formally as

$$C_{xy} = \int_{-\infty}^{\infty} \int_{-\infty}^{\infty} (x - \overline{x})(y - \overline{y}) P_{XY}(x, y)\, dx\, dy \tag{2.18}$$

If we have a stationary, ergodic, and zero mean process, then C_{xy} reduces to the crosscorrelation R_{xy} and can be expressed as a time (or space) average:

$$R_{xy} = \lim_{T \to \infty} \frac{1}{T} \int_0^T X(t)Y(t+\tau)\, dt \tag{2.19}$$

When the mean is not zero, then

$$C_{xy}(\tau) = R_{xy}(\tau) - \mu_x \mu_y \tag{2.20}$$

Similarly for the autocovariance

$$C_{xx}(\tau) = R_{xx}(\tau) - \mu_x^2 \tag{2.21}$$

assuming real data in all cases.

2.2 FUNDAMENTAL PROBABILITY DENSITY FUNCTIONS

Probability density functions are the asymptotically smooth estimates of the normalized frequency of occurrence (histograms) of data collected from experiments. As such, they are not really achieved in practice because for the least frequent occurrences, which make up the "tails" of the distribution, we must wait a long time and collect a lot of data to get a good estimate of the distribution. There are, however, several useful models for probability density functions, which seem to be appropriate for descriptions of noise in signal processing systems. Starting with the most fundamental of these, the binomial distribution, we can derive most of the more common of these distributions.

The binomial distribution is a discontinuous probability density function appropriate for describing the classic coin flip experiment, where only two possible outcomes can occur: heads or tails, with equal probabilities (1/2). For the roll of dice with six possible outcomes per die, it also applies. The binomial distribution is also useful in describing electron fluctuation statistics in image tubes, such as microchannel plate image intensifiers. It is also appropriate for calculating the probability of detection in radar systems for multiple radar scans. The probability of occurrence of an event is p, and the probability that a contrary event occurs is $q = 1 - p$. If the number of ways the event can occur is defined by the number of combinations we can form from n objects taken k at a time, then that quantity is

$$\frac{n!}{k!(n-k)!}$$

The probability of each outcome is $p^k q^{n-k}$, so that the probability distribution for a particular value of $k = r$ is

$$\rho(r) = \sum_{k=1}^{n} \frac{n!}{k!(n-k)!} p^k q^{n-k} \delta(r - k) \qquad (2.22)$$

The mean value for this distribution is given by $\mu_r = np$ and the variance is $\sigma_r^2 = npq$. In the limit of large n the binomial distribution becomes the Gaussian distribution defined by

$$p(x) = \frac{1}{\sqrt{2\pi}\sigma} \exp\left[-(x-\mu)^2/2\sigma^2\right] \qquad (2.23)$$

which is completely defined by its first and second moments, or mean (μ) and (equivalently) variance (σ^2), respectively. It is plotted in Fig. 2.5. It is the projection (or marginal distribution) for the bivariate Gaussian defined by

$$p(x, y) = \frac{1}{2\pi \sigma_x \sigma_y \sqrt{1-\rho^2}} \exp\left[\frac{\frac{x^2}{\sigma_x^2} + \frac{y^2}{\sigma_y^2} - \frac{2\rho x y}{\sigma_x \sigma_y}}{-2(1-\rho^2)}\right] \qquad (2.24)$$

where $\mu_x = \mu_y = 0$ for simplicity, and ρ is the correlation coefficient. [Recall the discussion in Section 2.1 leading up to Equation (2.13).]

The Gaussian distribution describes thermal noise in transistors and detectors, and it is a widely used model for noise in electronics and systems analysis. It is also the distribution of the sum of many random variables with arbitrary distributions (i.e., central limit theorem).

2.2 FUNDAMENTAL PROBABILITY DENSITY FUNCTIONS

In the limit of large $n (n \to \infty)$ and for $p \to 0$ such that the product $np = m$ is a constant, the binomial distribution approaches the Poisson distribution, defined by

$$p(x) = \sum_{k=1}^{n} \frac{m^k e^{-m}}{k!} \delta(x - k) \qquad (2.25)$$

and plotted in Fig. 2.6 for the continuous random variable, where the mean equals the variance and is defined as $m (> 0)$. This distribution describes random arrival event probabilities for photon detection and electron emission from hot cathodes, as well as reliability statistics for components (such as mean time between failures). From the standpoint of optical processing, the Poisson and Gaussian models are the most frequently exploited to characterize noise in photodetectors, where Poisson statistics underlie shot noise from incident photons and Gaussian statistics underlie thermal (Johnson) noise in resistors.

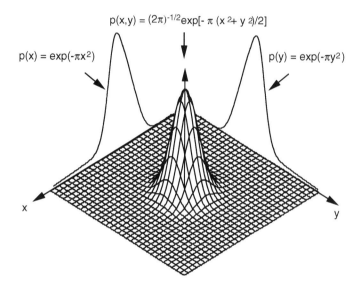

Figure 2.5 Bivariate Gaussian probability density function (asymptotic model of two-dimensional frequency of co-occurrence histogram for a Gaussian random process), showing marginal distributions along x and y-axes for $\mu_x = \mu_y = 0$, $\sigma_x = \sigma_y = 1$.

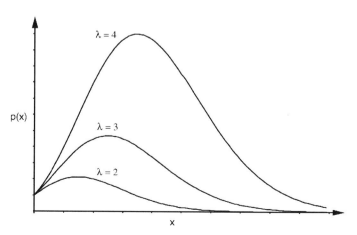

Figure 2.6 Continuous Poisson probability density function for varying parameter values ($m = 2, 3, 4$).

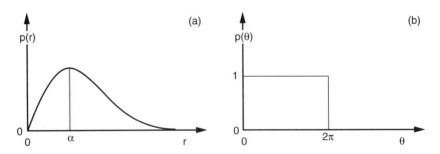

Figure 2.7 (a) Rayleigh probability density function, and b) uniform probability density function, which, when multiplied together, are equivalent to the bivariate Gaussian in polar coordinates under conditions specified on pages 44 and 45.

It is interesting to note that the model probability distributions change when the random variable (x) undergoes a single-valued transformation of the form $y = f(x)$. In general, the relationship between the probability distributions for x and y is

$$p(y) = p(x) \left| \frac{dx}{dy} \right| \tag{2.26}$$

If $y = f(x)$ is linear, then the form of the new probability distribution is left unchanged. For instance, if $y = ax + b$, then a Gaussian transforms to a Gaussian in which $\mu_y = a\mu_x + b$ and $\sigma_y^2 = a^2 \sigma_x^2$. For multiple random variables, we must replace the derivative in Equation (2.26) with the Jacobian of the transformation. That the Jacobian arises can be understood if it is remembered that the *probability* must be equal before and after the transformation; that is,

$$\int \int_A P_{XY}(x, y) \, dx \, dy = \int \int_{A'} P_{X'Y'}(x', y') \, dx' \, dy' \tag{2.27}$$

where the transformation is now a vector transformation in two dimensions $x' = f(x)$, where $\mathbf{x}' = (x', y')^T$ and $\mathbf{x} = (x, y)^T$. Thus the Jacobian, given by

$$J = \begin{vmatrix} \frac{\partial x}{\partial x'} & \frac{\partial x}{\partial y'} \\ \frac{\partial y}{\partial x'} & \frac{\partial y}{\partial y'} \end{vmatrix} \tag{2.28}$$

is used in Equation (2.27) to yield

$$\int \int_A P_{XY}(x, y) \, dx \, dy = \int \int_A P_{XY}(x, y) |J| \, dx \, dy \tag{2.29}$$

This result can be used to understand one of the fundamental cases of multidimensional probability distribution functions. For instance, the bivariate Gaussian distribution function [Equation (2.24)] can be reexpressed in polar coordinates such that $(x, y) \to (r, \theta)$. If we assume that the bivariate distribution is separable [i.e., $p(x, y) = p(x)p(y)$, which corresponds to assuming that $\rho = 0$] and we assume that $\sigma_x = \sigma_y$, then the new probability density function is $p(r, \theta) = p(r)p(\theta)$, where

$$p(r) = \frac{r}{\alpha^2} \exp(-r^2/2\alpha^2) u(r) \tag{2.30}$$

is the Rayleigh distribution with mean $= \mu_r = (\pi/2)^{1/2}\alpha$ and variance $= \sigma_r^2 = (2-\pi/2)\alpha^2$, and

$$p(\theta) = \frac{1}{2\pi} \text{rect}\left(\frac{\theta - \pi}{2\pi}\right) \quad (2.31)$$

is the uniform distribution with mean $= \mu_\theta = \pi$ and variance $= \sigma_\theta^2 = \pi^2/12$. In this case the magnitude of the Jacobian is

$$|J| = \begin{vmatrix} \cos\theta & r\sin\theta \\ -\sin\theta & r\cos\theta \end{vmatrix} = r \quad (2.32)$$

so that the differential element of area $dxdy$ becomes $rdrd\theta$.

These last two distributions are plotted in Figs. 2.7(a) and (b). The Rayleigh distribution describes the amplitude distribution of the vector sum of a large number of independent two-dimensional vectors in which it is not necessary that the amplitudes of the components of the vectors follow any specific distribution. It also describes the amplitude of the electromagnetic field scattered from a large number of scatterers as well as the amplitude of noise in an AM receiver when no signal is present. The uniform distribution function also happens to be a good model for phase noise in electronic systems.

2.3 MATCHED FILTER DERIVATION AND PROPERTIES OF CORRELATION

The most basic property of the matched filter (or correlator) is that it yields the maximum output signal-to-noise ratio (SNR) when the input signal is corrupted by additive white (spectrally flat) Gaussian noise. Other criteria besides output SNR can be used, such as maximum likelihood, to obtain the form of the matched filter, but they are beyond the scope necessary for our purposes. As it turns out, the matched filter will be optimum under more general conditions—that is, for colored (spectrally nonflat) and non-Gaussian noise.

Consider the derivation of the matched filter based on maximizing the output SNR, given by

$$\text{SNR} = \frac{s_0^2(t)}{\overline{n_0^2(t)}} \quad (2.33)$$

If we assume that the transfer function of the optimal filter is denoted by $H(\omega)$ and the input signal spectrum is $S(\omega)$, then

$$s_0(t) = F^{-1}\{H(\omega)S(\omega)\} \quad (2.34)$$

$$s_0(t) = \frac{1}{2\pi}\int_{-\infty}^{\infty} H(\omega)S(\omega) \exp(j\omega t)\, d\omega \quad (2.35)$$

Assuming a noise model whereby stationary white noise with spectrum $S_n(\omega) = N/2$ is added to the input signal, then the variance of the output noise is

$$\overline{n_0^2(t)} = \frac{1}{2\pi}\int_{-\infty}^{\infty} |H(\omega)|^2 \left(\frac{N}{2}\right) d\omega \quad (2.36)$$

when no signal is present. Therefore,

$$\text{SNR} = \frac{\left|\int_{-\infty}^{\infty} H(\omega)S(\omega)\exp(j\omega t)\, d\omega\right|^2}{\pi N \int_{-\infty}^{\infty} |H(\omega)|^2\, d\omega} \qquad (2.37)$$

Now let's use the Schwarz ineqality:

$$\left|\int_{-\infty}^{\infty} F_1(\omega)F_2(\omega)\exp(j\omega t)\, d\omega\right|^2 \leq \int_{-\infty}^{\infty} |F_1(\omega)|^2\, d\omega \int_{-\infty}^{\infty} |F_2(\omega)|^2\, d\omega \qquad (2.38)$$

Then, if $F_1(\omega) = H(\omega)$ and $F_2(\omega) = S(\omega)\exp(j\omega t)$, we obtain

$$\left|\int_{-\infty}^{\infty} H(\omega)S(\omega)\exp(j\omega t)\, d\omega\right|^2 \leq \int_{-\infty}^{\infty} |H(\omega)|^2\, d\omega \int_{-\infty}^{\infty} |S(\omega)|^2\, d\omega \qquad (2.39)$$

Therefore,

$$\text{SNR} \leq \frac{1}{\pi N} \int_{-\infty}^{\infty} |S(\omega)|^2\, d\omega \qquad (2.40)$$

The SNR is maximum only if the Schwarz *equality* holds, which is possible only if

$$H(\omega) = KS^*(\omega)\exp(-j\omega t_0) \qquad (2.41)$$

or

$$H(\omega) = KS(-\omega)\exp(-j\omega t_0) \qquad (2.42)$$

where K is an arbitrary constant. Therefore we obtain

$$h(t) = \text{FT}^{-1}\{H(\omega)\} = \text{FT}^{-1}\{KS(-\omega)\exp(-j\omega_0 t)\} = Ks^*(t - t_0) \qquad (2.43)$$

Thus the matched filter impulse response, $h(t)$, is a scaled, shifted, and time-reversed replica of the input signal, hence the name "matched filter." The schematic diagram of the matched filter process (with an additive noise model) is shown in Fig. 2.8. With this figure in mind we can interpret the process of matched filtering with the following discussion. Given that the input signal is $r(t) = s(t) + w(t)$, where $w(t)$ is additive white noise and the impulse response $h(t) = s(-t)$, then the output $y(t)$ is given by $y(t) = R_{rs}(t)$ or

$$\begin{aligned} R_{rs}(t) &= r(t) \otimes s(-t) \\ &= \int_{-\infty}^{\infty} s(-\tau) r(t - \tau)\, d\tau \\ &= \int_{-\infty}^{\infty} s(\tau') r(t + \tau')\, d\tau' \\ &= \int_{-\infty}^{\infty} s(\tau') [s(t + \tau') + w(t + \tau')]\, d\tau' \end{aligned} \qquad (2.44)$$

where $\tau' = -\tau$. Therefore, $R_{rs}(t) = R_{ss}(t) + R_{nn}(t)$, where $R_{ss}(t)$ is the correlation peak and $R_{nn}(t)$ are the sidelobes, as shown in Fig. 2.9.

To determine whether a signal is present or not, we must look at the statistics of the output (correlation) signal and apply simple detection theory. Even if the signal is not present, noise with a Gaussian distribution will be present, which we will assume has zero mean and unit variance for simplicity. (Real correlation values are unipolar, not bipolar, as assumed here, although real signals can be bipolar.) Under these circumstances the received signal is $r(t) = 0 + w(t)$. The "null" signal hypothesis would be denoted H_0 such that $r(t) = w(t)$ with conditional probability:

$$P\{r|H_0\} = \frac{1}{\sqrt{2\pi}} \exp\left(\frac{-r^2}{2}\right) \tag{2.45}$$

where $P\{r|H_0\}$ is the conditional probability and reads "probability of r given the hypothesis that only noise is present." Likewise, if the signal is also present, the conditional probability is

$$P\{r|H_1\} = \frac{1}{\sqrt{2\pi}} \exp\left[-\frac{(r-1)^2}{2}\right] \tag{2.46}$$

(which reads "probability of r given hypothesis that signal and noise are present"). Using the maximum likelihood criterion, the likelihood ratio is

$$\lambda(r) = \frac{P\{r|H_0\}}{P\{r|H_1\}} = \exp\left[-\left(r - \frac{1}{2}\right)\right] \tag{2.47}$$

The decision rule is then: Choose hypothesis H_1, if $\exp(r - \frac{1}{2}) > T$, where T is a threshold on the correlation function. For the log-likelihood ratio this becomes $r \geq \frac{1}{2} + \ln T$, where r is the measured (or received) signal. Above T we should choose the hypothesis H_1—that is, that the signal is present. Below T we choose H_0—that is, that there is no signal present.

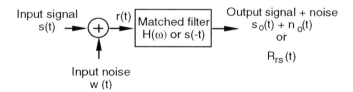

Figure 2.8 Schematic block diagram of matched filter.

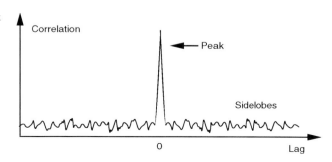

Figure 2.9 Output signal of a matched filter showing peak at location of signal and sidelobes corresponding to "noise".

Graphically the situation is depicted as shown in Fig. 2.10, where the classic Type II error (i.e., the probability of false alarm, P_{FA}) is the area under the tail of the curve $P\{h_0|H_0\}$ above the threshold T—that is, the probability that a signal is declared present when, in fact, only noise is actually present. The probability of missed detection is the area under the curve $P\{h_1|H_1\}$, below T. Clearly, T depends on the sidelobe variance; that is, $\sigma_{nn^2} = R_{nn}(\tau)$ for $\tau \neq 0$, as depicted in Fig. 2.9.

In completing this review of correlation we will summarize some of the basic properties of correlation.

1. $R_{xx}(\tau)$ is an even function of τ with a zero-lag value $R_{xx}(0) = \sigma_x^2$.
2. That $R_{xx}(\tau)$ is an even function follows from stationarity; that is, $R_{xx}(t, t+\tau) = R_{xx}(t+\tau, t) = R_{xx}(\tau)$.
3. Furthermore, $|R_{xx}(\tau)| \leq R_{xx}(0)$, which follows from the fact that, given stationarity, the mean and mean-square values of $x(t)$ and $x(t+\tau)$ are all equal and that $R_{xx}(\tau) \leq 1$, always (assuming unity variance).
4. If $x(t)$ has a periodic component, so does $R_{xx}(\tau)$, but it will not contain any information about the phase.
5. If $x(t)$ has no periodic component, then

$$\lim_{\tau \to \infty} R_{xx}(\tau) \to 0$$

This result implies that $x(t+\tau)$ becomes completely uncorrelated with $x(t)$ for large τ. Since a constant is a special case of a periodic function, then $R_{xx}(\infty) = 0$ implies a zero mean process.

6. The Fourier transform of $R_{xx}(\tau)$ is real, symmetric, and non-negative. This follows from the even (symmetry) property of $R_{xx}(\tau)$.
7. For cross-correlation, $R_{xy}(\tau) = R_{yx}(-\tau)$ even though neither $R_{xy}(\tau)$ or $R_{yx}(\tau)$ are (by themselves) necessarily symmetrical (about lag $\tau = 0$). This is called the *skew-symmetric property*.

The relationship of correlation to power spectral density is expressed by the Fourier transform pair; that is,

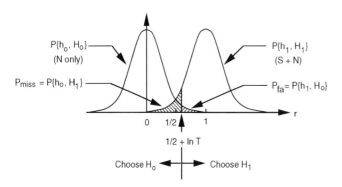

Figure 2.10 Probability distribution for noise alone, and signal plus noise, showing relationship to the threshold value chosen for detection and the corresponding shaded regions representing probability of missed detection and probability of false alarm.

$$R_{xx}(\tau) = \frac{1}{2\pi} \int_{-\infty}^{\infty} S_{xx}(\omega) \exp(j\omega\tau) \, d\omega \qquad (2.48)$$

and

$$S_{xx}(\omega) = \int_{-\infty}^{\infty} R_{xx}(\tau) \exp(-j\omega\tau) \, d\omega \qquad (2.49)$$

where Equations (2.48) and (2.49) constitute the Weiner–Khinchine theorem.

In practice the definition of the autocorrelation function most often used is

$$R_{xx}(\tau) = \lim_{\tau \to \infty} \frac{1}{T} \int_{-\infty}^{\infty} x(t) x(t+\tau) \, dt \qquad (2.50)$$

which is true for stationary ergodic processes. In general, correlation and power spectral density are equivalent descriptions of a signal or stochastic process. Under the special circumstances where we are dealing with wideband chirp signals they are equivalent in a computational sense, as well.

2.4 NONLINEAR TRANSFORMATIONS AND OPERATIONS

Any mathematical analysis for signal processing or optical systems would not be realistic or complete without taking into account nonlinear transformations. These transformations may be applied to coordinates or to amplitude (or intensity). In the case of nonlinear coordinate transformations, geometric distortion occurs in optics (and is often classified as an aberration). In addition, nonlinear transformations are inherent in the Abel transform, making it space-variant (see Chapter 3), which requires a quadratic transformation between coordinates to make it space-invariant. The Mellin transform, as we will see, can be made equivalent in form to the Fourier transform by virtue of an exponential coordinate transformation (or logarithmic if viewed in reverse). The advantage of the Mellin transform is that it can make certain signal processing functions, especially matched filtering, scale-invariant. In the context of coordinate transformations then, invariance to "distortions" (shift, scale, rotation, and others) is very desirable, since it usually improves the performance of desired signal processing functions, or makes them more analytically tractable to analyze using the well-known properties of linear systems.

Nonlinear transformations on signal amplitude or intensity occur for a number of reasons: Photodetectors and focal plane arrays possess detector elements that are square-law devices with respect to incident field strength. Detectors, as well as spatial light modulators, are saturating and nonlinear devices with limited dynamic range. They can be modeled as memoryless nonlinearities (for example, as binarizing transfer characteristics or linear characteristics with saturation) followed by linear filters. Nonlinear coordinate transformations are also useful in certain types of image restoration and enhancement algorithms, such as histogram equalization for contrast modulation. Of course, historically, photography was the earliest example of nonlinear image transfer.

Let us consider two memoryless nonlinearities of particular interest in optical signal processing: square-law detection (as already mentioned) and binarization. Focal plane arrays

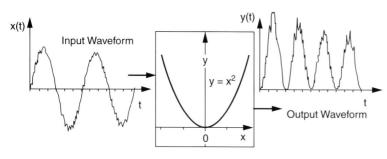

Figure 2.11 Square-law detection process, showing input and output waveforms, and non-linear transfer function.

or single detectors are square-law detectors of light amplitude [although the detected signal (voltage or current) is a linear function of incident light power]. In this discussion we will treat signals as stochastic in nature, since it is more realistic to do so and it is well-treated in the literature. If the statistics of the input signal or image are Gaussian, then, in general, the output statistics of a memoryless nonlinearity are transformed to some other form. For the sake of simplicity, let us deal with one-dimensional time-domain signals. The signal transformation is given by $g(t) = F[f(t)]$, where $f(t)$ is the input signal and $g(t)$ is the output signal. For a square-law detector $g(t) = K[f(t)]^2$, where K is an appropriate scale factor. This is illustrated in Fig. 2.11.

If a stationary Gaussian distributed signal with zero mean and variance σ_f^2 is passed through this square-law detector, then the output correlation $R_{yy}(\tau)$ is given by

$$R_{yy}(\tau) = R_{xx}^2(0) + 2R_{xx}^2(\tau) \tag{2.51}$$

or

$$R_{yy}(\tau) = \sigma_x^4 + 2R_{xx}^2(\tau) \tag{2.52}$$

where the output variance is σ_y^2. This result is useful when interpreting the readout of an optical correlator from a focal plane array or a single detector. The power spectrum is often calculated after detection in order to evaluate the effect of subsequent electrical filtering on the detected signal. Assuming that the spectrum of the input signal (optical field distribution) is denoted by $S_{xx}(\omega)$, the square-law detection process yields the spectrum of the focal plane readout as

$$S_{yy}(\omega) = 2\pi \sigma_x^4 \delta(\omega) + S_{xx}(\omega) * S_{xx}^*(\omega) \tag{2.53}$$

Consider the following examples.

EXAMPLE 2.1 Square-Law Detection of Simple Low-Pass Spectrum The ideal low-pass spectrum is given by

$$S_{xx}(\omega) = \begin{cases} 1, & |\omega| \leq \omega_c \\ 0, & |\omega| > \omega_c \end{cases}$$

with autocorrelation:

$$R_{xx}(\tau) = \frac{\sin(\omega_c \tau)}{\pi \tau}$$

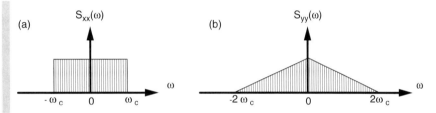

Figure 2.12 (a) Power spectrum of ideal low-pass signal and (b) power spectrum after square-law detection.

where ω_c is the cutoff frequency of the low-pass filter. $S_{xx}(\omega)$ is illustrated in Fig. 2.12(a). The spectrum of the detected output is given by

$$S_{yy}(\omega) = \frac{2}{\pi}\omega_c^2\delta(\omega) + \frac{1}{\pi}\text{tri}\left(\frac{\omega}{2\omega_c}\right)$$

which is shown in Fig. 2.12(b). The factor in front of the delta function actually corresponds to its area. As can be readily appreciated, the square-law device spreads the spectrum of the input signal. This is typical of nonlinear devices.

EXAMPLE 2.2 Effect of Binary Signal Transfer Characteristic on Autocorrelation Function Another case of interest in optical signal processing is a binarizing spatial light modulator, in which the transfer characteristic is described by

$$y(t) = \begin{cases} 1, & x(t) \geq 0 \\ -1, & x(t) < 0 \end{cases}$$

and as illustrated in Fig. 2.13.

Given that the input to this transfer characteristic is Gaussian distributed and has the correlation function $R_{xx}(\tau)$, the output correlation $R_{yy}(\tau)$ is given by

$$R_{yy}(\tau) = \frac{2}{\pi}\sin^{-1}[R_{xx}(\tau)]$$

This result is useful when binarizing the output power spectrum from a Fourier transform architecture, since the power spectrum and correlation function are related by the Weiner–Khinchine theorem. It is also useful in interpreting the correlation function when it is calculated between binary signals as seen in binary correlators.

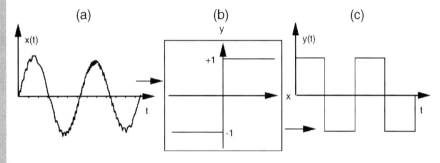

Figure 2.13 Binary transfer characteristic and its effect on a arbitrary (stochastic) signal.

We can also look at the effect of a nonlinear transfer characteristic on the Fourier transform of an input signal, since this is an important process in optical processing. Therefore, consider the following example.

EXAMPLE 2.3 Effect of Sigmoidal Transfer Characteristic on Fourier Transform of a Signal As a final example of the effects of nonlinear transfer on input signal characteristics, consider the effect of a nonlinearity on the power spectrum of the input signal. Let us consider a sigmoidal transfer characteristic, which is useful as a model of a saturating spatial light modulator. This transfer characteristic is defined by

$$f(x) = \frac{\alpha - \beta}{1 + \exp[-\gamma(x - \mu_x)]} + \beta$$

where α = saturation level, β = noise floor, x = modulated input signal, μ_x = mean value of input signal, and γ = slope of linear region. This is plotted in Fig. 2.14 for $\alpha = 1$, $\beta = 0$, $\mu_x = 0$, and $\gamma = 1$.

The output signal is simply $f(x)$ for a particular input amplitude x. The limiting values of $f(x)$ can be determined by choosing

$$\exp[-\gamma(x - \mu_x)] \gg 1 \text{ and } \ll 1$$

corresponding to $x \to -\infty$ and $x \to \infty$, respectively, such that $f(-\infty) = \beta$ and $f(\infty) = \alpha$, respectively.

The slope at the center of the linear region is

$$\gamma = \frac{4f'(x = \mu_x)}{\alpha - \beta}$$

where $f'(x)$ is the first derivative of the transfer characteristic with respect to x. Consider this nonlinear response as representing a three-port (spatially modulating) device as shown in Fig. 2.15. The electrical modulating signal is given by

$$I(x) = I_0 + I_1 \cos(kx)$$

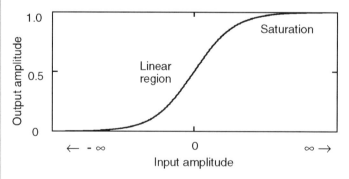

Figure 2.14. Sigmoidal non-linear transfer characteristic as a general model for a spatial light modulator input-output relationship.

2.4 NONLINEAR TRANSFORMATIONS AND OPERATIONS 53

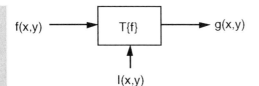

Figure 2.15 Three-port device (representing a spatial light modulator) in which the input-output relationship is non-linear and where the transmitted light is controllable by an intermediate modulating input (electrical or optical).

The mean value of $I(x)$ is I_0 if an integral number of cycles spans the spatial extent of the modulating device. We will assume for simplicity that the input $f(x, y)$ is a normally incident plane wave so that $f(x, y) = 1$ over the input aperture. The resulting output signal spectrum is shown in Fig. 2.16(b), corresponding to the input signal spectrum in Fig. 2.16(a). This particular example applies to the case where we desire to convert $f(x, y)$, which may have been a temporally modulated electrical signal, into a spatially modulated optical signal.

An important example of nonlinear signal transformation treated by Javidi is associated with optical correlation via Fourier transformation in which the two functions to be correlated are first added and then Fourier-transformed, subjected to a nonlinear transform (like the one described above), and then inverse Fourier-transformed. This extends the discussion of the previous example. In this process, one function is shifted relative to the other, to actually represent their relative positions in the input plane of the correlator as shown in block diagram form in Fig. 2.17.

An important advantage of employing a nonlinearity in the Fourier plane of this correlator is that the output signal-to-noise ratio (SNR) or peak-to-sidelobe ratio (PSR) is enhanced. An important requirement is that the phase of the cross-term be preserved, so that unwanted distortions will not destroy the correlation peak. We will focus our attention, therefore, on the intermediate (Fourier) plane and the output of the nonlinearity employed in the Fourier plane—that is, the so-called joint power spectrum. We will follow closely the development by Javidi.

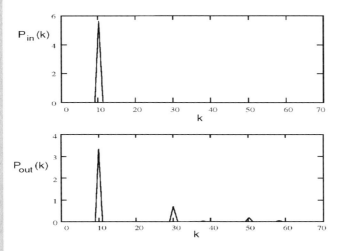

Figure 2.16 (a) Input and (b) output spectrum of a sinusoidal signal passed through a sigmoidal spatial light modulator transfer characteristic. (Note the increased high frequency content in the output spectrum.)

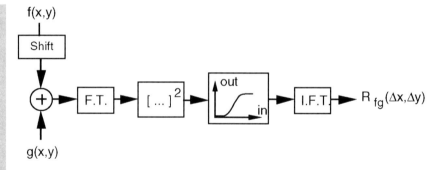

Figure 2.17 Overall correlation process implemented by Fourier transformation. A nonlinear process is introduced at the intermediate Fourier plane, before the final inverse Fourier transform is carried out.

Given that the two functions are $f(x + x_0, y)$ and $g(x - x_0, y)$, the first step of the process (Fourier transform) yields

$$I(u, v) = F(u, v) \exp[j\phi_F(u, v)] \exp[-jx_0u] \\ + G(u, v) \exp[j\phi_G(u, v)] \exp[jx_0u] \quad (2.54)$$

After Fourier transformation a square-law detection process is employed (representing a CCD detector). The resulting intensity distribution is given by

$$|I(u, v)|^2 = F^2(u, v) + G^2(u, v) \\ + 2F(u, v)G(u, v) \cos[\phi_F(u, v) - \phi_G(u, v) + 2x_0u] \quad (2.55)$$

To implement pure linear correlation, this result would be subsequently Fourier-transformed. In practice, the intermediate result is used to electrically modulate a nonlinear device, known as a spatial light modulator. This device can be described in general by a transformation of the form $t(I)$. By again exploiting the Fourier transform, we can write

$$T(\omega) = \int_{-\infty}^{\infty} t(I) \exp(-j\omega I) \, dI \quad (2.56)$$

where $T(\omega)$ is a Fourier amplitude variable with argument ω related to the input amplitude I, and

$$t(I) = \frac{1}{2\pi} \int_{-\infty}^{\infty} T(\omega) \exp(j\omega I) \, d\omega \quad (2.57)$$

Thus, in general terms, the output of the nonlinear transformation is

$$t(I) = \frac{1}{2\pi} \int_{-\infty}^{\infty} T(\omega) \exp\{j\omega [F^2(u, v) + G^2(u, v)]\} \\ \exp\{j2\omega F(u, v)G(u, v) \cos[\phi_F(u, v) - \phi_G(u, v) + 2x_0u]\} \, d\omega \quad (2.58)$$

Consider as a specific case the general kth-law nonlinearity. The kth law nonlinear transfer characteristic is defined as

$$t(I) = \frac{I^k}{I_0}, \quad 0 \leq k \leq 1 \quad (2.59)$$

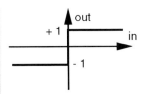

Figure 2.18 Binary (hard clipping) non-linear transfer function (kth law non-linear device for $k = 0$).

where $k \to 1$ corresponds to a linear device and $k \to 0$ corresponds to a hard-clipping (binarizing) device, as shown in Fig. 2.18. The corresponding Fourier transform of $t(I)$ is given by

$$T(\omega) = \frac{2}{(j\omega)^{k+1}} \Gamma(k+1) \tag{2.60}$$

where $\Gamma(k+1)$ is the gamma function of order $k+1$ such that $\Gamma(k+1) = k!$ The resulting output of the nonlinearity is then

$$t_{nk}(I) = \sum_{\substack{n=1 \\ (n \text{ odd})}}^{\infty} \frac{\varepsilon_n \Gamma(k+1)[F(u,v)G(u,v)]^k}{2^k \Gamma\left(1 - \frac{n-k}{2}\right) \Gamma\left(1 + \frac{n+k}{2}\right)} \cos[n(\phi_F - \phi_G + 2x_0 u)] \tag{2.61}$$

Each harmonic term (indexed by n) in this result is phase-modulated by $n(\phi_F - \phi_G)$, and higher orders are displaced by $2nx_0$ (which is obtained via diffraction in an actual optical correlator). The envelope of each term is proportional to $[F(u,v)G(u,v)]^k$. If the input signals have zero phase difference, then the correlation is just the inverse Fourier transform of $[FG]^k$. Of course, for $k = 1$ linear correlation is the result. Then for the $n = 1$ term the resulting joint power spectrum is

$$t_{11}(I) = 2F(u,v)G(u,v)\cos[\phi_F(u,v) - \phi_G(u,v) + 2x_0 u] \tag{2.62}$$

For $k = 0$ (hard-clipping nonlinearity) and $n = 1$ the joint power spectrum is

$$t_{10}(I) = \frac{4}{\pi} \cos[\phi_F(u,v) - \phi_G(u,v) + 2x_0 u] \tag{2.63}$$

Of course, if the two input functions have the same phase, then

$$t_{10}(I) = \frac{4}{\pi} \cos(2x_0 u) \tag{2.64}$$

and the corresponding output signal is the autocorrelation, which is an impulse function of magnitude $2/\pi$ located at $2x_0$.

2.5 MIXING AND MODULATION

In many instances, temporal modulation of signals is employed in optical signal processing—for example, for amplitude modulation (AM) of laser sources as well as for amplitude and frequency modulation (FM) of acousto-optic (AO) cells. In the case of AO cells, mixing is also required to heterodyne the input AM or FM up to the center frequency of the AO cell. Of course, two-dimensional spatial light modulators (SLMs) can be thought of as mixers as well (actually as a matrix of mixers), but they are usually analyzed spatially. Temporal properties of

SLMs—or, for that matter, detector arrays—become important when their response times are on the order of expected frame rates, because residual signal spatial modulation can then carry over to the next frame of spatial modulation as an undesirable distortion or artifact. Modulation can be carried out using a passive device such as a mixer, which is nonlinear and generates new frequencies by virtue of the product operation. It is a three-port device as shown in Fig. 2.19, in which the two input waveforms are denoted by $m(t)$ and $x(t)$. The output is a product of $m(t)$ and $x(t)$ scaled by some conversion gain G such that $y(t) = Gm(t)x(t)$. A low-pass filter often follows the mixer to reject high-frequency mixing components, as will be seen later. In the case of heterodyning we have $x(t) = \cos \omega_c t$, where ω_c is the carrier frequency such that $y(t) = m(t) \cos \omega_c t$, assuming $G = 1$. The Fourier transform of this output waveform is

$$Y(\omega) = \frac{1}{2}[M(\omega - \omega_c) + M(\omega + \omega_c)] \tag{2.65}$$

indicating that the resulting signal has upper and lower sidebands located equally about the carrier frequency ω_c. However, on reception of this signal in a communications receiver system, a second heterodyning operation is performed in order to recover the original modulation signal, $m(t)$. On reception the same carrier is applied to $y(t)$ to obtain

$$Z(\omega) = \frac{1}{2}[Y(\omega - \omega_c) + Y(\omega + \omega_c)] \tag{2.66}$$

Substituting Equation (2.65) into Equation (2.66) yields

$$Z(\omega) = \frac{1}{2}M(\omega) + \frac{1}{4}[M(\omega - 2\omega_c) + M(\omega + 2\omega_c)] \tag{2.67}$$

As already mentioned, a low-pass filter may be employed to eliminate the higher-frequency terms. Then $Z(\omega) \to M(\omega)/2$ or $z(t) \to m(t)/2$ so that even ideally, G is only 1/2. Implicit in obtaining this result is that the transmitter and receiver (or modulator and demodulator) have the same carrier signals in terms of frequency and phase. Hence, they are called synchronous and homodyned. This system is also known as double sideband (DSB) modulation. If the phase difference is not zero between transmit and receive, then the system can be characterized by describing the receive signal as

$$z(t) = m(t) \cos(\omega_c t) \cos(\omega_c t + \phi) \tag{2.68}$$

or

$$z(t) = m(t) \cos^2(\omega_c t) \cos \phi - m(t) \sin(\omega_c t) \cos(\omega_c t) \sin \phi \tag{2.69}$$

Rewriting this result as

$$z(t) = \frac{1}{2}m(t)\{[1 + \cos(2\omega_c t)] \cos \phi - \sin(2\omega_c t) \sin \phi\} \tag{2.70}$$

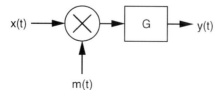

Figure 2.19 Three-port modulation device for the mixing of two signals by the product operation (followed by a fixed gain).

it is clear that subsequent low-pass filtering yields $z(t) \to \frac{1}{2} m(t) \cos \phi$, which is a useful result for at least two reasons. First, only receivers that modulate synchronously with their transmitters receive an unattenuated signal; and, second, synchronous detection is phase-sensitive, which is useful in photon signal detection.

If we introduce an amplitude offset (or bias), then we can define the modulation as $1 + m(t)$, which ensures the presence of a carrier (whereas before we tacitly suppressed it by writing the modulation as just $m(t)$). In addition, if $m(t) < 1$ and it is low bandwidth compared to the carrier, we will generate an envelope on the carrier that is a faithful replica of only the modulation. The traditional DSB system is therefore a suppressed carrier modulation and is usually implemented as a balanced modulator, the name deriving from the fact that in practice the carrier term mentioned above is balanced-out or suppressed.

Frequency modulation can be expressed as a phase modulation in the general expression for a signal, namely, $x(t) = x_0 \cos \phi(t)$ where $\phi(t) = \omega_c t + \phi_0 + k f(t)$, where ω_c is the carrier frequency, ϕ_0 the initial phase, and k the proportionality constant between phase and the modulating signal $f(t)$. In fact, there is no significant difference between phase and frequency modulation.

For the case where the instantaneous frequency ω_i is given by $\omega_i = \omega_c + \Delta \omega \cos \omega_m t$, where $\Delta \omega \ll \omega_c$, the phase variation is

$$\phi(t) = \omega_c t + \frac{\Delta \omega t}{\omega_m} \sin \omega_m t + \phi_0 \qquad (2.71)$$

so that the modulated carrier is

$$x(t) = x_0 \cos \left(\omega_c t + \frac{\Delta \omega t}{\omega_m} \sin \omega_m t + \phi_0 \right) \qquad (2.72)$$

where the modulation index $\Delta \omega / \omega_m$ determines whether the FM system is narrowband or broadband. Narrowband FM occurs for $\Delta \omega / \omega_m \ll T/2$. In this case the resulting FM modulator output is

$$x(t) = x_0 \left[\cos(\omega_c t) - \frac{\Delta \omega t}{\omega_m} \sin(\omega_m t) \sin(\omega_c t) \right] \qquad (2.73)$$

which is similar to the output of a product modulator in AM. For an arbitrary waveform $f(t)$ a similar result ensues:

$$z(t) = x_0 \left[\cos(\omega_c t) - \frac{\Delta \omega t}{\omega_m} f(t) \sin(\omega_c t) \right] \qquad (2.74)$$

For the more general case of broadband FM, Equation (2.73) can be rewritten as

$$x(t) = x_0 \left\{ \cos(\omega_c t) \cos \left[\frac{\Delta \omega}{\omega_m} \sin(\omega_c t) t \right] - \sin(\omega_c t) \sin \left[\frac{\Delta \omega}{\omega_m} \sin(\omega_c t) t \right] \right\} \qquad (2.75)$$

The two terms on the right-hand side of Equation (2.75) are the real and imaginary parts of the term $\exp[j(\Delta \omega / \omega_m) \sin \omega_m t]$, which can be written as a Fourier series:

$$\exp \left[j \frac{\Delta \omega}{\omega_m} \sin(\omega_m t) t \right] = \sum_{n=-\infty}^{\infty} J_n \left(\frac{\Delta \omega}{\omega_m} \right) \exp(j n \omega_m t) \qquad (2.76)$$

where $J_n(x)$ is the nth-order Bessel function of the first kind defined here as

$$J_n\left(\frac{\Delta\omega}{\omega_m}\right) = \frac{\omega_m}{2\pi}\int_{-\pi/\omega_m}^{\pi/\omega_m} \exp\left[j\frac{\Delta\omega}{\omega_m}\sin(\omega_m t)t\right]\exp(-jn\omega_m t) \qquad (2.77)$$

The final form of Equation (2.75) can then be rewritten as

$$x(t) = x_0 \sum_{n=-\infty}^{\infty} J_n\left(\frac{\Delta\omega}{\omega_m}\right)\cos[(\omega_c - n\omega_m)t] \qquad (2.78)$$

Thus, the resulting signal is spread significantly in frequency about the carrier (and its image frequency). To see this consider the following example.

EXAMPLE 2.4 FM Spectrum of a Signal For the signal

$$x(t) = x_0 \cos\left(\omega_c t + \frac{\Delta\omega t}{\omega_m}\sin\omega_m t\right)$$

assume that $x_0 = 1$ and that $\Delta\omega/\omega_m = 2$. Then using a table of Bessel function values we obtain $J_0(2) \approx 0.22$, $J_1(2) \approx 0.58$, $J_2(2) \approx 0.35$, $J_3(2) \approx 0.13$, and so on. The corresponding plot of the magnitude spectrum $|X(\omega)|$ is shown in Fig. 2.20.

Both AM and FM (narrowband and broadband) are used to modulate acousto-optic cells for communication and radar signal processing. We will discuss this further in Chapter 8.

An important signal used in optical signal processing, as well as radar and communication systems in general, is the chirp signal, which is both AM and FM. One of the simplest expressions for this signal is

$$x(t) = x_0 \operatorname{rect}\left(\frac{t}{\tau}\right)\cos\left(\omega_c t + \frac{1}{2}\alpha t^2\right) \qquad (2.79)$$

where the instantaneous frequency is $\omega_i = \omega_c + \alpha t$ and is plotted in Fig. 2.21(a) for a positive value of chirp slope α. The corresponding plot of $x(t)$ is shown in Fig. 2.21(b). The corresponding spectrum of this signal is

$$X(\omega) = \int_{-\tau/2}^{\tau/2} \cos\left(\omega_c t + \frac{1}{2}\alpha t^2\right) dt \qquad (2.80)$$

which can be written as

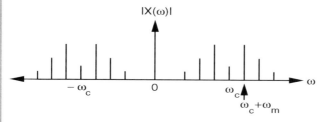

Figure 2.20 Amplitude spectrum of the FM signal with sinusoidal phase.

$$X(\omega) = \frac{1}{2}\sqrt{\frac{\pi}{\alpha}} \exp[-j(\omega_c - \omega)^2/2\alpha][C(k_1) + jS(k_1)C(k_2) + jS(k_2)] \quad (2.81)$$

where

$$C(k) = \int_0^k \cos\left(\frac{\pi}{2}k'^2\right) dk' \quad \text{and} \quad S(k) = \int_0^k \sin\left(\frac{\pi}{2}k'^2\right) dk' \quad (2.82)$$

are the Fresnel cosine and sine integrals (discussed further in Chapter 3) and

$$k_{1,2} = \frac{\alpha\tau/2 \pm (\omega_c - \omega)}{\sqrt{\pi\alpha}} \quad (2.83)$$

The corresponding magnitude of $X(\omega)$ is plotted in Fig. 2.22. This result is important in understanding not only chirp radar, but the focusing properties of lenses on coherent light and similar properties of acousto-optic Bragg cells modulated with broadband FM signals. In fact, it is fundamental to understanding the one-to-one correspondence between Fourier transforms and correlation, especially for those special class of signals that have a linear FM relationship. Fourier transforms can be expressed entirely in the time domain as correlations between two signals under these circumstances, making them amenable to generation solely by physical (time-domain) devices, such as chirp generators and pulse compressors, rather than by algorithms implemented as digital time-to-frequency transformations (FFTs). In fact, the notion of pulse compression can be understood as a correlation of the chirp signal with itself, which occurs in radar when the received signal from each impulse reflector is convolved (loosely speaking) with a filter impulse response with the same functional shape as the received signal. The result of this process is the corresponding autocorrelation (or matched filter output) given by

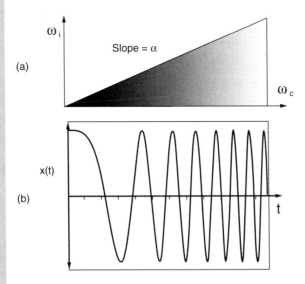

Figure 2.21 (a) Linear chirp frequency characteristic (upchirp), and (b) corresponding signal waveform.

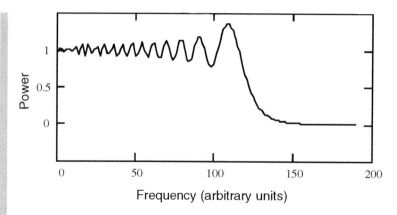

Figure 2.22 Amplitude spectrum of linear chirp FM signal.

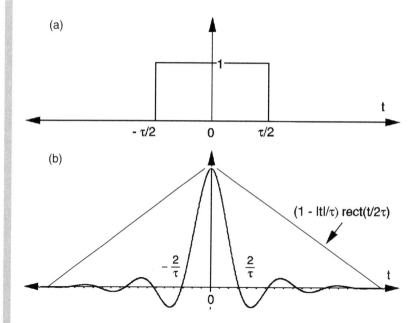

Figure 2.23 (a) Matched filter (received) compressed pulse derived by pulse compression from the original (transmitted) pulse shown in (b).

$$R_{xx}(t) = \frac{\sin(\alpha t/2)}{(\alpha t/2)} \left(1 - \frac{|t|}{\tau}\right) \text{rect}\left(\frac{t}{2\tau}\right) \cos(\omega_c t) \qquad (2.84)$$

where t is a lag variable (not time). This is plotted (without the carrier) in Fig. 2.23(b) to compare it with the original pulse shown in Fig. 2.23(a). By comparing the mainlobe width of $R_{xx}(t)$ with the overall width of $x(t)$, we can show that the process of calculating $R_{xx}(t)$ is equivalent to pulse compression.

PROBLEM EXERCISES

1. By using the rectangular to polar coordinate transformation, show that the bivariate Gaussian with zero means ($\mu_x = \mu_y = 0$) and equal variances ($\sigma_x = \sigma_y = \sigma$) converts to the product of a Rayleigh and uniform distribution.

2. In the limit of large n ($n \to \infty$) and for $p \to 0$ such that the product $np = m$ is a constant, show that the binomial distribution approaches the Poisson distribution. Without the restriction of $np = m =$ constant, but allowing m to be very large and continuous, show how the Gaussian distribution is derived.

3. Using the result of Example 2.2,

$$R_{yy}(\tau) = \frac{2}{\pi}\sin^{-1}[R_{xx}(\tau)]$$

   show that $R_{yy}(\tau)$ can be expressed as a series using the Fourier expansion for $\sin^{-1}(\ldots)$:

$$R_{yy}(\tau) = \frac{2}{\pi}\sum_{n=1}^{\infty}\frac{1}{n}[J_0(n\pi) - (-1)^n]\sin[n\pi R_{xx}(\tau)]$$

   Then calculate the power spectrum of the output using the Wiener–Khintchine theorem.

4. In the previous problem assume that the input waveform is a sinusoid; that is,

$$x(t) = A\cos(\omega t + \phi)$$

   What is the output spectrum $S_{yy}(\omega)$? How does it compare to $S_{xx}(\omega)$?

5. The equivalent noise bandwidth of a linear filter with a transfer function $H(\omega)$ is given by the normalized mean-square value of the output signal $y(t)$, assuming that the input process is white (as required in the definition of equivalent noise bandwidth) with power spectral density $N/2$. Given that the filter impulse response is $h(t) = A\exp(-\alpha t)u(t)$, what is the equivalent noise bandwidth and how does it compare to the 3-dB bandwidth?

6. Show explicitly that the marginal distribution for the bivariate Gaussian with $\mu_x = \mu_y = 0$ is Gaussian.

7. Given the periodic function

$$f(x) = \cos(kx + \phi)$$

   where ϕ is an arbitrary constant (or phase shift), calculate the corresponding autocorrelation function and plot your result.

8. For a system that measures a signal $s(x, y)$ with additive white uncorrelated noise $n(x, y)$ of zero-mean and unit variance, show that the autocorrelation of the output signal $r(x, y) = s(x, y) + n(x, y)$ is given by

$$R_{rr}(x', y') = R_{ss}(x', y') + \delta(x', y')$$

   If the input signal is shifted by an amount (x_0, y_0) what happens to the location of $R_{rr}(x', y')$? Sketch the correlation surface before and after the input signal is shifted.

9. If the signal in Problem 8 is defined by

$$s(x, y) = \exp[-(a + jb)(x^2 + y^2)]$$

determine the result for $R_{rr}(x', y')$.

10. If the 2-D random process defined by $n(x, y)$ is white, then its autocorrelation is

$$R_{nn}(x', y') = \delta(x - x')\delta(y - y')$$

If this 2-D random process is multiplied by a circular function representing a circular aperture of radius r_0 centered on the origin and defined by $\text{circ}(r/r_0)$, then

$$n(x, y) = \text{circ}(r/r_0)n(x, y)$$

where $r = (x^2 + y^2)^{1/2}$. Show that the new autocorrelation function is

$$R_{nn}(x, y; x', y') = \text{circ}(r'/r_0)\delta(x - x')\delta(y - y')$$

where $r' = (x'^2 + y'^2)^{1/2}$. Then show that the autocorrelation of the Fourier transform of $n(x, y)$, given by $N(u, v)$ is

$$R_{nn}(x, y; x', y') = 2\pi r_0 \frac{J_1\left[r_0\sqrt{(x - x')^2 + (y - y')^2}\right]}{(x - x')^2 + (y - y')^2}$$

11. For a stationary Gaussian input to a square-law detector as discussed on pp. 49–51, prove that Equations (2.51) and (2.52) are true by using the relation

$$R_{yy}(\tau) = \overline{y(t)y(t + \tau)} = \overline{x^2(t)x^2(t + \tau)}$$

and the fact that if another function $z(t) = x(t) + y(t)$, then

$$\overline{z(t)z(t + \tau)} = \overline{x(t)x(t + \tau)} + \overline{y(t)y(t + \tau)} + \overline{y(t)x(t + \tau)} + \overline{x(t)y(t + \tau)}$$

that is, the average of the sums is the sum of the averages.

12. Phase-sensitive detection is often used to obtain an enhanced ability (over square-law detectors) to detect signals buried in noise. Typically a square-wave reference signal is mixed with the input signal at the same fundamental frequency ($f = \omega/2\pi$). One way to do this is to simply switch the input signal along two different paths, one with unity gain and one with inverting unity gain, and add the two paths before low-pass filtering. (This is similar to optical chopping.) The reference signal can be used to do the switching at frequency f, which is equivalent to mixing the bipolar (± 1) reference square wave and the input sinusoid. Typical output waveforms for variations in the phase of the input signal with respect to the reference are shown. If the input signal is given by

$$V = V_0 \cos(\omega t + \phi)$$

expand the square wave in a Fourier series and show that after mixing and low-pass filtering, any frequencies $> 2\omega$ yield

$$V_{\text{out}} = \frac{1}{2} V_0 \cos \phi$$

Will this approach still work if the input and reference signals are at different frequencies (ω_i and

ω_r respectively)? What is the form of the output voltage then? What happens if $\omega_i = 3\omega_r$? Look at the output power spectrum.

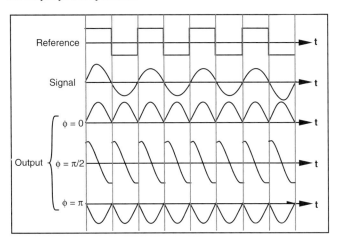

13. Verify that Equations (2.62) and (2.63) are true for $k = 1$ and $k = 0$ (with $n = 1$), respectively.

14. Consider the nonlinear joint transform correlator for the case of a hard-clipping nonlinearity ($k = 0$). Calculate the correlation plane output for noisy images containing white Gaussian noise and a narrow rectangular pillbox. Evaluate the results in terms of the ratio of peak (zero-lag) correlation to output correlation plane variance at all other points (analogous to the correlation or matched filter signal-to-noise ratio). Compare your results to linear correlation ($k = 1$). Use the MATLAB tools in Appendix C.

15. For the architecture of Fig. 2.17 assume that $f(x, y)$ is a 2-D rectangular function and that $g(x, y)$ is also the same 2-D rect function without noise. Calculate directly using MATLAB: (a) the first Fourier transform, (b) the result of square-law detection, (c) the result of passing the detected signal through a hard-clipping nonlinearity as shown in Fig. 2.18, and (d) the result of inverse Fourier transforming. Interpret your results in terms of the analytical results of Equations (2.55) and (2.64) for zero noise variance.

16. Derive the result of Equation (2.84) and model the pulse compression (autocorrelation of the chirped pulse) using MATLAB.

BIBLIOGRAPHY

B. Javidi, "Analysis of Nonlinear Optical Correlation," SPIE Proceedings, Vol. 977, *Real Time Signal Processing XI*, pp. 307–323 (1988).

B. Javidi, "Comparison of the Nonlinear Joint Transform Correlator and the Nonlinearly Transformed Matched Filter Based Correlator for Noisy Input Scenes," *Opt. Eng.* **29**, 1013 (1990).

L. Maisel, *Probability, Statistics and Random Processes*, Simon and Schuster, New York (1971).

P. Z. Peebles, Jr., *Probability, Random Variables, and Random Signal Principles*, 2nd ed., McGraw-Hill, New York (1987).

C. Ruhla, *The Physics of Chance*, Oxford University Press, New York (1992).

N. S. Tzannes, *Communication and Radar Systems*, Prentice-Hall, Englewood Cliffs, NJ (1985).

Chapter 3
MATHEMATICAL TRANSFORMS USED IN OPTICAL SIGNAL PROCESSING

3.1 OVERVIEW

In addition to the Fourier and Hankel transforms, there are other integral transforms that are useful in optics and optical signal processing, which will be introduced here from a purely mathematical perspective. These transforms will be useful when we discuss the physical theory of diffraction, as well as image processing and pattern recognition applications of optics. These transforms consist of the

- Fresnel transform
- Hilbert transform
- Radon transform
- Mellin transform
- Wavelet transform

We will first introduce and briefly describe each of these transforms in order to familiarize ourselves with their appearance, notation, and area of application.

The Fresnel transform is given by

$$F(x, y) = -ja \int_{-\infty}^{\infty} \int_{-\infty}^{\infty} f(x', y') \exp\{j\pi a[(x - x')^2 + (y - y')^2]\} dx'dy' \quad (3.1)$$

and arises from consideration of the diffraction of light through an aperture in the so-called near-field (or Fresnel zone), which is close to the aperture compared to its size. x and y are spatial coordinates, as are x' and y'. In the limit as the plane of observation, defined by (x, y), physically approaches the input aperture plane (x', y'), the observed function $F(x, y)$ approaches $f(x, y)$, which can be considered an image projection operation. In the other extreme, when the (x, y) plane recedes far away, $F(x, y)$ approaches the Fourier transform of $f(x, y)$.

The Hilbert transform, given by

$$F_H(x) = \frac{1}{\pi} \int_{-\infty}^{\infty} \frac{f(x')\,dx'}{x' - x} \tag{3.2}$$

(shown in one dimension), arises when considering single-sided (causal and/or positive frequency) signals. In optics it is used to analyze the results of Schlieren imaging. In solid-state physics and the study of the optical properties of materials the Hilbert transform pair are often called the Kramers–Krönig relations, which relate the real and imaginary parts of electrical conductivity and the real and imaginary parts of the index of refraction, respectively.

The Radon transform is the mathematical basis for tomographic image processing. It was derived in 1917 by Johann Radon. He proved that the complete set of one-dimensional (1-D) projections of two-dimensional (2-D) or three-dimensional (3-D) functions contain all the information in the original function under suitable conditions. By integrating a 2-D function over sets of parallel lines at varying aspect angles, the required projections can be derived. Applications in various fields are numerous, including crystallography, radio astronomy, diagnostic radiology, nuclear magnetic resonance, radiative scattering, and geophysics. The Radon transform is also used in synthetic aperture radar signal processing. It can reduce the computations required to obtain the 2-D Fourier transform under certain conditions. The Radon transform is given by

$$F_R(p, \phi) = \int_{-\infty}^{\infty} \int_{-\infty}^{\infty} f(x, y)\delta(p - x\cos\phi - y\sin\phi)\,dx\,dy \tag{3.3}$$

The Abel transform, which is a special case of the Radon transform, occurs when considering circularly symmetric distributions in two dimensions projected into one dimension. It has application to the frequency analysis of optical imaging systems, especially those exhibiting circular symmetry and employing raster scanning (i.e., television systems).

The Mellin transform is useful in pattern recognition and is given by

$$F_M(s) \int_0^{\infty} f(x) x^{s-1}\,dx \tag{3.4}$$

in one dimension. The Mellin transform is related to the Laplace transform; hence s is used to denote the Mellin space, suggestive of the s-plane associated with the Laplace transform. It is used to calculate the moments of a function and enables scale-invariant optical matched filtering. In fact the Mellin transform can be calculated optically, as will be discussed in the chapter on pattern recognition (Chapter 12).

Finally, the newest addition to mathematical transforms used in optics and signal processing is the wavelet transform. It can be viewed as a windowed Fourier transform, in which the window and exponential kernel of the original Fourier transform together combine to become the new kernel of the wavelet transform. Thus, the wavelet transform looks like this:

$$W(a, b) = \frac{1}{a} \int_{-\infty}^{\infty} s(x) h\left(\frac{x-b}{a}\right) dx \tag{3.5}$$

in which $h(x)$ can take on many forms, depending on the originator and application, as will be seen in Section 3.6, and where b is a shift parameter and a is a scale parameter.

Of course, the wavelet transform carries more significant information than simply a windowed Fourier transform. The wavelet transform contains information on a signal parametrized with respect to scaling and shifting of the kernel. It also is more consistent with realistic signals in which both spatial and spatial frequency extent (or "regions of support") are finite in extent. That is, wavelet transforms are useful in characterizing nonstationary signals, a fact which overcomes a major pitfall of Fourier transforms, which is that they produce a global rather than local description of signals. The wavelet transform for a 1-D signal is also a transformation that generates a 2-D parameter space. Hence, 2-D signals must be transformed to a 4-D parameter space. There are many choices for the kernel (or analyzing wavelet) $h(x)$, as discussed later.

3.2 FRESNEL TRANSFORM

Although the Fresnel transform emerges naturally as part of physical optics, we will treat it here to introduce it as a mathematical concept in its own right. In optical diffraction a quadratic phase factor appears in the propagation of light between surfaces in an optical system. This factor is of the form

$$h(x) = \exp(\pm j\pi a x^2) \tag{3.6}$$

in one dimension. If we take the Fourier transform of this function [using the positive sign in Equation (3.6)], we obtain in a formal sense

$$H(k) = \int_{-\infty}^{\infty} \exp(j\pi a x^2) \exp(-j2\pi k x) \, dx \tag{3.7}$$

where k is spatial frequency, having units of inverse length, and a is a parameter known as the "chirp rate," having units of inverse length squared. Completing the square in the argument of the integrand, we obtain

$$H(k) = \exp(-j\pi k^2/a) \int_{-\infty}^{\infty} \exp[j\pi a(x - k/a)^2] \, dx \tag{3.8}$$

By changing variables (i.e., $x' = x - k/a$), this expression simplifies to

$$H(k) = \exp(-j\pi k^2/a) \int_{-\infty}^{\infty} \exp(j\pi a x'^2) \, dx' \tag{3.9}$$

If the limits of integration are finite, this integral must be evaluated by expanding the integrand into $\cos(\pi x'^2) + \sin(\pi x'^2)$ and using tabulated numerical values in a standard table of integrals or by using a nomographical technique such as the Cornu spiral seen in many optics texts. In any case, Equation (3.9) is the Fresnel *integral* (not the Fresnel transform). If the limits are kept at $\pm\infty$, we can still expand the above integrand and get an analytical (closed-form) result. First we change variables such that $ax'^2 = z^2/2$. Then

$$H(k) = \exp(-j\pi k^2/a) \sqrt{\frac{2}{a}} \int_{-\infty}^{\infty} [\cos(\pi z^2/2) + j \sin(\pi z^2/2)] \, dz \tag{3.10}$$

3.2 FRESNEL TRANSFORM

which can be evaluated in closed form to yield

$$H(k) = \left(\frac{1+j}{\sqrt{2a}}\right) \exp(-j\pi k^2/a) \tag{3.11}$$

Likewise, if we choose $f(x) = exp(-j\pi ax'^2)$, then

$$H(k) = \left(\frac{1-j}{\sqrt{2a}}\right) \exp(j\pi k^2/a) \tag{3.12}$$

Denoting $\exp(j\pi ax'^2)$ by $h_1(x')$ and $\exp(-j\pi ax'^2)$ by $h_2(x')$, it is then clear that

$$H_1(k)H_2(k) = 1/a \tag{3.13}$$

Therefore, using the convolution theorem, we obtain

$$h_1(x) * h_2(x) = \delta(x)/a \tag{3.14}$$

By extension in two dimensions we can write

$$H_1(k_x, k_y)H_2(k_x, k_y) = 1/a^2 \tag{3.15}$$

and

$$\exp[j\pi a(x^2 + y^2)] * \exp[-j\pi a(x^2 + y^2)] = \delta(x, y)/a^2 \tag{3.16}$$

or in vector notation:

$$\exp(j\pi a\mathbf{x}^2) * \exp(-j\pi a\mathbf{x}^2) = \delta(\mathbf{x})/a^2 \tag{3.17}$$

Now that we have introduced the Fresnel integral and some of its properties, we can continue the discussion by considering the Fresnel *transform* of a function $f(x, y)$, defined by

$$F(x, y) = -ja \int_{-\infty}^{\infty} \int_{-\infty}^{\infty} f(x', y') \exp\{j\pi a[(x - x')^2 + (y - y')^2]\} dx'dy' \tag{3.18}$$

where x and y are spatial variables, which may take on a different interpretation (i.e., spatial frequency) when distances are large compared to the input plane extent, as in Fraunhofer diffraction. By inspection we can rewrite this integral formally as a convolution between the function $f(\mathbf{x})$ and the kernal $\exp(j\pi a\mathbf{x}^2)$:

$$F(\mathbf{x}) = -ja \exp(j\pi a\mathbf{x}^2) * f(\mathbf{x}) \tag{3.19}$$

where bold $\mathbf{x}$ denotes a vector. If we now convolve both sides of Equation (3.19) with $ja \exp(-j\pi a\mathbf{x}^2)$, we obtain

$$ja \exp(-j\pi a\mathbf{x}^2) * F(\mathbf{x}) = ja \exp(-j\pi a\mathbf{x}^2) * -ja \exp(j\pi a\mathbf{x}^2) * f(\mathbf{x}) \tag{3.20}$$

Using the result of Equation (3.17) again, Equation (3.20) reduces to

$$f(\mathbf{x}) = ja \exp(-j\pi a\mathbf{x}^2) * F(\mathbf{x}) \tag{3.21}$$

which is the inverse Fresnel transform. Thus $f(\mathbf{x})$ can be recovered from $F(\mathbf{x})$. This means that the Fresnel transform is unique, and thus Equations (3.19) and (3.21) constitute the Fresnel transform pair.

If a Fourier transform is now taken of either Equation (3.19) or Equation (3.21) and the convolution theorem is used along with the relation

$$\text{FT}\{\exp[\mp j\pi a(x^2 + y^2)]\} = \pm \frac{j}{a} \exp[\mp j\pi(k_x^2 + k_y^2)/a] \tag{3.22}$$

(see Table 1.5 in Appendix A for the Gaussian example), we obtain

$$\text{FT}\{F(\mathbf{x})\} = -\exp[j\pi|k|^2/a]\,\text{FT}\{f(\mathbf{x})\} \tag{3.23}$$

This result means that the Fresnel transform in the spatial domain (or the inverse Fresnel transform) represented by the convolution operation of Equation (3.19) [or Equation (3.21), respectively] can be viewed as simply the result of multiplying a positive (or negative) multiplicative quadratic phase factor in the spatial frequency domain times the Fourier transform of the original function. In other words, the effect of the Fresnel transform on a function $f(\mathbf{x})$ can be viewed as the convolution of $f(\mathbf{x})$ with a chirp function in the space domain. In the Fourier domain, however, the effect of the Fresnel transform can be viewed as a multiplication (or mixing) of a chirp signal with the Fourier transform of $f(\mathbf{x})$. As an example of the Fresnel transform of a familiar object, consider a rectangular aperture.

EXAMPLE 3.1 Fresnel Transform of a Rectangular Aperture The Fresnel transform of a rectangular aperture illustrated in Fig. 3.1 is given by the Fourier transform of a real chirp signal over the extent of the window, namely,

$$F(k) = \int_{-x_0}^{x_0} \cos\left(\frac{1}{2}ax^2\right) \exp(-jkx)\,dx$$

$$F(k) = \frac{1}{2}\int_{-x_0}^{x_0} \exp\left[j\left(\frac{1}{2}ax^2 - kx\right)\right] dx + \frac{1}{2}\int_{-x_0}^{x_0} \exp\left[-j\left(\frac{1}{2}ax^2 + kx\right)\right] dx$$

$$= \frac{1}{2}\sqrt{\frac{\pi}{a}} \exp(jk^2/2a) \left[C\left(\frac{ax_0/2 - k}{\sqrt{\pi a}}\right) + jS\left(\frac{ax_0/2 - k}{\sqrt{\pi a}}\right)\right]$$

$$+ \frac{1}{2}\sqrt{\frac{\pi}{a}} \exp(jk^2/2a) \left[C\left(\frac{ax_0/2 + k}{\sqrt{\pi a}}\right) + jS\left(\frac{ax_0/2 + k}{\sqrt{\pi a}}\right)\right]$$

where

$$C(x) = \int_0^x \cos \pi z^2/2\,dx \quad \text{and} \quad S(x) \int_0^x \sin \pi z^2/2\,dz$$

are the Fresnel integrals discussed earlier. These integrals are tabulated since they do not have a closed-form expression. The magnitude of $F(k)$ is given by

$$|F(k)| = \frac{1}{2}\sqrt{\frac{\pi}{a}}\{[C(x_1) + C(x_2)]^2 + [S(x_1) + S(x_2)]^2\}^{1/2}$$

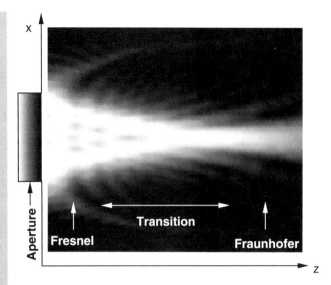

Figure 3.1 Rectangular aperture along one axis and corresponding Fresnel transform along the same axis, at various ranges (near, intermediate and far). Note that the far-range pattern closely resembles the sinc function, whereas the near range pattern approximates the aperture.

where

$$x_1 = \frac{ax_0/2 - k}{\sqrt{\pi a}} \quad \text{and} \quad x_2 = \frac{ax_0/2 + k}{\sqrt{\pi a}}$$

This general result corresponds to the images shown in Fig. 3.1. Values for $C(x)$ and $S(x)$ are given in standard optics texts.

3.3 HILBERT TRANSFORM

The Hilbert transform is important in signal processing because it expresses the fact that real signals are causal and real frequencies are only positive. It also relates the real (or energy dissipative) and imaginary (or energy storage) components of physically realizable signals or system responses. It is also important in understanding Schlieren imaging in optics where a knife-edge is used to block-off one-half of the spatial frequency plane in a coherent optical system, a technique useful in real-time image enhancement.

The Hilbert transform in one dimension is given by

$$F_{\mathrm{H}}(x) = \frac{1}{\pi} \int_{-\infty}^{\infty} \frac{f(x')\,dx'}{x' - x} \tag{3.24}$$

which can be written as a convolution:

$$F_H(x) = \frac{-1}{\pi x} * f(x) \qquad (3.25)$$

Using the convolution theorem, we can immediately write down the Fourier transform of the Hilbert transform:

$$\begin{aligned} \text{FT}\{F_H(x)\} &= j\,\text{sgn}(s)\text{FT}\{f(x)\} \\ &= \begin{cases} j\text{FT}\{f(x)\}, & s > 0 \\ 0, & s = 0 \\ -j\text{FT}\{f(x)\}, & s < 0 \end{cases} \end{aligned} \qquad (3.26)$$

(By now it should be obvious that whenever we introduce a new mathematical transform we first subject it to a Fourier transform. This is not because we know of nothing better to do, but because we often need to know how the mathematical operation represented by such a transform is viewed in the Fourier plane, something achieved easily with optics and so important for characterizing signals.)

Thus in the Fourier domain the Hilbert transform does not modify the magnitude of a signal, but it does affect the phase by shifting it by $\pm\pi/2$ according to the sign of s. Thus two successive Hilbert transforms will reverse all the phases and make the original signal negative. This last step corresponds to reconstructing (or "recovering") the original signal. We can express this as

$$f(x) = \frac{1}{\pi x} * F_H(x) \qquad (3.27)$$

or

$$f(x) = \frac{-1}{\pi} \int_{-\infty}^{\infty} \frac{F_H(x')\,dx'}{x' - x} \qquad (3.28)$$

Equations (3.24) and (3.28) therefore constitute the Hilbert transform pair. The Hilbert transform of a real function is a real function. Cosine components transform into negative sine components, and sine components transform into cosine components. In fact, the Hilbert transform of an even function is odd, and that of an odd function is even.

In order to gain a deeper understanding of the origins and meaning of the Hilbert transform, let us consider the analysis of a real physical signal in the time and frequency domain. (Extension to the space and spatial frequency domains is trivial.) We do this also because many readers may have more familiarity with time-domain signals, including their Hilbert transforms.

Given a real signal $v_R(t)$ we assume its Fourier transform exists such that

$$v_R(t) = \frac{1}{2\pi} \int_{-\infty}^{\infty} V_R(\omega)\,\exp(j\omega t)\,d\omega \qquad (3.29)$$

that is, $v_R(t)$ is simply given by the inverse Fourier transform of "some" function $V_R(\omega)$. The Fourier transform is therefore

$$V_R(\omega) = \int_{-\infty}^{\infty} v_R(t)\,\exp(-j\omega t)\,dt \qquad (3.30)$$

3.3 HILBERT TRANSFORM

$$= \int_{-\infty}^{\infty} v_R(t) \cos(\omega t)\, dt - j \int_{-\infty}^{\infty} v_R(t) \sin(\omega t)\, dt \qquad (3.31)$$

where the first integral is odd with respect to ω and the second integral is even. Since $v_R(t)$ is real and $V_R(\omega)$ is (in general) complex, we can also note that

$$V_R{}^*(-\omega) = V_R(\omega) \qquad (3.32)$$

Thus $V_R(\omega)$ is symmetric, with a real even and odd imaginary part. Now, if we write $V_R(\omega)$ in phasor notation, we obtain

$$V_R(\omega) = |V_R(\omega)| \exp[j\phi(\omega)] \qquad (3.33)$$

Equation (3.29) can then be written

$$v_R(t) = \frac{1}{2\pi} \int_{-\infty}^{\infty} |V_R(\omega)| \exp[j\phi(\omega)] \exp(j\omega t)\, d\omega \qquad (3.34)$$

or

$$v_R(t) = \frac{1}{2\pi} \int_{-\infty}^{0} |V_R(\omega)| \exp\{j[\phi(\omega) + \omega t]\}\, d\omega \\ + \frac{1}{2\pi} \int_{0}^{\infty} |V_R(\omega)| \exp\{j[\phi(\omega) + \omega t]\}\, d\omega \qquad (3.35)$$

If we then reverse the sign of ω in the first integral, use the symmetry property [Equation (3.32)] and the trigonometric identity $\cos x = [\exp(jx) + \exp(-jx)]/2$, we obtain

$$v_R(t) = \frac{1}{\pi} \int_{0}^{\infty} |V_R(\omega)| \cos[\phi(\omega) + \omega t]\, d\omega \qquad (3.36)$$

where we have restricted the integral to non-negative frequencies only.

Thus we conclude that all information necessary to reconstruct $v_R(t)$ lies in the positive-frequency domain (including zero). This result suggests that it should be possible to construct a complex representation of $v_R(t)$ that has Fourier components at positive (and zero) frequencies but that is zero for negative frequencies. We can establish the consequences of such a signal by defining a function of frequency $V(\omega)$ which is single-sided:

$$V(\omega) = \sqrt{2} \cdot V_R(\omega) u(\omega) \qquad (3.37)$$

where

$$u(\omega) = \begin{cases} 1, & \omega \geq 0 \\ 0, & \omega < 0 \end{cases} \qquad (3.38)$$

is the unit step function in frequency. Since Equation (3.37) is a product in the frequency domain, it implies that $v(t)$ is a convolution in time of $v_R(t)$ and $\mathrm{FT}^{-1}[u(\omega)]$. Therefore,

$$v(t) = \sqrt{2} \cdot v_R(t) * \mathrm{FT}^{-1}[u(\omega)] \qquad (3.39)$$

where the inverse Fourier transform of $u(\omega)$ is given by

$$\text{FT}^{-1}\{u(\omega)\} = \text{FT}^{-1}\left\{\frac{1}{2} + \frac{1}{2}\operatorname{sgn}(\omega)\right\} \tag{3.40}$$

$$= \frac{1}{2}\left[\delta(t) + \frac{1}{j\pi t}\right] \tag{3.14}$$

Thus (3.39) becomes

$$v(t) = \frac{\sqrt{2}}{2}v_R(t) + j\frac{\sqrt{2}}{2}\left(\frac{1}{-\pi t}\right) * v_R(t)$$

$$v(t) = \frac{\sqrt{2}}{2}\int_{-\infty}^{\infty} v_R(z)\delta(t-z)\,dz + \frac{\sqrt{2}}{2\pi j}\int_{-\infty}^{\infty} \frac{v_R(z)}{t-z}\,dz \tag{3.42}$$

where z is a dummy variable corresponding to time (in this case), and where the second integral on the right-hand side of (3.42) is, in fact, the Hilbert transform of $v_R(t)$. This simplifies to

$$v(t) = \frac{1}{\sqrt{2}}\left[v_R(t) + \frac{j}{\pi}\int_{-\infty}^{\infty}\frac{v_R(z)}{z-t}\,dz\right] \tag{3.43}$$

where the second term on the right-hand side can be identified as the imaginary component of $v(t)$. Then

$$v(t) = \frac{1}{\sqrt{2}}[v_R(t) + jv_I(t)] \tag{3.44}$$

Recalling Equation (3.32) and noting that $v(t)$ can be derived from $V(\omega)$ by using the complex exponential kernal, it is clear that

$$v_I(t) = \frac{-1}{\pi}\int_0^{\infty} |V_R(\omega)|\sin[\phi(\omega) + \omega t]\,d\omega \tag{3.45}$$

$v(t)$ is the so-called "analytic signal" associated with the real signal $v_R(t)$. Its spectrum is single-sided and contains all the information necessary to reconstruct $v(t)$.

The general properties of the Hilbert transform resemble those of the Fourier transform in some respects. These properties are summarized in Table 3.1 in Appendix A, and a list of common Hilbert transform pairs is given in Table 3.2.

As an example of the significance of the Hilbert transform for causal signals, consider the general causal signal defined by

$$h(t) = h(t)u(t) = \begin{cases} h(t), & t \geq 0 \\ 0, & t > 0 \end{cases} \tag{3.46}$$

where $u(t)$ is the unit step function; that is, $h(t)$ can only be causal. The Fourier transform of this signal is

$$H(\omega) = \int_{-\infty}^{\infty} h(t)u(t)\exp(-j\omega t)\,dt \tag{3.47}$$

3.3 HILBERT TRANSFORM

Splitting $h(t)$ into even $h_e(t)$ and odd $h_o(t)$ parts, we require, of course, that

$$h(t) = h_e(t) + h_o(t) \tag{3.48}$$

Referring to the construction in Fig. 3.2, it is clear that

$$h(t) = \frac{1}{2}[h(t) + h(-t)] + \frac{1}{2}[h(t) - h(-t)] \tag{3.49}$$

Note that we can express $h_o(t)$ as

$$h_o(t) = \text{sgn}(t)h_e(t) \tag{3.50}$$

and using Table 1.5 in Appendix A we obtain

$$\text{FT}\{\text{sgn}(t)\} = -2j/\omega \tag{3.51}$$

Therefore,

$$h(t) = [1 + \text{sgn}(t)]h_e(t) \tag{3.52}$$

and

$$H(\omega) = H_e(\omega) + j(-2/\omega) * H_e(\omega) \tag{3.53}$$

Figure 3.2 Simple graphical construction illustrating principle that all functions have an even and odd part.

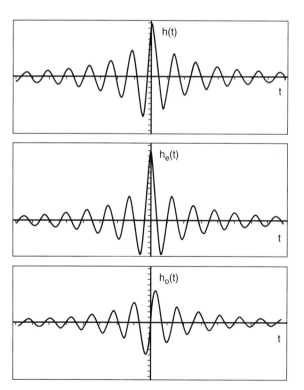

Note that the second term of Equation (3.53) is identical in form to (3.25). Thus the causal signal Fourier transform [or transfer function, if $h(t)$ represents an impulse response] can be expressed as

$$H(\omega) = G(\omega) + jB(\omega) \tag{3.54}$$

where $B(\omega)$ is the Hilbert transform of $G(\omega)$. Thus

$$B(\omega) = \frac{1}{\pi} \int_{-\infty}^{\infty} \frac{G(\omega')}{\omega' - \omega} d\omega' \tag{3.55}$$

and

$$G(\omega) = \frac{-1}{\pi} \int_{-\infty}^{\infty} \frac{B(\omega')}{\omega' - \omega} d\omega' \tag{3.56}$$

are a Hilbert transform pair.

EXAMPLE 3.2 Hilbert Transform Pair for an RC Circuit To apply the Hilbert transform to a specific case consider:

$$G(\omega) = \frac{1}{1 + (\omega \tau)^2}$$

(where $\omega = 2\pi f$) which corresponds to the in-phase (dissipative) component of the impedance of an RC circuit—that is, its conductance. The Hilbert transform of $G(\omega)$ is the susceptance $B(\omega)$ given by

$$B(\omega) = \frac{1}{\pi} \int_{-\infty}^{\infty} \frac{1}{1 + (\omega' \tau)^2} \frac{1}{\omega' - \omega} d\omega'$$

Therefore,

$$B(\omega) = \frac{-\omega \tau}{1 + (\omega \tau)^2}$$

$G(\omega)$ and $B(\omega)$ are plotted in Figs. 3.3(a) and 3.3(b) for comparison. Clearly $G(\omega)$ is even and $B(\omega)$ is odd, as expected.

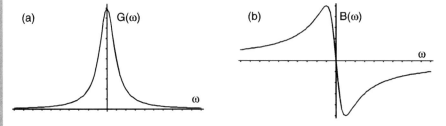

Figure 3.3 (a) Conductance (in-phase) and (b) susceptance (quadrature) components of an RC filter response.

3.4 RADON TRANSFORM

In the traditional (2-D) image processing situation the Radon transform can be viewed as a collection of 1-D projections of a 2-D function $f(x, y)$. A single projection located in a plane parallel to the x axis, as shown in Fig. 3.4, can be derived by integration along lines parallel to the y axis for an object rotated by an angle $\phi = \phi_0$ with respect to the x axis. The 1-D function thus generated has as its independent variable the perpendicular distance of the integration line from the origin measured along the x axis, denoted by the scalar value p comprising along with ϕ the vector $\mathbf{p} = (p, \phi)$. The composite of all such projections can be denoted by $F_R(p, \phi)$, implying that a new 2-D function has been created. However, since operations of interest usually act only on individual projections, we can view the projection data as 1-D but parameterized by ϕ. The projections can also be considered as resulting from integration over lines parallel to the y' axis when the object is fixed, where the (x', y') coordinate system is rotated by angle ϕ with respect to the (x, y) coordinate system. Mathematically, the projection can be viewed as a 1-D integral transform in which the kernal is a 1-D Dirac delta function that selects the projection's azimuth angle ϕ. If we define a position on the function $f(\mathbf{r}) = f(x, y)$ as $\mathbf{r} = (r, \theta)$ or $\mathbf{r} = r\cos\theta\mathbf{i} + r\sin\theta\mathbf{j}$, where $\mathbf{i}$ and $\mathbf{j}$ are unit vectors along the x and y axes, respectively, and we define the unit vector $\mathbf{n}$ as $\mathbf{n} = \cos\phi\mathbf{i} + \sin\phi\mathbf{j}$, then the coordinate p is defined by

$$p = r\cos(\theta - \phi) = \mathbf{r} \cdot \mathbf{n} \tag{3.57}$$

Then the kernal of the Radon transform can be defined as $\delta(p - \mathbf{r} \cdot \mathbf{n})$. The Radon transform is therefore defined as

$$F_R(p, \phi) = \int_{-\infty}^{\infty}\int_{-\infty}^{\infty} f(\mathbf{r})\delta(p - \mathbf{r} \cdot \mathbf{n})\, d^2\mathbf{r} \tag{3.58}$$

Figure 3.4 Projections of a two-dimensional function $f(x, y)$ onto the x-z and y-z planes, corresponding to the Radon transform samples at $\phi = 0$ and $\pi/2$.

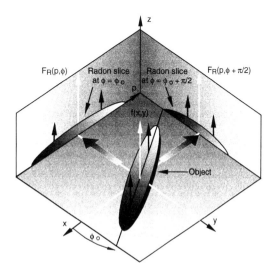

The new coordinates (p, ϕ) constitute Radon space, and a plot of $F_R(p, \phi)$ is termed a sinogram, because a single point in Cartesian space maps to a sinusoid in Radon space. The Radon transform for the object in Fig. 3.4 is shown on the side panel as a function of p. Clearly when $\phi = 0$ the Radon transform is taken along the longitudinal axis of symmetry, whereas when $\phi = \pi/2$ the two peaks coincide. To picture this qualitatively, imagine a flashlight illuminating the 3-D function as a physical object in the -x and -y directions onto the "walls" defined by the y–z and x–z planes, respectively. The shadows roughly resemble their projections.

An important theorem associated with the Radon transform is the projection-slice theorem, namely,

$$\text{FT}_{\text{2-D}}[f(\mathbf{r})] = \text{FT}_{\text{1-D}}\{F_R[f(\mathbf{r})]\} \tag{3.59}$$

which says that the 2-D Fourier transform of a function $f(x, y)$ can be obtained by first taking the Radon transform and then taking a 1-D Fourier transform on the projection data. Thus one can get to 2-D Fourier space by two different routes. In pattern recognition the 2-D Fourier transform is often desired, but the processing overhead for calculating it by brute force may be too great to be acceptable in some practical situations. The important consequence of the projection-slice theorem is that by calculating the Radon transform first and then using a 1-D Fourier transform, the computational burden for obtaining the 2-D Fourier transform can be reduced. The price paid is that nonuniform (polar) sampling may alias the image frequency content. The degree of aliasing depends on the actual frequency content of the image and how fine the image is sampled in the azimuth angle ϕ. (The computational burden is often reduced because not all Radon slices are needed to adequately reconstruct the desired object; that is, not all polar samples are needed.)

A slice of the Radon transform at $\phi = \phi_0$ is given by

$$F_R(p, \phi_0) = F_R\{f(x, y)\}|_{\phi=\phi_0} \tag{3.60}$$

$$= \int_{-\infty}^{\infty} \int_{-\infty}^{\infty} f(x, y)\delta(p - x\cos\phi_0 - y\sin\phi_0)\,dx\,dy \tag{3.61}$$

The delta function "sifts out" of the integral the function $f(x, y)$ along the locus of points defined by $p = x\cos\phi_0 + y\sin\phi_0$. This expression is equivalent to the 1-D Fourier transform of $f(x, y)$ evaluated at zero spatial frequency ($v = 0$) along an axis perpendicular to the Fourier axis, when the space coordinates in $f(x, y)$ are coordinate-transformed by a rotation matrix $\mathbf{R}$ such that $\mathbf{x}' = \mathbf{R}\mathbf{x}$. Thus

$$F_1(x') = F_{1D}\{f(x', y')\}|_{v=0} \tag{3.62}$$

$$= \int_{-\infty}^{\infty} f(x', y')\exp(-j2\pi yv)\,dy|_{v=0} \tag{3.63}$$

The rotation transformation is explicitly given by Equation (1.27) for $\phi = \phi_0$:

$$\begin{bmatrix} x' \\ y' \end{bmatrix} = \begin{bmatrix} \cos\phi_0 & \sin\phi_0 \\ -\sin\phi_0 & \cos\phi_0 \end{bmatrix}\begin{bmatrix} x \\ y \end{bmatrix} \tag{3.64}$$

Therefore, by inspection of Equation (3.64) we obtain

$$x' = x \cos \phi_0 + y \sin \phi_0 = p \tag{3.65}$$

which is consistent with the meaning of the delta function as given by its argument in Equation (3.61). Equation (3.62) can be helpful in implementing the Radon transform in an optical Fourier transform architecture.

Now let's consider a simple example of the Radon transform calculated using Equation (3.58).

EXAMPLE 3.3 Radon Transform of an Impulse Array Consider the impulse array defined by

$$f(x, y) = \delta(x - 1)\delta(y) + \delta(x - 2)\delta(y - 1) + \delta(x - 2)\delta(y + 1)$$

and illustrated in Fig. 3.5(a). Applying the Radon transform given by

$$F_R(p, \phi) = \int_{-\infty}^{\infty} \int_{-\infty}^{\infty} f(x, y) \delta(p - x \cos \phi - y \sin \phi) \, dx dy$$

we obtain

$$F_R(p, \phi) = \delta(p - \cos \phi) + \delta(p - 2 \cos \phi - \sin \phi) + \delta(p - 2 \cos \phi + \sin \phi)$$

which is illustrated in Fig. 3.5(b). Note that at angles $\phi = 0$, $\pi/4$, $3\pi/4$, $5\pi/4$, and $7\pi/4$, the array points line up, causing a discontinuous jump in the "serpentine walls" of the sinogram, denoted by heavy dots in the sinogram.

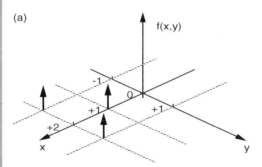

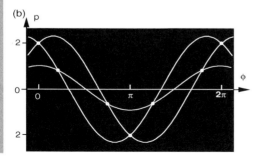

Figure 3.5 (a) An array of impulse functions whose Radon transform is the sinogram shown in (b) in plan view.

In principle we can obtain an exact reconstruction of the original image $f(x, y)$ by using the inverse Radon transform, which is given by

$$f(\mathbf{r}) = \frac{1}{2\pi^2} \int_0^\pi \int_0^\infty \frac{\frac{dF_R(p, \phi)}{dp}}{r \cos(\phi - \theta)} d\phi \qquad (3.66)$$

This result, however, is often not exactly realized in practice. To understand this consider the following example.

EXAMPLE 3.4 Reconstruction of a Point Object by Back-Projection Consider the case where $f(\mathbf{r})$ is simply given by $f(\mathbf{r}) = \delta(x)\delta(y)$ corresponding to a 2-D impulse function (or "thumbtack"). Using Equation (3.58),

$$F_R(p, \phi) = \int_{-\infty}^\infty \int_{-\infty}^\infty \delta(x)\delta(x)\delta(p - \mathbf{r} \cdot \mathbf{n}) d^2\mathbf{r}$$

we find that

$$F_R(p, \phi) = \frac{1}{\pi r}$$

This result is termed the point spread function of the system representing the Radon transform. The back-projection of an arbitrary function $f(\mathbf{r})$ is then derived from the expression

$$f'(\mathbf{r}) = f(\mathbf{r}) * \frac{1}{\pi r}$$

which can be interpreted to mean that $f'(\mathbf{r})$ is a blurred replica of $f(\mathbf{r})$.

To improve on this result the back-projections can be filtered with some transfer function or filter $H(\boldsymbol{\rho})$ in the frequency domain $\boldsymbol{\rho}$. To see this, recall the previous equation and note that by the convolution theorem we can write

$$F'(\boldsymbol{\rho}) = \mathbf{F}(\boldsymbol{\rho}) F\left\{\frac{1}{\pi r}\right\} = \mathbf{F}(\boldsymbol{\rho}) \frac{1}{\pi \rho} \qquad (3.67)$$

in the frequency domain, where bold $\mathbf{F}$ denotes a Fourier transform. To get an accurate reconstruction we must construct a filter that cancels the blurring effect of pure back-projection. A filter of the form $H(\boldsymbol{\rho}) = \pi \rho A(\boldsymbol{\rho})$ will do this, where $\pi \rho$ is the so-called rho filter and $A(\boldsymbol{\rho})$ is an apodizing function that is necessary to ensure that the rho filter is finite and bounded. It plays the same role as a pupil function in describing a lens. The filtered reconstruction, $F'_f(\boldsymbol{\rho})$, then becomes

$$F'_f(\rho) = H(\rho) F'(\rho) = \pi \rho A(\rho) \mathbf{F}(\rho) \frac{1}{\pi \rho} = A(\rho) \mathbf{F}(\rho) \qquad (3.68)$$

Therefore, $f'_f(\mathbf{r}) = a(\mathbf{r}) * f(\mathbf{r})$, which, when written out explicitly, is

$$f'_f(\mathbf{r}) = \int_0^\pi \int_{-\infty}^\infty h(p) F_R(p - \mathbf{r} \cdot n) d\phi dp \qquad (3.69)$$

where $h(p)$, in general, can be bipolar.

There is an interesting connection between the Radon transform and the Abel transform worth noting here. If the original function $f(x, y)$ is circularly symmetric, then the Radon transform reduces to the Abel transform. Since $f(x, y)$ would then depend only on $r = (x^2 + y^2)^{1/2}$ and not on the polar angle, all profiles would be equivalent. To see this, note that if we let $\phi_0 = 0$ in the definition of the Radon transform (3.58), then $F_R(p, \phi_0) = F_R(p)$ and

$$F_R(p) = \int_{-\infty}^{\infty} \int_{-\infty}^{\infty} f(\mathbf{r}) \delta(p - x) \, dx \, dy \tag{3.70}$$

Then

$$F_R(p) = \int_{-\infty}^{\infty} \int_{-\infty}^{\infty} f\left[(x^2 + y^2)^{1/2}\right] \delta(p - x) \, dx \, dy$$

$$= 2 \int_{-\infty}^{\infty} f\left[(p^2 + y^2)^{1/2}\right] dy$$

By changing variables such that $r^2 = p^2 + y^2$, we obtain

$$F_R(p) = 2 \int_{|p|}^{\infty} \frac{f(r) r \, dr}{\sqrt{(r^2 - p^2)}} = F_A(p) \tag{3.71}$$

which is the Abel transform, where r is the radius and x is the abscissa in the image plane. The Abel transform is useful when circularly symmetric image distributions are projected into one dimension. The Abel transform is important in optical image scanning, especially in understanding the conversion of the edge response (spatial step function) of an optical system to the corresponding point spread function (impulse response). It is also helpful in analyzing television (raster) displays.

3.5 MELLIN TRANSFORM

The Mellin transform in one dimension is given by

$$F_M(s) = \int_0^{\infty} f(x) x^{s-1} \, dx \tag{3.72}$$

and is equivalent to the Laplace transform when a change of variables is made given by $x = \exp(-t)$. Substituting $\exp(-t)$ for x in Equation (3.72) yields the Laplace transform:

$$F_M(s) = \int_0^{\infty} g(t) \exp(-st) \, dt \tag{3.73}$$

where $g(t) = f[\exp(-t)]$. $F_M(s)$ can also be regarded as the $(s - 1)$th moment of $f(x)$, as discussed in the following example.

EXAMPLE 3.5 Mellin Transform of a Triangle Function Given the function $g(t) = f[\exp(-t)] \to f(x)$ defined as

$$f(x) = \text{tri}\left(\frac{x - 1}{1}\right) = \begin{cases} x, & 0 \le x \le 1 \\ -x + 2, & 1 < x < 2 \\ 0, & \text{elsewhere} \end{cases}$$

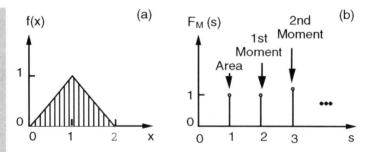

Figure 3.6 (a) Triangle function and (b) corresponding Mellin transform showing moments.

illustrated in Fig. 3.6(a), its Mellin transform is

$$F_M(s) = \int_0^\infty \text{tri}\left(\frac{x-1}{1}\right) x^{s-1} \, dx$$

The result corresponding to $f(x)$ is

$$F_M(s) = \frac{2(2^s - 1)}{s(s+1)}$$

which is shown in Fig. 3.6(b). Note that $F_M(1) = 1$ is the area of $f(x)$, $F_M(2) = 1$ is the first moment, $F_M(3) = 7/6$ is the second moment, and so forth. This property is useful when applying the calculation of moments by optical means to the problem of feature extraction for pattern recognition.

It also interesting to note that the Mellin transform is related to the Laplace transform in the same way as the z transform is to the Laplace transform. Instead of using $x = \exp(-t)$, we use $z = \exp(-s)$ to derive the z transform. Thus, if we write down the inverse Mellin transform,

$$f(x) = \frac{1}{2\pi j} \int_{c-j\infty}^{c+j\infty} F_M(s) x^{-s} \, ds \tag{3.74}$$

we see that it's analogous to the z transform given by

$$F(z) = \int_{-\infty}^{\infty} f(t) z^{-t} \, dt \tag{3.75}$$

except that t is real, whereas s may be complex. Likewise x is real, but z may be complex. Some examples of Mellin transform pairs are included in Table 3.3 in Appendix A.

Let us consider the last and perhaps most important property of the Mellin transform for our purposes: scale invariance. The scale invariance associated with the Mellin transform can be explained easily if we define the Mellin transform as

$$F_M(\omega) = \int_0^\infty f(x) x^{j\omega - 1} \, dx \tag{3.76}$$

where $-j\omega$ replaces s in Equation (3.72). Essentially we are restricting our attention to values of s along the imaginary ($s = j\omega$) axis. In this case if we let $x = \exp(-t)$, then

$$F_M(\omega) = \int_0^\infty f[\exp(t)] \exp(-j\omega t)\, dt \qquad (3.77)$$

Thus $F_M(\omega)$ becomes the Fourier transform of the function $g(t) = f[\exp(t)]$ whose independent variable t is related to x logarithmically. We can now show that if two functions exist given by $f_1(x)$ and $f_2(x) = f_1(ax)$ [i.e., $f_2(x)$ is a scaled version of $f_1(x)$], then their respective Mellin transforms $F_{1M}(\omega)$ and $F_{2M}(\omega)$ differ only by a phase factor, so that $|F_{1M}(\omega)| = |F_{2M}(\omega)|$. To see this note that

$$F_{2M}(\omega) = \int_0^\infty f_1(ax) x^{j\omega - 1}\, dx \qquad (3.78)$$

and redefine variables: $x' = ax$. Then

$$F_{2M}(\omega) = a^{-j\omega} \int_0^\infty f_1(x') x'^{j\omega - 1}\, dx' \qquad (3.79)$$

$$F_{2M}(\omega) = a^{-j\omega} F_{1M}(\omega) \qquad (3.80)$$

The factor $a^{-j\omega}$ can also be written in phasor notation as $\exp(-j\omega \ln a)$ to emphasize its role as a phase factor. If $F_{1M}(\omega)$ and $F_{2M}(\omega)$ happen to be inputs to a matched filter [recalling that Equation (3.77) is cast in the form of a Fourier transform], then their product is

$$F_{1M}(\omega) F_{2M}^*(\omega) = |F_{1M}(\omega)|^2 \exp(-j\omega \ln a) \qquad (3.81)$$

and the corresponding inverse Fourier transform necessary to achieve the cross-correlation is given by

$$f_1[\exp(t)] * f_1[\exp(t)] * \delta(x - \ln a) \qquad (3.82)$$

Hence, it is clear that the correlation peak has not been degraded by the difference in scale, but, by virtue of the logarithmic coordinate scaling, has merely been shifted to a new position $x = \ln a$. Thus scale invariance has been achieved by the use of the Mellin transform by converting the scale factor a to a coordinate shift. The following example illustrates the effect of logarithmic scaling inherent in the Mellin transform applied to an object function in two dimensions.

EXAMPLE 3.6 Logarithmic Scaling of a Rectangular Pillbox In Fig. 3.7(a) we see two squares of different sizes in the xy plane, representing an image of two pillboxes. When the logarithmic transformation is applied to the spatial coordinates (i.e., $x' = \ln x$ and $y' = \ln y$), we obtain the result of Fig. 3.7(b). [The above equation can also be expressed as a nonlinear vector transformation: $\mathbf{x}' = \ln \mathbf{x}$, where $\mathbf{x}'$ is the vector (x', y') and $\mathbf{x}$ is the vector (x, y).] It is obvious that the scale of the two pillboxes is now the same. This property is important in image correlation when two different sized pillboxes are located in *different* image planes, but they are meant to represent the same object which happens to be viewed with different scales. Then their peak cross-correlation will be degraded over the case when they have the same scale. Thus the logarithmic transformation inherent in the Mellin transform will enable them to have effectively the same scale and thus a peak correlation value.

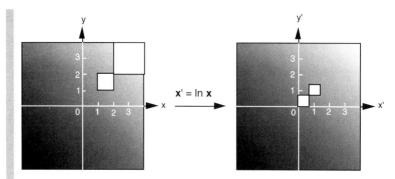

Figure 3.7 Two rectangular pillboxes of different sizes can be brought to equal size by the logarithmic scaling of the Mellin transform.

3.6 WAVELET TRANSFORM

The wavelet transform is linear and square-integrable like the classic Fourier, Hilbert, Radon, and other transforms. Its kernels are called wavelets, which are not fixed, but are an infinite set derivable from a "mother" kernel $h(x)$ by scale changes (denoted by a) and shifts (denoted by b). The general 1-D wavelet is defined as

$$\psi(x) = \frac{1}{\sqrt{a}} h\left(\frac{x-b}{a}\right) \quad (3.83)$$

where $a^{-1/2}$ is a normalization factor. Usually the form of $h(x)$ is $h(x) = w(x)f(x)$, where $w(x)$ is a window function (usually Gaussian) and $f(x)$ is a modulation term usually of the form $\exp(jk_0 x)$. Although wavelet transforms can be discrete or continuous, in optics they may in practice by hybrid: discrete in scale and continuous in shift variables.

For every input coordinate there are two output coordinates, hence a 1-D signal yields a 2-D wavelet transform, and a 2-D signal yields a 4-D wavelet transform. One reason that optical implementations are of interest is that they can perform the wavelet transform in parallel. Another, less obvious advantage of optics is that Fourier optics can map shift continously into complex phase information, which is invariant under square-law detection.

Wavelet transforms are useful for information processing of signals because they inherently sample regions of low spatial frequency content at low rates and sample high spatial frequency content at high rates. This is expressed by the fact that wavelet transforms preserve the ratio of frequency to frequency resolution (i.e., a measure of constant relative bandwidth), $f/\Delta f$, which is also known as the Q (quality factor) of the signal. Thus wavelets should be very effective at bandwidth reduction and for nonstationary signals—that is, signals whose frequency content varies with time.

Another advantage of the wavelet transform is that it provides information on both the space (or time) domain and the corresponding frequency domain. The Wigner distribution and ambiguity function, described in Chapter 11, also provide both space and frequency domain information but are of second-order in the signal, whereas the wavelet transform is first-order. This means that the wavelet transform accommodates linear superposition.

The wavelet transform of a 2-D image $s(x, y)$ is a 4-D function defined by

$$W(a_x, a_y, b_x, b_y) = (a_x a_y)^{-1/2} \int_{-\infty}^{\infty} \int_{-\infty}^{\infty} s(x, y) h\left(\frac{x - b_x}{a_x}, \frac{x - b_y}{a_y}\right) dx dy \quad (3.84)$$

which is the convolution of $s(x, y)$ with the scaled-window $h(x, y)$, or correlation if $s(x, y)$ is shifted and $h(x, y)$ is only scaled. It turns out that Equation (3.84) can be implemented optically by a shadow-casting system in the space domain, which would ignore the phase factor associated with translation in space. The frequency-domain representation of the wavelet transform given by Equation (3.84) is

$$W(a_x, a_y, b_x, b_y) = (a_x a_y)^{1/2} \int_{-\infty}^{\infty} \int_{-\infty}^{\infty} S(u, v) H(a_x u, a_y v)$$
$$\times \exp[-j2\pi(b_x u + b_y v)] du dv \quad (3.85)$$

where $S(u, v)$ and $H(u, v)$ are the Fourier transforms of $s(x, y)$ and $h(x, y)$. Thus Equation (3.85) indicates that the wavelet transform of $s(x, y)$ is the inner product of its Fourier transform and that of the wavelets, which can be implemented optically by spatial filtering in the Fourier domain. Table 3.5 in Appendix A summarizes the most common wavelets and their Fourier transforms in one dimension. We will discuss some of these below.

The wavelet transform can be considered as a correlation of a signal $s(x, y)$ with a set of dilated and shifted wavelets $h(x, y)$. In the Fourier domain the wavelet transform can be viewed as a set of wavelet transform filters. In two dimensions each pair of scale and shift varaibles, of course, may differ, leading to the possibility of rotation as well as dilation and translation. The fact that the wavelet transform in general depends on four parameters means that it is four-dimensional, thus requiring a 4-D matrix, implemented, for instance, in the form of a multichannel correlator.

Historically, the wavelet transform was anticipated by the work of Haar, who first constructed a complete orthonormal set of bipolar step functions by dilations and contractions (scale changes). Of course, Gabor developed elementary filter functions, which are one of the antecedents of more recent wavelets. These are defined as a Fourier transform pair with the same functional form in both domains:

$$g(x, y) = \exp\{-\pi[(x - x_0)^2 \alpha^2 + (y - y_0)^2 \beta^2]\}$$
$$\exp\{-2\pi j[u_0(x - x_0) + v_0(y - y_0)]\} \quad (3.86)$$

$$G(u, v) = \exp\left\{-\pi\left[\frac{(u - u_0)^2}{\alpha^2} + \frac{(v - v_0)^2}{\beta^2}\right]\right\}$$
$$\exp\{-2\pi j[x_0(u - u_0) + y_0(v - v_0)]\} \quad (3.87)$$

These functions contain the properties of spatial localization, orientation selectivity, spatial frequency selectivity, and quadrature phase relationship by virtue of the parameters $x_0, y_0, u_0, v_0, \alpha$, and β. If $\alpha \neq \beta$, then a further degree of freedom of rotation is incorporated in addition to position, modulation, and scale. The Gabor filter is interesting because it achieves the theoretical lower bound on the joint uncertainty between the two domains:

84 MATHEMATICAL TRANSFORMS USED IN OPTICAL SIGNAL PROCESSING

spatial (x, y) and spatial frequency (u, v). In addition, because of their above-mentioned selectivities, Gabor filters are a good model for visual perception, as well as neural network weighting functions. The real part of Equation (3.86) and its Fourier transform (3.87) are plotted in Fig. 3.8.

Subsequently, wavelet transform theory was developed by Morlet and co-workers and by Daubechies. There are several types of wavelets that have been developed, including the Haar, Morlet, and Mexican-hat wavelets. The Haar wavelet is defined as a bipolar step function:

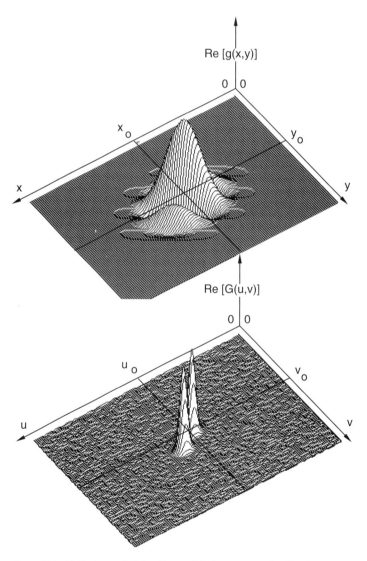

Figure 3.8 (a) Real part of Gabor filter and (b) its corresponding Fourier transform oriented at 45° with respect to the (x, y) and (u, v) axes.

$$h(x) = \text{rect}\left[2\left(t - \frac{1}{4}\right)\right] - \text{rect}\left[2\left(t - \frac{3}{4}\right)\right] \tag{3.88}$$

which illustrated in one-dimension in Fig. 3.9(a). Its corresponding Fourier transform is

$$H(\omega) = 2j \, \exp(-j\omega/2) \left[\frac{1 - \cos(\omega/2)}{(\omega/2)}\right] \tag{3.89}$$

and is shown in Fig. 3.9(b). Since $h(x)$ is antisymmetric about $x = 1/2$, then $H(\omega)$ has a phase factor $\exp(-j\omega/2)$. The modulus is positive, even, and symmetric about $\omega = 0$, and its square (power spectrum) converges very slowly because of the high-frequency content inherent in $h(x)$. As an image filter it will therefore tend to emphasize high-frequency content.

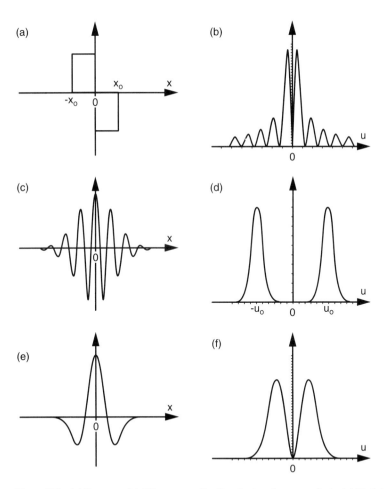

Figure 3.9 (a) Haar wavelet, (b) corresponding Fourier transform magnitude, (c) Morlet wavelet, (d) corresponding Fourier transform magnitude, (e) Mexican-hat wavelet, and (f) corresponding Fourier transform magnitude. Note the variation in frequency content. Since the Haar wavelet is antisymmetric, the magnitude of its Fourier transform inverts the negative frequency side of the real component.

The Morlet wavelet, first used to analyze sound patterns, is given, in general, by

$$h(x) = \exp(jk_0 x) \exp(-x^2/2) \tag{3.90}$$

where its real part is even and shown in Fig. 3.9(c). Its Fourier transform is

$$H(\omega) = 2\pi \{\exp[-(\omega - \omega_0)^2/2] + \exp[-(\omega + \omega_0)^2/2]\} \tag{3.91}$$

which is shown in Fig. 3.9(d). It is not, strictly speaking, an admissible function, because it does not have a zero mean and hence zero (DC) Fourier component. However, in the limit of larger ω_0 the value of $H(0)$ is very close to zero and hence satisfactory for numerical computation. Notice its similarity to the generalized Gabor filter described earlier. Gabor also introduced the Mexican-hat wavelet by taking the Laplacian of a Gaussian function, which yields

$$h(x) = (1 - |x|^2) \exp(-|x|^2/2) \tag{3.92}$$

as shown in Fig. 3.9(e) with Fourier transform:

$$H(\omega) = \omega^2 \exp(-\omega^2/2\pi) \tag{3.93}$$

as shown in Fig. 3.9(f). Both $h(x)$ and $H(\omega)$ are even and real-valued. As an image filter the Gabor wavelet is similar to the human visual response. More recently, there have been other forms of wavelet invented, including those by Meyer and Daubechies.

EXAMPLE 3.7 Modified Gabor (or Mexican Hat) Wavelet Filtering of a Simple Letter "E"
As an example of the effect of a wavelet filter on an image, consider the use of a modified Gabor (or Mexican hat) wavelet on the letter "E". In this case the impulse response of the wavelet is given by

$$h(x, y) = [1 - (ax)^2][1 - (by)^2] \exp[-(ax)^2/2] \exp[-(by)^2/2]$$

This filter can be implemented as a two-dimensional convolution in the space domain or as an equivalent matched filter in the spatial frequency domain. The letter "E" is shown in Fig. 3.10(a). The result of filtering with the Gabor filter in which $a = 0$ and $b = 1$ is purely one-dimensional (along the y axis) as shown in Fig. 3.10(b), resulting in the emphasis of the horizontal edges of the letter "E". These are really "local" rather than "global" features of the letter "E", even though the matched filter implementation of this wavelet filter is considered a "global" operation. The result of filtering in which $a = b = 1$ is two-dimensional and corresponds to convolving a template given by:

$$\begin{bmatrix} 0 & 0 & -1 & 0 & 0 \\ 0 & 17 & -41 & 17 & 0 \\ -1 & -41 & 100 & -41 & -1 \\ 0 & 17 & -41 & 17 & 0 \\ 0 & 0 & -1 & 0 & 0 \end{bmatrix}$$

with the letter "E" as shown in Fig. 3.10(c). In this case the edges and corners are emphasized such that, after suitable thresholding, only the corners of the letter "E" survive, thus showing how even more local features can be emphasized.

Figure 3.10 Spatial filtering of (a) the letter "E" by a Gabor wavelet filter of varying scale parameters: (b) $a = 0, b = 1$, and (c) $a = b = 1$.

PROBLEM EXERCISES

1. Use Equation (3.22) to find the Fresnel transform of $f(x, y) = \sin(2\pi ux) \sin(2\pi vy)$.

2. Calculate the Hilbert transform of (a) $\sin(\pi ax)/(\pi ax)$ and (b) $\sin^2(\pi ax)/(\pi ax)^2$.

3. If $G(f) + jB(f)$ is the transfer function of a causal filter, show that

$$B(f) = -\frac{2f}{\pi} \int_0^\infty \frac{G(f')}{f^2 - f'^2} df'$$

and

$$G(f) = \frac{2}{\pi} \int_0^\infty \frac{f'B(f')}{f^2 - f'^2} df'$$

4. Verify the projection-slice theorem for the Gaussian function.

5. Calculate the Radon transform of a square pillbox. Delineate three sectors for ϕ as follows: $0 \leq \phi < \pi/4, \pi/4$, and $\pi/4 \leq \phi < \pi/2$. For each range of ϕ there will be more than one expression for $F(p, \phi)$, depending on the value of p.

6. Calculate the sinogram for the following two arrays:

   (a) $f(x, y) = \delta(x - x_0, y) + \delta(x + x_0, y) + \delta(x, y - y_0) + \delta(x, y + y_0)$ and

   (b) $f(x, y) = \delta(x - x_0, y - y_0) + \delta(x + x_0, y - y_0) + \delta(x - x_0, y + y_0) + \delta(x + x_0, y + y_0)$.

   Plot your result in "plan view" and describe how they differ.

7. Show that the Abel transform (or circularly symmetric Radon transform) of the sinc function is the jinc function. (If the jinc function is the point spread function of an optical system, then the sinc function is the line spread function.)

8. Show that the Abel transform of a Gaussian is a Gaussian.

9. The inverse Abel transform is defined by

$$f(r) = -\frac{1}{\pi} \int_r^\infty \frac{\frac{dF_A(x)}{dx}}{\sqrt{x^2 - r^2}} dx$$

Show that if $F_A(x)$ is given by $\cos(2\pi a x)/\pi a$, then $f(r) = J_0(2\pi a r)$. Graph both $F_A(x)$ and $f(r)$ to visualize and interpret your result.

10. Calculate the Mellin transform of the following functions: (a) $\exp[-(x-1)^2]$, (b) $1/(1+x)$, (c) $\exp(-ax)$, and (d) $\text{rect}[(x-1/2)/1]$. Plot your results and for the appropriate values of the variable s, specify the area, mean, and variance.

11. Given $f_1(x) = \text{rect}(x/x_0)$ calculate the Mellin transform of $f_1(ax)$. Determine the cross-correlation function and plot it. How does it differ from the autocorrelation of $f_1(x)$?

12. Given the line spread function $f(x)$:

$$f(x) = \begin{cases} f(x), & |x| < a \\ 0, & |x| > a \end{cases}$$

determine the corresponding point spread function $h(r)$.

13. Show that the arbitrary separable signal defined by

$$f(x, y) = f_1(x) f_2(y)$$

can be reconstructed exactly from two projections. How should the projection angles be related?

14. Given the arbitrary rectangular function $f(x, y)$ illustrated below and defined by

$$f(x, y) = \text{rect}\left[\frac{x - \left(\frac{x_1 + x_2}{2}\right)}{\frac{x_2 - x_1}{2}}\right] \text{rect}\left[\frac{y - \left(\frac{y_1 + y_2}{2}\right)}{\frac{y_2 - y_1}{2}}\right]$$

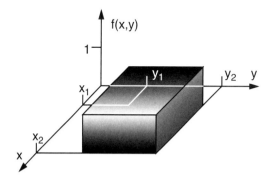

show explicitly that the magnitude of its Mellin transform $M(s)$ is scale invariant. Plot the magnitude of $M(s)$.

15. Calculate the Mellin transform of the circular pillbox $f(r) = \text{circ}(r/r_0)$ and show that it is independent of r_0.

16. Calculate the Abel transform of the jinc function given by

$$\text{jinc}(r) = \frac{J_1(2\pi \rho r)}{\pi r/\rho}$$

17. Starting with the expression for the two-dimensional Fourier transform of a function $f(x, y)$,

$$F(u, v) = \int_{-\infty}^{\infty} \int_{-\infty}^{\infty} f(x, y) \exp[-j2\pi(ux + vy)] \, dx \, dy$$

rewrite it as a triple integral incorporating a Dirac delta function, where the third integral is over the variable $t = qp$, such that $q \geq 0$. Using the scaling property of delta functions and the polar spatial frequency variables $u = q \cos \phi$ and $v = q \sin \phi$, show that the projection-slice theorem results.

18. Given the Hankel transforms

$$G(\rho) = \int_0^{\infty} f(r) J_n(\rho r) r \, dr$$

and

$$f(r) = \int_0^{\infty} G(\rho) J_n(\rho r) \rho \, d\rho$$

show that the Dirac delta function has a Bessel integral representation:

$$\delta(r - r') = r \int_0^{\infty} J_n(\rho r) J_n(\rho r') \rho \, d\rho$$

19. Show the following Mellin transforms:

$$\int_0^{\infty} x^{s-1} \sin(kx) \, dx = k^{-s}(s-1)! \sin\left(\frac{\pi s}{2}\right) \quad (-1 < s < 1)$$

and

$$\int_0^{\infty} x^{s-1} \cos(kx) \, dx = k^{-s}(s-1)! \cos\left(\frac{\pi s}{2}\right) \quad (0 < s < 1)$$

20. Given a family of lines (equally spaced in angle and equal in length) shown below, find the corresponding sinogram and sketch the result. What function satisfies the locus of points in Radon space?

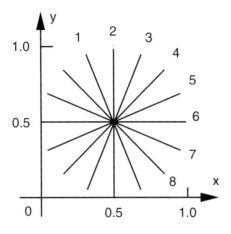

21. Use the projection slice theorem to show that a 2-D convolution of two functions $f(x, y)$ and $g(x, y)$ is equivalent to the 1-D convolution of their projections.

22. Use the projection slice theorem to derive the inverse Radon transform; that is, show that

$$f'(r) = \int_0^\pi F_R(p, \phi) \, d\phi$$

where $f(r)$ represents an estimate of the original image $f(x, y)$ in which the transformation $p = \mathbf{n} \cdot \mathbf{r}$ is termed *back-projection* and is satisfied for a set of vectors $\mathbf{r}$ in a plane. (The constraint $p = \mathbf{n} \cdot \mathbf{r}$ smears information in Radon space in a direction perpendicular to the p axis).

23. Find the Radon transform of

$$f(x, y) = \begin{cases} \exp[-(p_0 - x \cos\phi_0 - y \sin\phi_0)^2] & \text{for } -L < x < L, -L < y < L \\ 0, & \text{otherwise} \end{cases}$$

plot this function and its Radon transform using MATLAB.

24. Analyze the effect of correlating a rectangular pillbox with the Mexican-hat wavelet for varying scale relative to the pillbox. Use fast correlation to calculate your results and interpret the results for small, medium, and large wavelet scale parameters. Use MATLAB.

25. The 2-D wavelet transform given by

$$W(a_x, a_y, b_x, b_y) = \frac{-1}{\sqrt{a_x a_y}} \int_{-\infty}^{\infty} \int_{-\infty}^{\infty} s(x + b_x, y + b_y) h^*\left(\frac{x}{a_x}, \frac{x}{a_y}\right) dx dy$$

is in the form of a correlation [as opposed to convolution as in Equation (3.84)], where a_x and a_y are dilation and b_x and b_y are translation parameters. [Note that the filter $h(x/b_x, y/b_y)$ is only in the form of a dilated wavelet.] Write the general form of the corresponding Fourier-domain wavelet $W(a_x, a_y, b_x, b_y)$. For the Haar, Morlet, Mexican-hat, and Meyer's wavelets derive the form of $W(a_x, a_y, b_x, b_y)$ for a signal $s(x, y) = 1$.

26. A relative of the Fourier transform first presented by R. V. L. Hartley is the Hartley transform, given by

$$F_H(\omega) = \frac{1}{\sqrt{2\pi}} \int_{-\infty}^{\infty} f(t)(\cos \omega t + \sin \omega t) \, dt$$

with inverse given by

$$f(t) = \frac{1}{\sqrt{2\pi}} \int_{-\infty}^{\infty} F_H(\omega)(\cos \omega t + \sin \omega t) \, d\omega$$

An advantage of the Hartley transform is that it is entirely real. Show that the Hartley transform is related to the Fourier transform by the relation

$$F_H(\omega) = \text{Re}[F(\omega)] - \text{Im}[F(\omega)]$$

Compare the Hartley and Fourier transforms for the signal defined by

$$f(t) = \exp(-t/\tau)u(t)$$

and plot $F_H(\omega)$, $\text{Re}[F(\omega)]$, and $\text{Im}[F(\omega)]$.

27. Using the fundamental wavelet transform for a 1-D signal—that is,

$$W(a,b) = \frac{1}{\sqrt{|a|}} \int s(t) h^* \left(\frac{t-b}{a} \right) dt \quad (a \neq 0)$$

where $h(t)$ is selected from the set shown in Figs. 3.9(a), 3.9(c) and 3.9(e)—calculate $W(a,b)$ for each wavelet for the signal:

$$S(t) = \delta(t - t_1) + \delta(t - t_2) + \exp(j\omega_1 t) + \exp(j\omega_2 t)$$

Calculate the Fourier transform $S(\omega)$ for comparison. Plot both $S(\omega)$ and $W(a,b)$ versus time and frequency using MATLAB. How do the time and frequency resolutions compare for high and low frequencies?

28. Given the input signal defined by the function

$$f(x) = \begin{cases} 16 & 0 \leq x < 1/4 \\ -8 & 1/4 \leq x < 1/2 \\ 16 & 1/2 \leq x < 3/4 \\ 20 & 3/4 \leq x < 1 \end{cases}$$

represent $f(x)$ as a weighted linear superposition of discrete Haar wavelets using the inner product integral:

$$b_n = \int_{-\infty}^{\infty} \psi_{jk}(x) f(x) \, dx$$

where $\psi_{jk}(x)$ is given by $\phi(x)$ for $n = 0$, where $\phi(x)$ is a low-pass filter (analyzing function), and for $n = 1, 2, 3$ by

$$\psi_{00}(x) = \psi(x)$$

$$\psi_{10}(x) = \sqrt{2}\psi(2x)$$

$$\psi_{11}(x) = \sqrt{2}\psi(2x - 1)$$

where $\psi(x)$ is the "Mother" wavelet (a bandpass filter), $\psi(2x)$ is a dilated version, and $\psi(2x - 1)$ is a dilated and translated version. Plot $f(x)$ and the individual members of the Haar family defined above to represent $f(x)$. Then write out the representation of $f(x)$ in terms of a matrix vector product, that is,

$$\mathbf{y} = \mathbf{Mb}$$

where $\mathbf{y}$ is the column vector describing $f(x)$, $\mathbf{M}$ is the matrix representing the discrete Haar wavelet transform, and $\mathbf{b}$ is the vector representing the Haar wavelet coefficients. The wavelet transform matrix is analogous to the discrete Fourier transform matrix. The key difference is that the Haar wavelet coefficients come from n subintervals, whereas the Fourier coefficients come from values at n points.

BIBLIOGRAPHY

M. Abramowitz and L. A. Stegun, eds., *Handbook of Mathematical Functions*, National Bureau of Standards Applied Math Series, 55 (June 1964).

L. C. Andrews and B. K. Shivamoogi, *Integral Transforms for Engineers and Applied Mathematicians*, Macmillan, New York, (1988).

H. H. Barrett, "The Radon Transform and Its Applications," in *Progress in Optics XXI*, E. Wolf, ed., North-Holland, Amsterdam, pp. 217–286 (1984).

H. H. Barrett and W. Swindell, "Analog Reconstruction Methods for Transaxial Tomography," *Proc. IEEE* **65**, 89–107 (1977).

R. N. Bracewell, *The Fourier Transform and its Applications*, McGraw-Hill, New York (1978).

C. E. Cook, "Pulse Compression—Key to More Efficient Radar Transmission," *Proc. IRE* **48**, 310 (1960).

I. Daubechies, "Orthogonal Bases of Compactly Supported Wavelets," *Commun. Pure Appl. Math.* **XLI**, 909–996 (1988).

S. R. Deans, "The Radon Transform," in *Mathematical Analysis of Physical Systems*, R. E. Mickens, ed., Van Nostrand Reinhold, NY, pp. 81–133 (1985).

R. Duda and P. Hart, "Use of the Hough Transform to Detect Lines and Curves in Pictures," *Commun. ACM* **15**, 11–15 (1972).

D. Gabor, "Theory of Communication," *J. IEEE* **93**, 429–457 (1946).

A. F. Gmitro, J. E. Greivenkamp, W. S. Swindell, H. H. Barrett, M. Y. Chiu, and S. K. Gordon, "Optical Computers for Reconstructing Objects from Their X-Ray Projections," *Opt. Eng.* **19**, 260 (1983).

J. W. Goodman, *Introduction to Fourier Optics*, McGraw-Hill, New York (1968).

P. Goupillard, A. Grossman, and J. Morlet, "Cycle-octave and Related Transforms in Seismic Signal Analysis," *Geoexploration* **23**, 85–102 (1984/85).

I. S. Gradshteyn and I. M. Ryzhik, *Table of Integrals, Series and Products*, Academic Press, New York (1980).

A. Haar, "Zur Theorie der Orthogonalen Funktionensysteme," *Math. Ann.* **69**, 331–371 (1910).

R. V. L. Hartley, "A More Symmetrical Fourier Analysis Applied to Transmission Problems," *Proc. IRE* **30**, 144 (1942).

E. Hecht and A. Zajac, *Optics*, Addison-Wesley, Reading, MA (1979).

Y. Li and Y. Zhang, "Coherent Optical Processing of Gabor and Wavelet Expansions of One- and Two-Dimensional Signals," *Opt. Eng.* **9**, 1865 (1992).

Y. Meyer, *Ondelettes et Operateurs, Tome I: Ondelettes*, Hermann, Paris (1990).

A. Papoulis, *Probability, Random Variables and Stochastic Processes*, McGraw-Hill, New York (1965).

L. J. Pinson, *Electro-Optics*, John Wiley & Sons, New York, (1985).

H. L. Resnikoff, "Wavelets and Adaptive Signal Processing," *Opt. Eng.* **31**, 1229 (1992).

W. Steier and R. Shori, "Optical Hough Transform," *Appl. Opt.*, **25**, 2734–2738 (1986).

J. Villasenor and R. N. Bracewell, "Optical Phase Obtained by Analogue Hartley Transform," *Nature* **330**, 735 (1987).

L. G. Weiss, "Wavelets and Wideband Correlation Processing," *IEEE Signal Proc. Mag.*, pp. 13–32, Jan. 1994.

Chapter 4
FUNDAMENTAL PROPERTIES OF LIGHT AND GEOMETRICAL OPTICS

4.1 OVERVIEW

In this chapter we will focus our attention on the basic concepts of geometrical optics. Before we do so, however, we will look at a more fundamental level by describing the basic wave properties of light, including polarization. (Polarization is important in understanding many spatial light modulators.) Later in Chapter 5 we will cover physical optics, especially interference and diffraction. It might seem a little incongruous to start with basic wave properties only to return to them after an interruption for geometric optics. However, this is often the way the subject is taught, because it is more intuitive, and historically, geometric optics was understood and applied first. Besides, there are numerous practical results that we can introduce that will be useful in designing optical processing architectures.

4.2 FUNDAMENTAL SCALAR AND VECTOR PROPERTIES OF LIGHT

An electromagnetic wave is made up of oscillating electric (**E**) and magnetic (**H**) fields which possess a relationship to one another expressed in terms of vectors that correspond to the strength and direction of the electric field and magnetic field at a specific point in space and a particular instant of time. The direction of these vector quantities is given in reference to the direction of propagation of the electromagnetic wave, thus forming an orthogonal triad. Such a picture of electromagnetic waves (Fig. 4.1) constitutes the most general and fundamental description of the propagation of light in a vacuum.

For either the electric or magnetic field the magnitude is often defined as the scalar quantity $u(x, y, z, t)$, which satisfies the wave equation

$$\nabla^2 u = \frac{1}{v_p^2} \frac{\partial^2 u}{\partial t^2} \tag{4.1}$$

whose solution for propagation along the positive z axis (customary for the conventional propagation direction) is given by

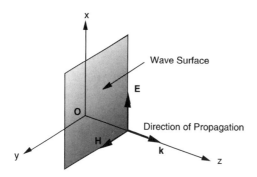

Figure 4.1 Fundamental schematic description of electromagnetic waves showing vector triad of electric (**E**), magnetic (**H**) and propagation direction (**k**).

$$u(z, t) = u_0 \cos(kz - \omega t) \quad (4.2)$$

which describes a traveling wave solution (see Fig. 4.2) since an arbitrary point z on the waveform (representing a point of known phase with respect to some fixed coordinate system) is given by setting the total phase equal to zero, namely, $kz - \omega t = 0$ such that $z = v_p t = \omega t/k$. Thus $v_p = \omega/\mathbf{k}$ is the so-called phase velocity of the wave. For arbitrary propagation direction we obtain

$$u(\mathbf{r}, t) = u_0 \cos(\mathbf{k} \cdot \mathbf{r} - \omega t) \quad (4.3)$$

where the wavevector **k** is in the direction of propagation, and u_o is the peak magnitude of the field ($|\mathbf{E}|$ or $|\mathbf{H}|$), $\mathbf{k} \cdot \mathbf{r} - \omega t = $ constant defines equiphase surfaces perpendicular to **k**, and **r** is an arbitrary radius vector. Spherical waves are also solutions to the wave equation and are given by

$$u(r, t) = u_o \frac{1}{r} \cos(k \cdot r - \omega t) \quad (4.4)$$

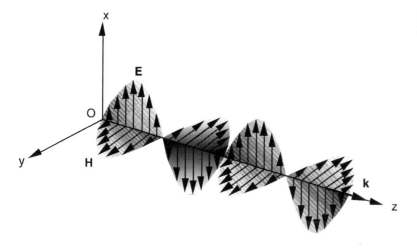

Figure 4.2 Plane harmonic electromagnetic wave.

where $r = (x^2 + y^2 + z^2)^{1/2}$. More general solutions of the wave equation can be constructed by linear superposition of two or more harmonic waves of different frequencies ($\omega + \Delta\omega$ and $\omega - \Delta\omega$) described by

$$u = 2u_o \exp[j(kz - \omega t)] \cos(\Delta k z - \Delta \omega t) \qquad (4.5)$$

This superposition acts like a packet of waves with wavenumber and frequency dispersion (width) of Δk and $\Delta \omega$, respectively, such that the packet of waves possesses a group velocity $v_g = \Delta\omega/\Delta k$. In the continuum limit we obtain $v_g = d\omega/dk$. The so-called dispersion (spatial spreading due to multiple frequencies spanning the finite width $\Delta\omega$) corresponds to the fact that the source that creates the wavepacket cannot support the presence of a single frequency because all practical sources have finite duration, which automatically implies a finite Fourier spectral distribution. This type of wave is depicted in Fig. 4.3.

Another type of dispersion occurs even if the source is otherwise nondispersive (i.e., infinite in duration). Dispersion in this case can be caused by propagating through a medium other than vacuum where the index of refraction (n) depends on frequency. The index of refraction (n) is defined as the ratio (c/v) of the velocity of light in vacuum (c) to the velocity of light in the medium (v). For a dispersive medium, n can also be expressed as a function of wavelength λ or wavevector (magnitude) $k = 2\pi/\lambda$, given by $n = n(k)$. Hence, in a dispersive medium the group velocity becomes

$$v_g = \frac{d}{dk}\left(\frac{ck}{n}\right) = v_p \left(1 - \frac{k}{n}\frac{dn}{dk}\right) \qquad (4.6)$$

where $v_p = c/n$ is the phase velocity. The type of dispersion discussed so far is the traditional one, which is a property of propagation of waves in a material medium. Another type of dispersion—that is, a shift of frequency (or wavelength)—occurs over time, resulting from the motion of the source. If the source of electromagnetic waves moves with respect to an observer (or detector) fixed in inertial space, the frequency is shifted according to

$$f' = f(1 \pm v/c) \qquad (4.7)$$

(+ for approaching and − for receding). Thus the relative frequency (or Doppler) shift is

$$\Delta f/f = \pm v/c \qquad (4.8)$$

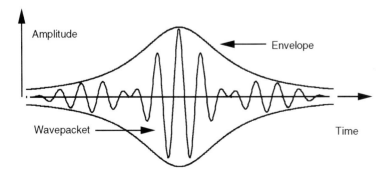

Figure 4.3 Superposition of multiple frequency plane waves that form a finite wavepacket.

4.3 POLARIZATION

As mentioned earlier, the propagation of light is fully described as the propagation of oscillating electric and magnetic field vectors perpendicular to the direction of propagation. These fields are described by the relations

$$\mathbf{E} = \mathbf{E}_0 \exp[j(\mathbf{k} \cdot \mathbf{r} - \omega t)] \quad (4.9)$$

and

$$\mathbf{H} = \mathbf{H}_o \exp[j(\mathbf{k} \cdot \mathbf{r} - \omega t)] \quad (4.10)$$

where $\mathbf{E}$ and $\mathbf{H}$ are the electric and magnetic fields and $\mathbf{k} \cdot \mathbf{r}$ is a scalar dot product. The polarization direction of such a wave in an isotropic medium is given by the wavevector $\mathbf{k}$ and the average energy flow by the Poynting vector $\mathbf{S}$ defined by

$$\langle \mathbf{S} \rangle = \langle \mathbf{E}^* \times \mathbf{H} \rangle = \frac{1}{2} \mathbf{E}_o \times \mathbf{H}_o = \frac{1}{2\mu\omega} |\mathbf{E}_o|^2 \mathbf{k} \quad (4.11)$$

where μ = magnetic permeability of the medium such that $\mathbf{H}_o = \mathbf{E}_o/\mu v_p = \varepsilon v_p \mathbf{E}_o$, where ε is the permittivity of the medium. The symbol $\langle \ldots \rangle$ denotes a time average in this case. Thus the magnitude of the average flux (or irradiance) is

$$I = \langle \mathbf{S} \rangle = \frac{1}{2\mu v_p} |\mathbf{E}_o|^2 \quad (4.12)$$

Given the expression for the electric field shown in Equation (4.9), we can now consider the superposition of two electric fields, which together represent an electromagnetic wave propagating along the z axis with (elliptic) polarization. This wave is defined by

$$\mathbf{E} = \mathbf{i} E_x + \mathbf{j} E_y = \mathbf{i} E_0 \exp[j(kz - \omega t)] + \mathbf{j} E_0 \exp[j(kz - \omega t + \phi)] \quad (4.13)$$

where $\mathbf{i}$ and $\mathbf{j}$ are unit vectors along the x and y axes, respectively, and where ϕ is the arbitrary phase difference between E_x and E_y. When the phase difference $\phi = \pm\pi/2$ we have circular polarization. Then

$$\mathbf{E} = E_o(\mathbf{i} \pm \mathbf{j}) \exp[j(kz - \omega t)] \quad (4.14)$$

When looking toward the source (viewing against the direction of propagation), if the electric vector rotates clockwise, the wave is right circularly polarized ($\phi = \pi/2$), and if the electric vector rotates counterclockwise, the wave is left circularly polarized ($\phi = -\pi/2$), as illustrated in Figs. 4.4 (a) and 4.4 (b).

In many instances, light is partially polarized, in which case the degree of polarization is given by the ratio $I_p/(I_p + I_u)$, where I_p is the flux density of polarized light and I_u is the flux density of unpolarized light.

Two important considerations in polarization are the materials and devices that actually produce or alter a particular polarization state. Generally, anisotropic materials are used. Optically anisotropic materials have an index of refraction that varies with direction in the material, and they are termed *birefringent*. Note that for such materials, different indices of

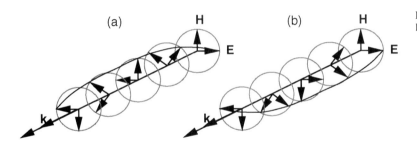

Figure 4.4 (a) Left circularly polarized light and (b) right circularly polarized light.

refraction result from different phase velocities, that is, $n_1 = c/v_{p1}$ and $n_2 = c/v_{p2}$ along two orthogonal axes. Other materials whose absorption constant varies with polarization are termed *dichroic materials*. In dichroic polarizers, one component of polarization is absorbed more than the other (transverse) component. Polaroid sheets are dichroic polarizers.

The two images produced by a birefringent material such as calcite are orthogonally polarized. Except for crystals with cubic symmetry, which behave optically like glass, all other crystals are optically anisotropic. Although tensor calculus is required to fully explain the propagation of light in these materials, we will give a simpler empirical account of specific practical devices.

Two orthogonal polarizations that make up an incident wave would ordinarily have different propagation velocities in a birefringent material. If, however, the incident wave propagates along the optical axis, then the velocities are the same. Crystals with one such axis are termed *uniaxial*, and those with two are termed *biaxial*.

Devices that change the relative phase between the two incident polarizations are termed *retarders*. The two most common types of retarders are quarter-wave and half-wave plates. Linear polarization can be converted to circular polarization via the quarter-wave plate, and it can be rotated by a fixed angle using a half-wave plate. To see this, consider a uniaxial birefringent plate as shown in Fig. 4.5.

An incident polarized wave making an angle θ with respect to the fast axis (higher propagation velocity axis) has both ordinary and extraordinary projections onto the so-called slow and fast axes, respectively. When propagating through the material, the phase of the slow-wave lags the fast wave. Such a birefringent plate creates a phase difference given by

$$\delta = 2\pi d(n_e - n_o)/\lambda \tag{4.15}$$

where n_e = index of refraction for extraordinary wave, n_o is the index of refraction for ordinary wave, and d is the thickness of plate. If $\delta = \pi/2$, we have the quarter-wave plate, and if $\delta = \pi$, we have the half-wave plate. Mica and quartz are usually used for these devices.

4.4 RECTILINEAR GLASS STRUCTURES AND THEIR PROPERTIES

Despite their simplicity, reflective and refractive glass structures that do not have any curved surfaces are very interesting and useful devices for propagating and manipulating optical wavefronts and images. These structures include flat plates, wedges, prisms, and multifaceted refractive elements. There are really only two basic laws that govern the directionality of light

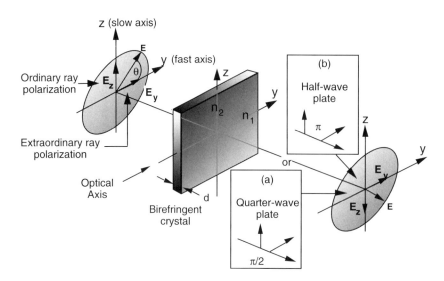

Figure 4.5 Uniaxial birefringent plate used as (a) quarter-wave and (b) half-waveplate.

when interacting with the so-called "rectilinear" glass surfaces (or any surfaces): the law of reflection and the law of refraction. (Although, as we shall see, absorption can also occur.)

The law of reflection simply states that the angle of incidence (θ) equals the angle of reflection (θ'), where both are measured with respect to the normal to the glass surface (which serves as the interface between two media) as shown in Fig. 4.6. The law of refraction is Snell's law, given by $n \sin \theta = n' \sin \theta'$, where n is the index of refraction of medium of the incident ray (e.g., air or vacuum), n' is the index of refraction of medium of the refracted ray (e.g., glass or water), and θ'' is the angle of refraction with respect to the normal.

Snell's law results from the requirement that the phase of electromagnetic waves at the medium (air–glass) interface match, or equivalently that the momentum of the light wave is conserved. Refer to Fig. 4.6 and note that for $n' > n$ the angle of refraction (θ') is less than the angle of incidence (θ').

In addition to direction, the intensity of light interacting with a media interface can be determined. The equations governing the amount of light reflected from and transmitted through a media interface (air–glass) are the Fresnel equations. Consider again the diagram of Fig. 4.6, where incident (E, H), reflected (E', H'), and transmitted (E'', H'') electric and

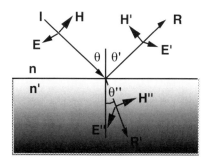

Figure 4.6 Reflection and refraction of light at a plane dielectric boundary including polarization effects. (Think of the **E** vectors as pointing out of the page.)

magnetic field strengths are defined, along with their corresponding wavevectors ($\mathbf{k}$, $\mathbf{k}'$, and $\mathbf{k}''$) and angles (θ, θ', and θ'').

Boundary conditions must be imposed on the electric and magnetic fields to yield the Fresnel equations. The requirement is that the tangential components of the E and H fields be continuous at the boundary—that is, $E + E' = E''$ for transverse electric (TE) polarization, and $H - H' = H''$ for transverse magnetic (TM) polarization.

The TE polarization corresponds to the case where the electric vector is perpendicular to the plane of incidence (but parallel to the media interface) as shown in Fig. 4.6. The TM polarization corresponds to the case where the magnetic vector is perpendicular to the plane of incidence (but parallel to the media interface). Applying these boundary conditions yields the Fresnel equations for the TE and TM polarizations. For the TE polarization,

$$\frac{E'}{E} = \frac{-\sin(\theta - \theta')}{\sin(\theta + \theta')} \tag{4.16}$$

is the amplitude reflectance ratio, and

$$\frac{E''}{E} = \frac{2\cos\theta\sin\theta'}{\sin(\theta + \theta')} \tag{4.17}$$

is the amplitude transmittance ratio. For the TM polarization,

$$\frac{E'}{E} = \frac{-\tan(\theta - \theta')}{\tan(\theta + \theta')} \tag{4.18}$$

is the amplitude reflectance ratio, and

$$\frac{E''}{E} = \frac{2\cos\theta\sin\theta'}{\sin(\theta + \theta')\cos(\theta - \theta')} \tag{4.19}$$

is the amplitude transmittance ratio. The reflected amplitude transmittance ratios are plotted as a function of angle of incidence (θ) in Figs. 4.7(a) and 4.7(b) (for both TE and TM polarizations) for the case where $n < n'$ (external reflection) and the case where $n > n'$ (internal reflection). By plotting the power ratios versus θ (i.e., reflectance R given by $R = |E'/E|^2$) as shown in Figs. 4.7(c) and 4.7(d), it is clear that there is an angle at which the TM polarization reflectance is zero, termed *Brewster's angle*, a characteristic used to make Brewster's windows for lasers. Furthermore, it is clear that at a slightly larger angle (the critical angle) there is total internal reflection. These angles are defined by $\theta_B = tan^{-1}(n'/n)$ for Brewster's angle and by $\theta_C = \sin^{-1}(n'/n)$ for the critical angle (where $n' > n$).

For the TE polarization there is an interesting special case. Consider the reflection coefficient R, which can be written

$$R = \left| \frac{n\cos\theta' - \cos\theta}{n\cos\theta' + \cos\theta} \right|^2 \tag{4.20}$$

For $\theta = 0$ we obtain

$$R = \left| \frac{n - 1}{n + 1} \right|^2 \tag{4.21}$$

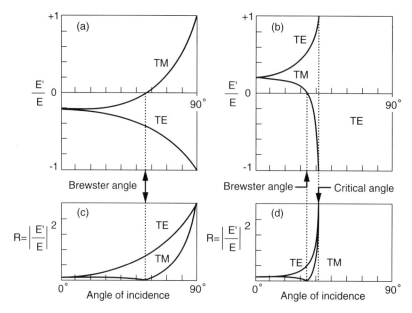

Figure 4.7 Reflected amplitude ratios for TE and TM polarizations for (a) external and (b) internal reflection and corresponding power ratios (or reflection coefficients) (c, d).

which corresponds to normal incidence. The transmittance is given by

$$T = 1 - R = \frac{4n \cos\theta \cos\theta'}{(n \cos\theta' + \cos\theta)^2} \tag{4.22}$$

For $\theta = \theta' = 0$ we obtain

$$T = \frac{4n}{(n+1)^2} \tag{4.23}$$

For instance, if $n = 1.5$ (typical for glass) we obtain $R = 0.04$ (4%). Therefore at normal incidence glass optical components have a loss of 4% per surface. This leads to significant loss in complex optical systems with many surfaces, unless antireflection coatings are used.

For polarized light the power reflection coefficient for TE and TM polarizations is given in compact form as

$$R_p = \frac{\tan^2(\theta - \theta')}{\tan^2(\theta + \theta')} \tag{4.24}$$

where the subscript p denotes p polarization (TM) and

$$R_s = \frac{\sin^2(\theta - \theta')}{\sin^2(\theta + \theta')} \tag{4.25}$$

where the subscript s denotes the s polarization (TE). Often incident light is unpolarized, in which case

$$R_u = \frac{1}{2}(R_p + R_s) \tag{4.26}$$

gives the power reflection coefficient assuming a uniform distribution of arbitrary polarizations. When $\theta + \theta' = \pi/2$, R_p is zero. Then the reflected light becomes completely polarized normal to the plane of incidence. Thus we see that a simple glass surface cannot only deflect light, but attenuate and polarize it.

By considering a parallel glass plate (as an example of two optical surfaces) we can consider further useful effects that rectilinear glass structures provide. Consider a simple parallel glass plate with index of refraction n and thickness t. Light of wavelength λ is normally incident on the plate. We can compare the phase of the light wave passing through the glass and traveling over a total distance AB with light simply propagating over the same distance without the glass plate, as shown in Fig. 4.8. In the first case (no glass plate), the number of waves in the distance AB is AB/λ, which yields a phase shift of $2\pi AB/\lambda$. In the second case, the number of waves in the distance AB is the sum of the number in the air, given by $(AB - t)/\lambda$, and the number in the glass, given by t/λ', where λ' is the wavelength in the glass medium, given by $\lambda' = \lambda/n$. The phase difference between these two cases is

$$\Delta\phi = \frac{2\pi(AB - t)}{\lambda} + \frac{2\pi t}{\lambda'} - \frac{2\pi AB}{\lambda} \tag{4.27}$$

or

$$\Delta\phi = 2\pi t \left(\frac{1}{\lambda'} - \frac{1}{\lambda}\right) \tag{4.28}$$

Since $\lambda' = \lambda/n$, then

$$\Delta\phi = \frac{2\pi t}{\lambda}(n - 1) \tag{4.29}$$

For a particular case, consider the following example. For a typical situation ($n = 1.5$, $t = 1$ mm, and $\lambda = 500$ nm) the phase shift is $\Delta\phi = 2\pi \times 10^3$ radians, a large number. Thus, it doesn't take much glass to retard the phase of an incident wave by many cycles. This will become interesting when we discuss spatial light modulators. In any case, it is obvious that a simple rectilinear glass structure can also be a phase shifter.

A simple flat plate can also be used to laterally deflect a laser beam as shown in Fig. 4.9.

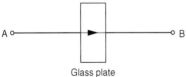

Figure 4.8 Propagation of light ray over an optical path AB (a) without glass plate and (b) with glass plate.

This effect must be considered when using flat-plate beamsplitters in an optical train. Using Snell's law it can be shown that the amount of lateral beam deflection (δ) is given by

$$\delta = t \sin\theta \left(1 - \frac{\cos\theta}{\sqrt{n^2 - \sin^2\theta}}\right) \quad (4.30)$$

where t is the plate thickness and n is the index of refraction.

In addition, in this configuration, the glass plate acts a polarizer if the angle of incidence is set to be Brewster's angle θ_B and the incident light is linearly polarized in the TM mode. Then no light is reflected from the first surface and no internal component is reflected either. Thus this so-called Brewster's window is "perfect."

If a simple parallel glass plate is darkened with an absorptive material, it becomes a neutral density filter. For normally incident unpolarized light the transmission coefficient is

$$T = \frac{\exp(-\alpha t)(1 - R)^2}{1 - R^2 \exp(-2\alpha t)} \quad (4.31)$$

which includes losses due to absorption (given by $\exp(-\alpha t)$, where α is the absorption coefficient) and losses from reflection (R). In the limit of no absorption (α approaching zero), this reduces to

$$T = \frac{2n}{n^2 + 1} \quad (4.32)$$

which is not the result of simply applying Equation (4.23) twice because Equation (4.23) applies to TE polarization (see Problem 17). Usually neutral density filters are characterized by optical density, defined as O.D. = $\log_{10}(10^{-D})$. Hence a filter with 10% transmission (10 dB loss) has an optical density of 1.

An important arrangement of rectilinear glass structures for use in lasers is the Fabry–Perot resonator. It consists of two plane parallel reflecting surfaces, as shown in Fig. 4.10. This arrangement can be used to make an interferometer by virtue of the multiple reflections between the two surfaces. The constructive interference condition is given by $2d \sin\theta = m\lambda$ (where m is an integer), which is very similar to the Bragg condition associated with acousto-optics as discussed later in the book.

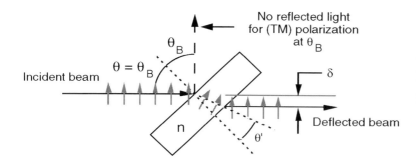

Figure 4.9 Flat glass plate used to laterally deflect a light ray. When oriented at Brewster's angle relative to the incident light beam it also acts to polarize initially unpolarized light. If the incident light is polarized in the TM mode, the reflected beam is extinguished, making the plate a perfect (Brewster's) window.

4.4 RECTILINEAR GLASS STRUCTURES AND THEIR PROPERTIES

It can be shown that for multiple reflections of an incident beam the sum of the transmitted beams is given by

$$E_T = \frac{E_o t^2}{1 - r^2 e^{i\delta}} \tag{4.33}$$

where r is the reflectivity, t is the transmittance of the surfaces, and $\delta = 4\pi d \cos\theta / \lambda$ is the phase difference between successive rays as shown in Fig. 4.10. The transmitted intensity, given by $I_T = |E_T|^2$, can be expressed as

$$I_T = I_o \frac{T^2}{(1-R)^2} \frac{1}{1 + \frac{4R}{(1-R)^2} \sin^2(\Delta/2)} \tag{4.34}$$

where $T = |t|$, $R = |r|$, and $\Delta = \delta + \delta_r$, where δ_r is the phase change for one reflection. The third factor in Equation (4.34), which explicitly depends on Δ, is plotted in Fig. 4.11 and is the so-called Airy function, shown for different values of reflection coefficient R.

By application of Fresnel's equations, the intensities of the maxima and minima are given by

$$I_{T_{max}} = I_o \frac{T^2}{(1-R)^2} \tag{4.35}$$

and

$$I_{T_{min}} = I_o \frac{T^2}{(1+R)^2} \tag{4.36}$$

The minimum of intensity approaches zero as $R \to 1$; and as R increases, the maximum sharpens as shown in Fig. 4.11.

When the Fabry–Perot arrangement is used in lasers, it serves as a resonator. After two reflections, it is necessary for the internal wave to superimpose upon itself to support oscillation. In resonators, the Q factor best characterizes the effectiveness as an oscillator; it is essentially the ratio of the power stored to the power extracted from the resonator and is given by

$$Q = \frac{r}{1-r} \tag{4.37}$$

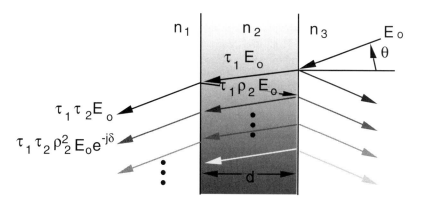

Figure 4.10 Fabry–Perot resonator structure.

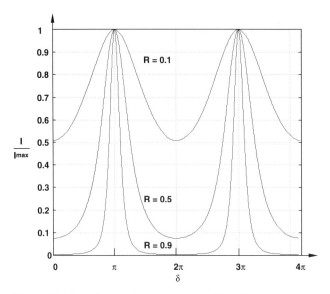

Figure 4.11 Intensity variation of output from Fabry–Perot resonator.

where r is the mirror reflectivity. If we take account of absorption in the medium between the two reflecting surfaces, as in a laser, then the Q value becomes

$$Q = \frac{|r_1 r_2 \exp(-2\alpha L)|}{1 - |r_1 r_2 \exp(-2\alpha L)|} \tag{4.38}$$

where r_1 is the first surface reflection coefficient, r_2 is the second surface reflection coefficient, and L is the length of cavity (between the two surfaces). This result will be used later when we discuss lasers.

The only other geometric parameter of rectilinear glass structures that we can change is to make the two faces through which light propagates nonparallel. Then we create wedges and prisms. The simplest of these is the glass wedge depicted in Fig. 4.12, where the apex angle is small. In this case the angle of deflection created in the transmitted beam is given by $\delta = (n-1)\theta$. This effect is important in precise alignment of an optical train with wedge-type beamsplitters (which avoid "ghosting"—that is, the creation of multiple secondary images characteristic of parallel-plate beamsplitters).

In addition to beam deflection, two prisms can be configured to expand a laser beam. Such an anamorphic prism pair is used to change an elliptical laser beam into a circular one and is depicted in Fig. 4.13. The angleof incidence is close to Brewster's angle so that the incident

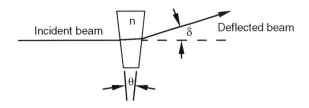

Figure 4.12 Glass wedge of narrow apex angle and resultant ray deflection.

beam (linear) polarization is maintained and reflection losses are minimized (especially if antireflection coatings are used). The degree of elliptical correction is determined by the angles between the prisms (see Problem 18).

If we increase the apex angle of the wedge in Fig. 4.12 significantly, we arrive at the standard prism shown in Fig. 4.14. Now the deflection angle is given by

$$\delta = \theta_1 + \theta_2 - \beta \tag{4.39}$$

In addition to deflecting light, a prism will disperse broadband light; for example, it will split incident white light into its component colors. The angle of deflection δ is then a function of wavelength, λ, given by

$$\delta = \theta_1 + \sin^{-1}[(n^2 - \sin^2 \theta_1)^{1/2} \sin \beta - \sin \theta_1 \cos \beta] - \beta \tag{4.40}$$

assuming n varies with wavelength. Thus a prism also acts to perform spectral analysis. This is the basis of spectroscopy, although modern instruments often use diffraction gratings in lieu of prisms.

If a right-angle prism is truncated as shown in Fig. 4.15(a) and light enters as shown, it will be refracted toward the prism base, where it is totally internally reflected. Upon leaving the exit face, the image of an object (as shown) is inverted. This is the so-called Dove prism. Furthermore, if the Dove prism is rotated about an axis along the direction of light propagation, the resulting inverted image rotates by twice the rotation angle of the prism. Hence rectilinear glass structures reverse, invert, and rotate an input image. An example of an array of Dove prisms, each at a different orientation angle, is shown in Fig. 4.15(b). Note that although the original array of images viewed through the array all have the same orientation, because each Dove prism is successively rotated by 12° starting with the top left-hand corner prism, each output image is rotated successively by 24°, thus spanning 360°.

By combining many wedges into an array of wedges as shown in Fig. 4.16, we can create an interesting discrete approximation to a thin lens. This rectilinear glass structure can be combined with an appropriate output lens to form multiple images as shown. Thus it serves as a simple means of multiplexing or replicating an input image for parallel processing. The same effect can be achieved by multiaperture lenslet arrays as discussed later.

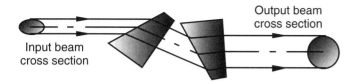

Figure 4.13 Anamorphic prism pair configured to expand a laser beam in one axis.

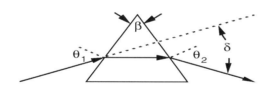

Figure 4.14 Standard prism showing ray deflection, where β is the apex angle.

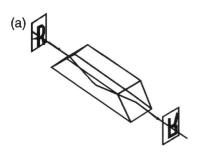

Figure 4.15 (a) Dove prism, showing ray trace and effect on image propagating through the prism; and (b) array of Dove prisms, which rotate identical replicated objects or images to different orientations.

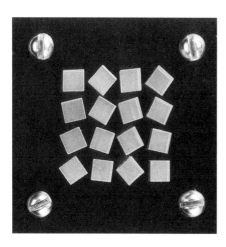

4.5 SIMPLE LENSES AND LENS COMBINATIONS

Although a full treatment of lenses and lens combinations requires an extensive discussion of ray tracing, aberrations, and the derivation of lens formulas, it is more expedient here to summarize some of the results in a more compact form. By doing so, we can capture the essentials needed to apply geometrical optics in a prescriptive manner to complex optical signal processing architectures. We will describe the effects of a simple thin spherical lens, distinguish positive and negative lenses, recite some of the key lens formulas and parameter definitions, and discuss two-lens combinations, including the telescope and telephoto lens. Finally, we will talk about a novel refractive system, called the lenslet array.

Consider first the simple thick lens depicted in Fig. 4.17. We are interested in its imaging properties; hence we show a simple object (arrow) in the object plane propagated through the lens by simple ray tracing to the inverted image in the focal plane. At once we see that a lens does at least two things: It magnifies and inverts the image of an object. This magnification in size is called lateral magnification for the lens system in a finite conjugate condition. In fact, the lateral magnification is given by $m = -s_i/s_o$, where s_i/s_o is the finite conjugate ratio—that

4.5 SIMPLE LENSES AND LENS COMBINATIONS   107

Figure 4.16 Multifaceted prism array viewed with another lens to form multiple images. The multifaceted prism is a discrete approximation to an ordinary lens.

is, the ratio of the image (s_i) to object (s_o) distances—measured from the so-called principal planes H_i and H_o, respectively, as shown in in Fig. 4.17. The principal planes coincide for a thin lens, in which case the edge thickness (t_e) is zero and the center thickness (t_c) approaches zero such that $t_c \ll D$ (diameter of lens aperture).

The radius of curvature of the first surface is denoted as r_1 and is positive if the center of curvature is to the right. The radius of curvature for the second surface is r_2 and is negative if the center of curvature is to the left. The object distance s_o is positive for an object to the left of H_o, and image distance s_i is positive for an image to the right of H_i (which hold true for real

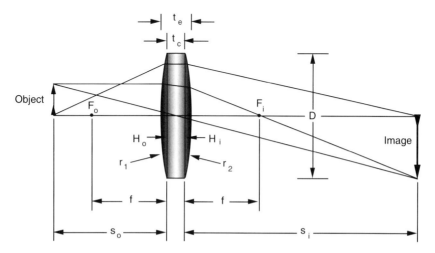

Figure 4.17 Thick lens and typical parameters.

or virtual objects). The focal length is defined as $F_o H_o$ or $F_i H_i$, which are equal in this case, and the angular subtense of the lens as seen from the object is defined as $\theta = \tan^{-1}(D/2s_o)$. In the case of a negative lens the object will form a virtual rather than real image. Various conditions in which simple positive and negative lenses can be used are shown in Table 4.1. Some of these are illustrated in Figs. 4.18(a)–(e).

Several important lens formulas are used in optical systems design, which are useful in developing layouts for optical signal processors. A key assumption underpinning these formulas is the paraxial approximation, in which rays are assumed to be very close to and nearly parallel to the optical axis. It is also assumed that lens surfaces are very nearly normal to the optical axis. These assumptions are good for large f numbers (denoted $f/\#$) defined by $f/\# = s_o/D$, where D is the lens diameter. A related parameter is numerical aperture (N.A.) defined by N.A. $= n \sin \theta$, where $n =$ object space refractive index (~ 1 for air) and $\theta = \tan^{-1}(D/2s_o)$. Thus $f/\# = 1/(2\text{N.A.})$ for large $f/\#$ values. The solid angle subtended by the lens for an observer situated at an on-axis object point is defined by $\Omega = 2\pi(1 - \cos \theta)$ measured in steradians, where θ is given above. If the observer is situated at the on-axis *image* point, then $\theta = \tan^{-1}(D/2s_i)$.

The basic paraxial lens formula for lens positioning is the Gaussian lens formula given by

$$\frac{1}{f} = \frac{1}{s_o} + \frac{1}{s_i} \tag{4.41}$$

which can also be written in the form

$$f = \frac{m}{m+1} s_o \tag{4.42}$$

where m is lateral magnification. This formula can be applied to any lens, positive or negative. A specification of the focal length of a thick lens is based on the general formula

Table 4.1
Image Characteristics for Real Objects Viewed by Thin Lenses

	Object		Image		
Figure	Location	Type	Location	Orientation	Relative Size
		Positive (Convex) Lens			
4.18(a)	$\infty > s_o > 2f$	Real	$f > s_i > 2f$	Inverted	Minified
4.18(b)	$s_o = 2f$	Real	$s_i = 2f$	Inverted	Same Size
4.18(c)	$f < s_o < 2f$	Real	$\infty > s_i > 2f$	Inverted	Magnified
4.18(d)	$s_o = f$	—	$\pm\infty$	—	—
4.18(e)	$s_o < f$	Virtual	$|s_i| > s_o$	Erect	Magnified
		Negative (Concave) Lens			
—	Anywhere	Virtual	$|s_i| > |f|$	Erect	Minified

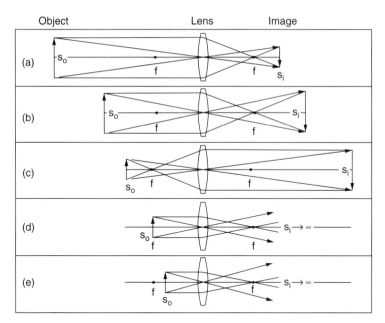

Figure 4.18 Various positive lens configurations for varying object distances: (a) $\infty > s_o > 2f$, (b) $s_o = 2f$, (c) $2f > s_o > f$, (d) $s_o = f$, and (e) $s_o < f$.

$$\frac{1}{f} = (n-1)\left(\frac{1}{r_2} - \frac{1}{r_2}\right) + \frac{(n-1)^2}{n}\frac{t_c}{r_1 r_2} \tag{4.43}$$

where r_1 and r_2 are the first and second surface radii of curvature, respectively (for light traveling from left to right), and t_c is the center thickness. This formula simplifies to the more common lensmaker's formula

$$\frac{1}{f} = (n-1)\left(\frac{1}{r_1} - \frac{1}{r_2}\right) \tag{4.44}$$

when we have a thin lens ($t_c = 0$).

For the case of two lenses combined, the paraxial lens combination formula is

$$\frac{1}{f} = \frac{1}{f_1} + \frac{1}{f_2} - \frac{d}{f_1 f_2} \tag{4.45}$$

which is true for all cases—that is, all values of lens separation d and all signs of f_1 and f_2. For fixed d the resultant value of f is invariant with respect to the interchange of lenses. For the case of two lenses a quantity of practical interest is the distance (s_{2i}) measured from the secondary principal point (H_{2i}) of the second element to the final combination focal point. This is given by

$$s_{2i} = \frac{f_2(f_1 - d)}{f_1 + f_2 - d} \tag{4.46}$$

As a special case of the combination of two lenses, consider a simple (Keplerian) telephoto lens combination as shown in Fig. 4.19(a). In this combination, one lens is smaller in diameter and shorter in focal length than the other. The particular configuration used illustrates the idea behind a laser beam expander, which is the reverse of a telephoto lens. It also assumes that the input beam is collimated so that the output beam is collimated if the lenses are separated by the sum of their focal lengths ($f_1 + f_2$). In this case the beam diameter is increased by the ratio of the focal lengths ($m = f_2/f_1$). In fact, for a slightly uncollimated light beam the divergence is reduced by m^{-1} in this arrangement. An even more special case occurs if the two lenses (assumed thin) have equal focal lengths ($f_1 = f_2$) and are in contact so that $d = 0$ [see Fig. 4.19(b)]. Then the combined focal length $f = f_2/2$ and the pair act a condenser lens, or low $f/\#$ collimator (which is good for collimating highly divergent light sources such as laser diodes).

As a final example of lens combinations, consider the standard astronomical telescope viewing a distant object that subtends a small angle at infinity. This situation is depicted in Fig. 4.20. The objective with focal length f_o forms an image $a'b'$ at a focal point F_o with light that comes from a distant object AB. The eyepiece has a much smaller focal length f_e, which is placed so that the intermediate image $a'b'$ falls at the minimum distance of distinct vision ($\delta \sim 25$ cm) within the focal length f_e from the eyepiece (i.e., between O and F_o) as shown. In this case the real image formed by the objective becomes a virtual object for the eyepiece, which is operated like the simple magnifier. Thus a virtual object $a'b'$ is formed at a distance δ from the eyepiece such that the angular subtense of the object is defined as

$$\alpha \approx \tan \alpha = \frac{a'b'}{f_o} \qquad (4.47)$$

and the angular subtense of the image is defined as

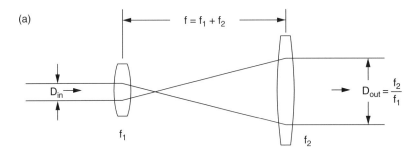

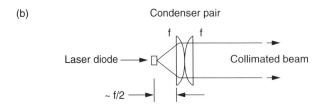

Figure 4.19 Two lenses (a) in combination for Keplerian telescope (or beam expander) and (b) as a condenser pair.

4.5 SIMPLE LENSES AND LENS COMBINATIONS

$$\beta = \tan \beta = \frac{a'b'}{f_e} \tag{4.48}$$

As a result the telescope presents to the observer an image whose subtense is greater than that of the real object viewed by the unaided eye by an amount given by the angular magnification (M):

$$M = \beta/\alpha = f_o/f_e \tag{4.49}$$

An important parameter of telescopes that will be derived later is the resolving power of the objective. It is given approximately by

$$\alpha \approx 1.22 \frac{\lambda}{D} \tag{4.50}$$

where λ is the wavelength of light and D is the objective diameter.

Another lens combination that can be used in optical signal processing is the multiple-lenslet array. In this case, small or microlenses are arranged side-by-side in 2-D arrays as shown in Figs. 4.21(a) and 4.21(b). Lenslet arrays typically consist of small-aperture (0.1- to 10-mm-diameter) lenses with focal lengths varying from 1 to 100 mm. Lenslets can be constructed by several techniques, including cementing lenslets to a common substrate or producing them on specially prepared substrates via a photothermal process. For application purposes we will consider only their optical characteristics.

There are two modes of operation in which lenslet arrays are generally used, corresponding to (1) close-up imaging of subareas and (2) image replication. Close-up imaging of subareas actually requires two sets of lenses (or one equivalent thick lens) as depicted in Fig. 4.22(a). In the first array, each lens images over a limited numerical aperture (by virtue of limiting stops or baffles placed between the object and lenslet, not shown). A second array in Fig. 4.22(a) erects each inverted subimage to form a composite in the final (output) image plane. Otherwise the image seen in the intermediate image plane would be scrambled since the overall image is not inverted (while each subimage is) by the first (inverting) array. An exit pupil for each lens is provided by a second baffle, also not shown. Note that in Fig. 4.22(a) the composite image will be formed by the overlap of each brightness profile. Each brightness

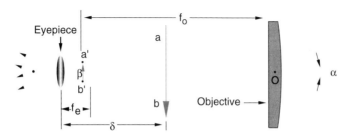

Figure 4.20 Standard astronomical telescope configuration.

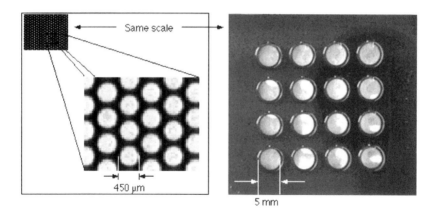

Figure 4.21 Two-dimensional lenslet arrays: (a) smaller hexagonally close-packed photoformed array (440-μm-diameter lenslets), and (b) larger machined array (5-mm-diameter lenslets in an 8 × 8 array).

profile derives from the vignetting of one circular stop in plane L_1 by the corresponding one in plane L_2, which is expressed as

$$I(y) = \left(\frac{2}{\pi}\right)\left(\frac{N\pi R_L^2}{t^2}\right)\left[\cos^{-1}(y/y_o) - (y/y_o)\sqrt{1-(y/y_o)^2}\right] \quad (4.51)$$

where R_L is the lens radius, N is the radiance of a Lambertian source, t is the one-to-one working distance, $y_o = 2nR_L t/T$, and T is the lens thickness. In this approach, each lenslet is modeled as a thick lens, where $T = n(\text{EFL} - s_o)$, EFL is the effective focal length, s_o is the working distance, and n is the index of refraction of lenslet material. This mode of operation can be useful for implementing compact image relay optics.

The second mode of operation of lenslet arrays, image replication, enables parallel channels to be created and information to be processed separately in each, as diagrammed in Fig. 4.22(b). An example of replication is shown in Fig. 4.23. This replication situation shown in Fig. 4.23 corresponds to hexagonally closed-packed lenslets with a diameter of 400 μm and glass thickness of 6 mm. The replicated images were created from a small TV display by replicating them onto a CCD focal plane array.

As a final section on simple lenses, we will summarize the basic image transfer properties of lenses in terms of light level. Refer to Fig. 4.24(a) for this discussion. The total flux P (in units of W) collected by each lens from a source of area dA, subtense (half-angle) α, and radiance B (in units of W · m^{-2}· Sr^{-1}) is given by

$$P = \pi B dA \sin^2 \alpha \quad (4.52)$$

The image irradiance (in W · m−2) produced by this lens at the focal plane can be computed using the relation

$$H = \frac{\pi BT}{m^2 + 4(m+1)^2(f/\#)^2} \quad (4.53)$$

4.5 SIMPLE LENSES AND LENS COMBINATIONS

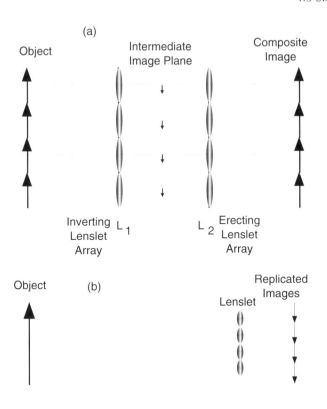

Figure 4.22 (a) Close-up subarea imaging mode and (b) replication mode for lenslet arrays.

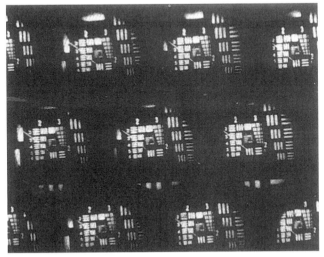

Figure 4.23 Replicated images produced by imaging a small television display using a hexagonally close-packed lenslet array. (N. F. Borelli, D. L. Morse, R. H. Bellman, and W. L. Morgan, "Photolytic technique for producing microlenses in photosensitive glass," Appl., Optics **24,** 2420 (1985). Reprinted with permission.)

where T is the lens transmission, m is the lens magnification, and f/# is the lens f number.

In addition to the above result, there is one other relationship worth noting here with regard to the transmission of light through lenses, and that is the cosine fourth law. The cosine fourth law says that if the distribution of irradiance in the object plane is uniform, then the

114 FUNDAMENTAL PROPERTIES OF LIGHT AND GEOMETRICAL OPTICS

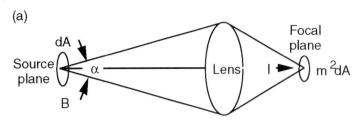

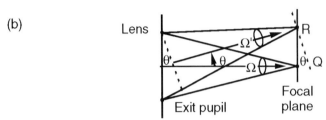

Figure 4.24 Defining geometry for: (a) total flux collected by lens, image irradiance produced by lens with source area dA, lens magnification m and associated f number, and (b) geometry for demonstrating cosine fourth law.

distribution in the image plane goes as the fourth power of the angle measured from the center of the exit pupil. By referring to Fig. 4.24(b), it should be clear that a factor of $\cos^2 \theta$ comes from the ratio of the solid angles Ω' and Ω subtended at the focal plane at an oblique position R and at the on-axis position Q, respectively. Another factor of $\cos \theta$ comes from the obliquity at the exit pupil, and the other factor comes from the obliquity at R with respect to Q.

PROBLEM EXERCISES

1. What must the magnification of a telescope be in order to make full use of the resolution of its objective lens? Assume that it is only diffraction-limited, but that the limit on the eyepiece is set by human visual acuity ($\sim 4 \times 10^{-4}$ radians).

2. Find the angle that polarizes natural (uniformly polarized) light reflected from glass with refractive index $n = 1.5$. Find the degree of polarization of the refracted light when it is incident at this angle.

3. Given two semitransparent parallel planes, the reflection and transmission coefficients at normal incidence are ρ_1 and τ_1 and ρ_2 and τ_2 for the first and second surfaces, respectively. The incident light is not monochromatic, so intensities add rather than amplitudes, as in ordinary interference. Find the reflection and transmission coefficients ρ and τ for the total system.

4. Prove that the expression [see Equation (4.4)]
$$\frac{1}{r} \cos(\mathbf{k} \cdot \mathbf{r} - \omega t)$$
is a solution to the wave equation [Equation (4.1)].

5. Derive the general expression for the group velocity of a wavepacket for a dispersive medium [Equation (4.6)] characterized by an index of refraction that varies with wavelength; that is, $n = n(k)$, where $k = 2\pi/\lambda$.

6. Assume that the index of refraction is given by the Cauchy formula
$$n = A + B\lambda^{-2}$$

Find the phase and group velocities for $\lambda = 633$ nm for a glass with $A = 1.4$ and $B = 2.5 \times 10^4$ nm^2.

7. If a light wave is traveling in glass with index of refraction $n = 1.5$, and the amplitude of the electric field is 100 V/m, what is the amplitude of the magnetic field and the Poynting vector?

8. Draw diagrams to show the type of polarization of the waves whose electric field amplitude is given by

   (a) $\mathbf{E} = \mathbf{i}E_o \cos(kz - \omega t) + \mathbf{j}E_o \cos(kz - \omega t - \pi/4)$

   (b) $\mathbf{E} = \mathbf{i}E_o \cos(kz - \omega t) + \mathbf{j}E_o \sin(kz - \omega t)$

   Express the amplitudes in complex vector form—that is, $\mathbf{E} = \mathbf{E}_o exp[j(kz - \omega t)]$ where $\mathbf{E}_o = (a\mathbf{i} + b\mathbf{j})E_o$.

9. Show that the major axis of the ellipse traced out by the magnitude of the electric field for elliptically polarized light makes an angle

$$\frac{1}{2} \tan^{-1}\left[\frac{2AB \cos(\phi)}{A^2 - B^2}\right]$$

with the x axis, where $\mathbf{E} = [A\mathbf{i} + B \exp(j\phi)\mathbf{j}] \exp[j(kz - \omega t)]$.

10. Show that the refractive index of a prism is given by

$$n = \frac{\sin[(\phi + \phi_m)/2]}{\sin(\phi/2)}$$

where ϕ is the apex angle and δ_m is the minimum angle of deviation of a ray from its initial direction as shown in the figure below. For small angles, show that $\delta_m \approx (n - 1)\phi$.

11. A telescope has an objective focal length f_o of 50 cm and an eyepiece with focal length f_e of 1 cm. What is the angular magnification of the telescope for the object and image located at infinity?

12. A thin biconvex lens with $n = 1.5$ has radii of curvature of 30 cm and 60 cm. If it is to form an image of an object half its size, what must the lens to object and lens to image plane distances be? Construct an appropriate ray diagram, which includes three rays from the object: one through the lens center, one collimated before the lens, and one collimated after the lens.

13. A compound lens consists of two thin biconvex lenses L_1 and L_2 of focal lengths 10 and 20 cm, separated by a distance of 8 cm. Determine the position of the image relative to the second lens if the input light to the first lens is collimated. What is the combined focal length (f) of the lenses?

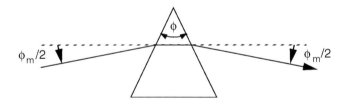

14. Consider two cylindrical lenses oriented orthogonally to each other, spaced apart by some distance d, and positioned along an optical axis such that both lenses image an input object of square shape onto a common focal plane, as illustrated below. If the focal length of lens L_1 is $f_1 = 100$ mm and the magnification in the corresponding axis is $m_y = -1$, what is the distance between the object and L_1 and the image and L_1? Lens L_2 has a focal length of $f_2 = 10$ mm. Where should L_2 be placed relative to object and image to focus light at the same plane as L_1? What is the resulting magnification m_x? If the input object is a square aperture of dimensions 10 mm × 10 mm, what are the corresponding dimensions of the image in the x' and y' directions?

15. Assume in the previous problem that the original input object is rotated about the optical axis by 45 degrees, forming a diamond shape. What is the exact shape of the image (i.e., sketch it showing relative dimensions)? If we let $m_x \to 0$, what happens to the image? If a linear detector array was then used to sample the resulting image, what would the readout look like? (Sketch the result and label axes and dimensions). What does this lens combination do; that is, what mathematical operation does it perform?

16. In practice, for sources of illumination not small compared to the distance to the point of observation, the irradiance does not obey the inverse square law. To what extent does the inverse square law err with respect to the exact irradiance versus distance from the source? Assume that the source is a circular disk of radius a uniformly illuminating the surroundings. At what range relative to a does the error reduce to 1%?

17. For unpolarized light propagating through a glass plate, show that Equation (4.31) is correct. Then show that Equation (4.32) holds true for no absorption (and normal incidence).

18. Using Equations (4.38) and (4.39), define the exact arrangement (facet normal orientation angles and apex angles) of the two prisms used in Fig. 4.13 to achieve a 1:1 aspect ratio exit beam from a 2:1 aspect ratio entrance beam. You may assume that the two prisms have the same apex angle β and the same index of refraction n, and you can constrain the entrance and exit beams, which are collimated, to be parallel to a common x axis.

19. Show that the result of Equation (4.40) is true by applying Snell's law to the prism shown in Fig. 4.14.

20. Consider a circular aperture representing a simple lens in a finite conjugate ratio condition (at unity magnification) imaging a diffuse object surface with irradiance H_o onto an image plane. For object points near the axis and for a lens diameter D small with respect to the object distance s_o, show that the image irradiance is given by

$$H_i = \frac{H_o}{4(f/\#)^2 + 1}$$

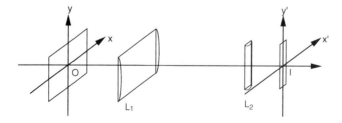

where $f/\# = f/D$ and H_o is the peak irradiance (on axis).

21. If a planoconvex lens has a radius of curvature of $R_2 = 30$ cm and an index of refraction of 1.52, what is its focal length? If an object is located 50 cm to the left of this lens, calculate the position of the image of the object and its lateral magnification.

22. A unit amplitude monochromatic plane wave propagating in an arbitrary direction with respect to the (x, y, z) coordinate system is defined by

$$u(x, y, z) = \exp(j\mathbf{k} \cdot \mathbf{r})$$

where $\mathbf{k} = k_x\mathbf{i} + k_y\mathbf{j} + k_z\mathbf{k}$ and $\mathbf{r} = x\mathbf{i} + y\mathbf{j} + z\mathbf{k}$. If the direction cosines are given by the triple (α, β, γ), then write down an expression for the lines of zero phase in the $z = 0$ plane and sketch the result for the $\alpha = 0.1, \beta = 0.1$ case. What is the slope of these lines and their spacing?

23. For a multiaperture lens system consisting of a 5×5 array of lenslets with 5-mm diameters on 6-mm spacings and with 20-mm focal lengths, determine the focal points (i.e., points of sharpest focus in the image plane) and the corresponding lateral magnification for all unique images of the total set of 25 images formed by the array of an object consisting of a square slide 25 mm on a side (with an arbitrary image on it) positioned at a distance of 60 mm from the center of the lenslet array. Carefully ray trace all unique peripheral and central rays from the object in order to determine the relative positions of central rays from the object where best focus occurs. Also determine the deviations from these points for best focus to those points of best focus for extremal points of the object (corners of the square slide) using the simple thin-lens formulae (i.e., determine depth of focus).

BIBLIOGRAPHY

M. Alonso and E. J. Finn, *Fundamental University Physics II: Fields and Waves*, Addison-Wesley, Reading, MA (1967).
R. H. Anderson, "Close-up Imaging of Documents and Displays with Lens Arrays," *Appl. Opt.* **18**, 477 (1979).
N. F. Borrelli and D. L. Morse, "Microlens Arrays Produced by a Photolytic Technique," *Appl. Opt.* **27**, 476 (1988).
N. F. Borrelli, D. L. Morse, R. H. Bellman, and W. L. Morgan, "Photolytic Technique for Producing Microlenses in Photosensitive Glass," *Appl. Opt.* **24**, 2520 (1988).
G. R. Fowles, *Introduction to Modern Optics*, Holt, Rinehart and Winston, New York (1968).
E. J. Hecht and A. Zajac, *Optics*, Addison-Wesley, New York (1974).
M. Hutley, R. Stevens, and D. Daly, "Microlens Arrays," *Phys. World*, pp. 27–32 (July 1991).
L. J. Pinson, *Electro-Optics*, John Wiley & Sons, New York (1985).

Chapter 5
SUMMARY OF PHYSICAL OPTICS

5.1 OVERVIEW

Except for our treatment of polarization (which is a fundamental vectorial property of electromagnetic waves), we have regarded the propagation of light in geometrical terms. Hence we have treated light as a form of energy propagated in straight lines. We now acknowledge three other properties of light that are explicitly dependent on its wave nature: coherence, interference, and diffraction.

5.2 COHERENCE AND INTERFERENCE

As described previously, the propagation of electromagnetic waves is described by a second-order partial differential equation [Equation (4.1)] so that the principle of linear superposition holds. We can immediately exploit this principle in describing coherence by considering two sources of plane harmonic linearly polarized waves. The fields produced by these two sources are given by

$$\mathbf{E}_1 = \mathbf{E}_{1o} \exp[j(\mathbf{k}_1 \cdot \mathbf{r} - \omega t + \phi_1)] \tag{5.1}$$

$$\mathbf{E}_2 = \mathbf{E}_{2o} \exp[j(\mathbf{k}_2 \cdot \mathbf{r} - \omega t + \phi_2)] \tag{5.2}$$

If the initial phase difference between these two sources, $\phi_1 - \phi_2$, is constant, then the sources are said to be mutually coherent. Under these circumstances if we examine the intensity of the total field at some point $\mathbf{r}$ from the sources we obtain

$$I = |\mathbf{E}|^2 = \mathbf{E} \cdot \mathbf{E}^* = (\mathbf{E}_1 + \mathbf{E}_2) \cdot (\mathbf{E}_1^* + \mathbf{E}_2^*) \tag{5.3}$$

$$I = |\mathbf{E}_1|^2 + |\mathbf{E}_2|^2 + 2\mathbf{E}_1 \cdot \mathbf{E}_2 \cos\theta \tag{5.4}$$

or

$$I = I_1 + I_2 + 2\mathbf{E}_1 \cdot \mathbf{E}_2 \cos\theta \tag{5.5}$$

where $\theta = (\mathbf{k}_1 - \mathbf{k}_2) \cdot \mathbf{r} + (\phi_1 - \phi_2)$, and $I_1 = |\mathbf{E}_1|^2$ and $I_2 = |\mathbf{E}_2|^2$.

5.2 COHERENCE AND INTERFERENCE

The interference term $2\mathbf{E}_1 \cdot \mathbf{E}_2 \cos\theta$ contains spatially periodic intensity variations because θ depends on $\mathbf{r}$. If the sources are mutually incoherent (and at the same wavelength), then $\phi_1 - \phi_2$ varies randomly in time. Then the mean value of the interference term ($\cos\theta$) is zero, and there is no visible interference. The degree of interference also depends on the polarization, but we will not consider it here.

In the classic Young's experiment that embodies these concepts a wavefront from a point source of light at infinity illuminates two slits (with separation h) in an otherwise opaque screen. A common source ensures that the screen sources are mutually coherent. After the plane wave passes through the two slits the resulting cylindrical wavefronts interfere creating loci of constructive (maximum) and destructive (minimum) interference. Their total intensity is measured at a remote point P a distance r from the screen (large with respect to the separation h). Then the fields measured at the screen can be adequately described as plane waves:

$$\mathbf{E}_1 = \mathbf{E}_{1o} \exp[j\mathbf{k}_1 \cdot \mathbf{r} - \omega t + \phi_1)] \quad (5.6)$$

$$\mathbf{E}_2 = \mathbf{E}_{2o} \exp[j\mathbf{k}_2 \cdot \mathbf{r} - \omega t + \phi_2)] \quad (5.7)$$

exactly the same as the fields described for introducing coherence.

If the normal distance z to the screen is large compared to h, then $\mathbf{k}_1 - \mathbf{k}_2 = \mathbf{j} kh/z$ so that $(\mathbf{k}_1 - \mathbf{k}_2) \cdot \mathbf{r} = khy/z$, where $\mathbf{k}_1 = \mathbf{k}_2 = \mathbf{k}$. Therefore,

$$I = |\mathbf{E}_1|^2 + |\mathbf{E}_2|^2 + 2\mathbf{E}_1 \cdot \mathbf{E}_2 \cos(khy/z - \phi) \quad (5.8)$$

where $\phi = \phi_1 - \phi_2$. If the slits are identical, then $\phi = 0$ and

$$I = 2I_0[1 + \cos(khy/z)] \quad (5.9)$$

where $I_0 = |\mathbf{E}_1|^2 = |\mathbf{E}_2|^2$. Bright fringes occur at $kyh/z = 2\pi n$ ($n = 0, 1, 2, \ldots$) in the corresponding interference pattern as shown Fig. 5.1. By using the fact that $k = 2\pi/\lambda$ and a trigonometric identity, we can put Equation (5.9) in the form

$$I = 4I_0 \cos^2(yh\pi/z\lambda) \quad (5.10)$$

In considering the total path length for each slit from the original source to the point of measurement, we can also treat the case where the amplitudes and phases of the two light waves along each path vary randomly in time. In calculating the total intensity (I) at the measurement point P, we then get

$$I = \langle \mathbf{E} \cdot \mathbf{E}^* \rangle = \langle |\mathbf{E}_1|^2 + |\mathbf{E}_2|^2 + 2\mathrm{Re}(\mathbf{E}_1 \cdot \mathbf{E}_2^*) \rangle \quad (5.11)$$

where $\langle \ldots \rangle$ denotes the time average defined by

$$\langle f(t) \rangle = \lim_{T \to \infty} \frac{1}{T} \int_0^T f(t)\, dt \quad (5.12)$$

Assuming the sources are stationary (temporally as well as spatially) and that polarization can be ignored, then

$$I = I_1^2 + I_2^2 + 2\mathrm{Re}(\mathbf{E}_1 \cdot \mathbf{E}_2^*) \quad (5.13)$$

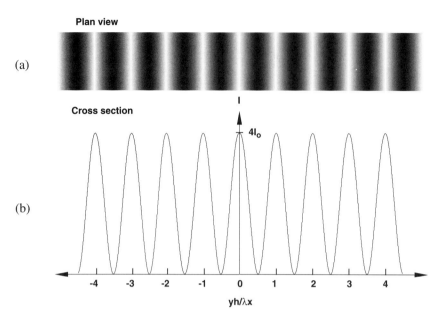

Figure 5.1 Interference pattern produced in Young's double slit experiment.

Looking at the two paths in Young's experiment, if the time to traverse path 1 is t and the time to traverse path 2 is $t + \tau$, then the interference term, defined as $2\text{Re}\Gamma_{12}(\tau)$, is given by

$$2\text{Re}\Gamma_{12}(\tau) = 2\text{Re}\langle \mathbf{E}_1(t) \cdot \mathbf{E}_2^*(t+\tau)\rangle \tag{5.14}$$

$\Gamma_{12}(\tau)$ is the so-called mutual coherence or correlation function. Note that $I_1 = \Gamma_{11}(0)$ and $I_2 = \Gamma_{22}(0)$. The normalized correlation function is then defined as

$$\gamma_{12}(\tau) = \frac{\Gamma_{12}(\tau)}{\sqrt{\Gamma_{11}(0)\Gamma_{22}(0)}} = \frac{\Gamma_{12}(\tau)}{\sqrt{I_1 I_2}} \tag{5.15}$$

where $\gamma_{12}(\tau)$ is periodic, and its absolute value $|\gamma_{12}(\tau)|$ is a measure of the degree of coherence, namely,

$$|\gamma_{12}(\tau)| = 1: \qquad \text{Complete coherence}$$
$$0 < |\gamma_{12}(\tau)| < 1: \qquad \text{Partial coherence}$$
$$|\gamma_{12}(\tau)| = 0: \qquad \text{Complete incoherence}$$

The degree of coherence is also related to the characteristics of the source, as discussed below.

An important consequence of the degree of coherence is the spectral resolution (or dispersion) of the wavetrain created by a source. Consider a point source that emits a single wavetrain of duration τ and radian frequency ω_o as shown in Fig. 5.2(a). This is a very realistic situation because all sources are of finite duration (they must be turned on and turned off). In complex notation this wavetrain is expressed as

$$f(t) = \begin{cases} \exp(-j\omega_o t) & -\tau/2 < t < \tau/2 \\ 0, & \text{otherwise} \end{cases} \tag{5.16}$$

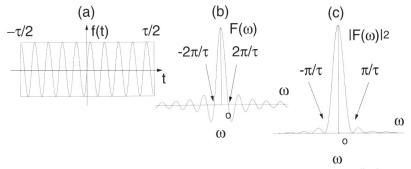

Figure 5.2 (a) Finite wavetrain emitted by a point source, (b) corresponding amplitude spectrum, and (c) power spectrum.

To determine the frequency spectrum of the source we must take the Fourier transform, which is

$$F(\omega) = \int_{-\tau/2}^{\tau/2} \exp[j(\omega - \omega_0)t] \, dt \quad (5.17)$$

or

$$F(\omega) = \tau \frac{\sin[(\omega - \omega_0)\tau/2]}{(\omega - \omega_0)\tau/2} \quad (5.18)$$

which is plotted in Fig. 5.2(b). The power spectrum $|F(\omega)|^2$ is shown in Fig. 5.2(c), from which we can determine the spectral width, which is $\Delta\omega = 2\pi/\tau$ or $\Delta f = 1/\tau$. For a sequence of wavetrains each lasting τ but occurring at random times, $|F(\omega)|^2$ is the same as the single pulse result given above. If, however, the pulses are not of the same duration—that is, τ varies from pulse-to-pulse—then the average τ given by $\langle\tau\rangle$ is used to obtain the spectral width $\Delta f = 1/\langle\tau\rangle$. The corresponding distance over which the resulting wavetrain is coherent is the coherence length defined by $l_c = c\langle\tau\rangle = c/\Delta f$. Using this and the fact that $\Delta f/f = \Delta\lambda/\lambda$, we obtain $l_c = \lambda^2 \Delta\lambda$ for the coherence length of a source in terms of wavelength dispersion. This feature of light propagation becomes important when considering lasers (particularly short pulse laser diodes) used in coherent optical processing architectures. It is essential that the coherence length be sufficient to support the creation of stable interference patterns and the accurate determination of their Fourier transforms.

As a last word on coherence, we should acknowledge the importance of spatial coherence, as opposed to temporal coherence, which was just discussed. If we have a point source at P, as depicted in Fig. 5.3, we can measure the fields at points P_1, P_2 and P_3. These are E_1, E_2 and E_3. Coherence between E_1 and E_2 corresponds to lateral (spatial) coherence. As we have seen, the degree of coherence between E_1 and E_3 is gauged by the distance r_{13} relative to the coherence length l_c (or r_{13}/c relative to τ) as determined by the temporal duration of the source. The degree of coherence between E_1 and E_2, on the other hand, depends on the extent of the source. E_1 and E_2 are completely mutually coherent for a point source; however, a very small laser source (e.g., laser diode) can have a distorted phase front due to rough facets or windows through which the emitted light passes. An example of light that is temporally coherent but spatially incoherent is light from a spectrally filtered gas discharge lamp. Spatially coherent but temporally incoherent light would come from a distant star (i.e., a point source that passes through a turbulent medium).

122 SUMMARY OF PHYSICAL OPTICS

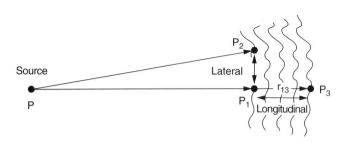

Figure 5.3 Simplified schematic illustrating geometry of lateral (or spatial) and longitudinal (or temporal) coherence.

5.3 SCALAR DIFFRACTION THEORY

If we recall the wave equation for electromagnetic wave propagation from Chapter 4 [Equation (4.1)], the solution for the field strength in spherical coordinates can be shown to be

$$U = U_o \frac{1}{j\lambda r} \exp[-j(kr - \omega t)] \qquad (5.19)$$

The dependence of the wave amplitude on $1/r$ is readily understood if it is realized that the power density must decrease as $1/r^2$ as the surface area increases as r^2, since the total power emanating from the source must remain constant. If, however, we displace the source from the origin by a distance r', the solution is

$$U = U_o \frac{1}{j\lambda |\mathbf{r} - \mathbf{r}'|} \exp[-j(k|\mathbf{r} - \mathbf{r}'| - \omega t)] \qquad (5.20)$$

This situation is illustrated in Fig. 5.4. The point r can be viewed as the point of observation. If there are many sources (continuously or discretely distributed) over some volume, the principle of superposition allows us to add all the fields as expressed by the integral

$$U(\mathbf{r}) = U_o \int_{-\infty}^{\infty} \int_{-\infty}^{\infty} \int_{-\infty}^{\infty} \frac{U(x', y', z')}{j\lambda |\mathbf{r} - \mathbf{r}'|} \exp[-j(k|\mathbf{r} - \mathbf{r}'| - \omega t)] \, d^3\mathbf{r}' \qquad (5.21)$$

where the integration is taken over all sources and where $U(x', y', z')$ is the field strength at the source point (x', y', z'). If, as in most optics problems, we confine all the sources to a plane perependicular to the optical axis, the above result reduces to

$$U(x, y, z) = U_o \int_{-\infty}^{\infty} \int_{-\infty}^{\infty} \frac{U(x', y')}{j\lambda |\mathbf{r} - \mathbf{r}'|} \exp[-j(k|\mathbf{r} - \mathbf{r}'| - \omega t)] \, dx' dy'$$

where $z = 0$. This is the field strength at the observation plane.

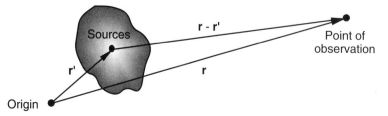

Figure 5.4 Schematic diagram of the general radiation problem for an arbitrary collection of sources displaced with respect to a coordinate system located at the origin shown and for an observation point removed accordingly from the centroid of the source distribution by the distance $|\mathbf{r} - \mathbf{r}'|$.

The above general result becomes more interesting if we consider what happens when the radiation described by it passes through an aperture or around some obstacle as suggested in Fig. 5.5. Rather than deriving this result from the boundary conditions, we will simply state the result, known as Huygen's approximation. Then the diffraction problem is expressed by the equation

$$U(x, y, z) = \int_{-\infty}^{\infty} \int_{-\infty}^{\infty} U_{\text{inc}} T(x', y', 0) \frac{1}{j\lambda |\mathbf{r} - \mathbf{r}'|} \exp[-j(k|\mathbf{r} - \mathbf{r}'| - \omega t)] \, dx' dy' \quad (5.22)$$

or

$$U(x, y, z) = \int_{-\infty}^{\infty} \int_{-\infty}^{\infty} U_{\text{trans}}(x', y') \frac{1}{j\lambda |\mathbf{r} - \mathbf{r}'|} \exp[-j(k|\mathbf{r} - \mathbf{r}'| - \omega t)] \, dx' dy' \quad (5.23)$$

where the transmission function $T(x', y', 0)$ describing the source plane includes any amplitude or phase factors. In other terms,

$$T(x', y') = \frac{U_{\text{trans}}(x', y', 0)}{U_{\text{inc}}(x', y', 0)} \quad (5.24)$$

where an obliquity term $\cos(\ldots)$ has been omitted for simplicity (since it is often close to unity). Equation (5.23) is the fundamental diffraction integral, and it is, in general, difficult to solve. Two approximations that we will consider are known as the Fresnel and Fraunhofer approximations, which roughly correspond to the near- and far-field, respectively.

In the far-field case the observation plane is far away with respect to the source plane or observation plane extent (in x, y)—that is, $z \gg x, y, x'$ or y'. Then

$$|\mathbf{r} - \mathbf{r}'| = [z^2 + (x - x')^2 + (y - y')^2]^{1/2}$$

$$= \left[1 + \frac{1}{2}\left(\frac{x - x'}{z'}\right)^2 + \frac{1}{2}\left(\frac{y - y'}{z}\right)^2 + \cdots \right] z \approx z \quad (5.25)$$

or

$$\frac{1}{|\mathbf{r} - \mathbf{r}'|} \approx \frac{1}{z} \quad (5.26)$$

The factor $k|\mathbf{r} - \mathbf{r}'|$ in the argument of the exponential cannot be approximated by kz because

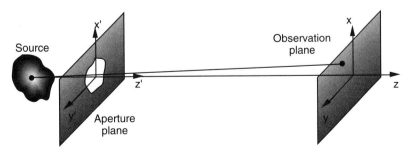

Figure 5.5 Schematic diagram of the general diffraction problem showing the source, diffracting aperture, and observation plane.

k is such a large number. In fact, important information is in this phase term. Thus, in the far-field approximation we obtain

$$U(x, y, z) = \frac{\exp(j\omega t)}{j\lambda z} \int_{-\infty}^{\infty} \int_{-\infty}^{\infty} U_{\text{trans}}(x', y') \exp[-j(k|\mathbf{r} - \mathbf{r}'| - \omega t)] \, dx' dy' \quad (5.27)$$

This result is still not simple enough to be readily solvable. A further approximation is necessary in the phase term. If we keep terms to second order in Equation (5.25), then

$$|\mathbf{r} - \mathbf{r}'| = z \left\{ 1 + \frac{1}{2} \left[\frac{(x - x')^2}{z^2} + \frac{(y - y')^2}{z^2} \right] \right\} \quad (5.28)$$

This is the so-called Fresnel approximation and leads to

$$U(x, y, z) = \frac{\exp[-j(kz - \omega t)]}{j\lambda z} \int_{-\infty}^{\infty} \int_{-\infty}^{\infty} U_{\text{trans}}(x', y')$$

$$\times \exp\left\{ -j\frac{k}{2z}[(x - x')^2 + (y - y')^2] \right\} dx' dy' \quad (5.29)$$

(where $\pi/\lambda z = k/2z$). This result is valid only when the third term in the series is

$$\exp\left\{ -j\frac{k}{8z^2}[(x - x')^2 + (y - y')^2]^2 \right\} \approx 1 \quad (5.30)$$

that is, when

$$z^3 \gg \frac{\pi}{4\lambda}[(x - x')^2 + (y - y')^2]_{\text{max}}^2 \quad (5.31)$$

where the maximum possible value of the right-hand side must be considered for points in the source and observation planes.

Equation (5.29) can be further simplified if we reexamine the phase term and note

$$\frac{\pi}{\lambda z}[(x - x')^2 + (y - y')^2] = \frac{\pi}{\lambda z}(x^2 + y^2) + \frac{\pi}{\lambda z}(x'^2 + y'^2) - \frac{2\pi}{\lambda z}(xx' + yy') \quad (5.32)$$

where the first term on the right-hand side is independent of the integration and the second term is negligible in the Fraunhofer approximation. Therefore under the constraint that

$$z \gg \frac{\pi}{\lambda}(x'^2 + y'^2)_{\text{max}} \quad (5.33)$$

we then have

$$U(x, y, z) = \frac{\exp[-j(kz - \omega t)]}{j\lambda z} \exp\left[\frac{\pi}{\lambda z}(x^2 + y^2)\right]$$

$$\times \int_{-\infty}^{\infty} \int_{-\infty}^{\infty} U_{\text{trans}}(x', y') \exp\left\{ -j\frac{2\pi}{\lambda z}[(xx' + yy')] \right\} dx' dy' \quad (5.34)$$

Consider Equations (5.31) and (5.33) corresponding to the constraints on (x', y') and (x, y) with respect to z for the Fresnel and Fraunhofer approximations, respectively. For a 10-cm limit on (x, y) and (x', y') at a wavelength of HeNe laser light ($\lambda = 633$ nm), we have $z_{\text{Fraunhofer}} \gg 99$ km and $z_{\text{Fresnel}} \gg 8$ m.

5.4 FRAUNHOFER DIFFRACTION

Diffraction is defined as the deviation of light from purely rectilinear propagation due to the presence of an obstruction. When a plane wave representing light impinges on an aperture, fringes are created along the outline of the aperture on a screen situated behind the aperture at close distances. Waves from one part of the aperture interfere with waves from another part of the aperture. Hence there is, in principle, little or no distinction between diffraction and interference. Every point on an impinging wavefront can serve as a source of spherical secondary wavelets (Huygens wavelets) of the same frequency as the primary incident wave, as illustrated in Fig. 5.6. The optical field at any point beyond the obstruction is the linear superposition of all such wavelets reaching that point. In the near-field (close to the aperture with respect to its size) the diffraction is termed *Fresnel diffraction*, whereas in the far-field (far from the aperture) it is termed *Fraunhofer diffraction*.

Before considering a few examples of Fraunhofer and Fresnel diffraction, we should consider from first principles the creation and propagation of radiation from a so-called coherent line source, as illustrated in Fig. 5.7. This is the simplest approach to explaining diffraction mathematically. The points along the y axis can be considered discrete emitters of Huygens wavelets. If we consider a point P in the plane of observation, which is far away from this source, then the rays from these point sources are essentially parallel, and all waves arriving at P from the various points have essentially the same amplitude (modulus). Under these circumstances the electric field strength at point P is given by

$$E_p = E_o(r)\exp[j(kr_1 - \omega t)] + E_o(r)\exp[j(kr_2 - \omega t)] + \cdots \\ + E_o(r)\exp[j(kr_N - \omega t)] \quad (5.35)$$

or

$$E_p = E_o(r)\exp[j(kr_1 - \omega t)]\{1 + \exp[j(kr_2 - r_1)] + \exp[j(kr_3 - r_1)] + \cdots \\ + \exp[j(kr_N - r_1)] \quad (5.36)$$

Figure 5.6 Huygen's model of diffraction showing production of secondary wavelets at the diffracting plane.

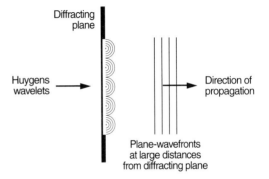

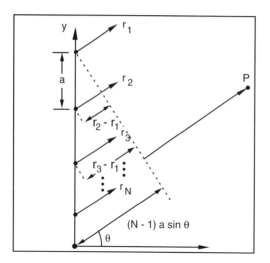

Figure 5.7 Coherent line source of Huygens wavelets.

Since the phase difference between adjacent source points viewed at an angle θ (with respect to the x axis) is $\delta = ka \sin\theta$ for a spacing a, we obtain

$$E_p = E_o(r) \exp[j(kr_1 - \omega t)]\{1 + \exp(j\delta) + [\exp(j\delta)]^2 + [\exp(j\delta)]^3 + \cdots \\ + [\exp(j\delta)]^{N-1}\} \quad (5.37)$$

In closed form Equation (5.37) becomes

$$E_p = E_o(r) \exp[j(kr_1 - \omega t)] \exp[j(N-1)\delta/2] \frac{\sin(N\delta/2)}{\sin(\delta/2)} \quad (5.38)$$

using the geometric series formula:

$$\sum_{n=0}^{N-1} a^N = \frac{1 - a^{-N}}{1 - a^{-1}} \quad (5.39)$$

Defining R as the distance from the center of the array to point P, we obtain

$$R = \frac{1}{2}(n-1)a \sin\theta + r_1 \quad (5.40)$$

Therefore,

$$E_p = E_o(r) \exp[j(kR - \omega t)] \frac{\sin(N\delta/2)}{\sin(\delta/2)} \quad (5.41)$$

so that the intensity is given by

$$I_p = I_o \frac{\sin^2(N\delta/2)}{\sin^2(\delta/2)} \quad (5.42)$$

which is plotted in Fig. 5.8 for cases of $N = 10, 20, 30, 45, 60$, and 100. Note that as N increases, the width of the mainlobe of each sinc function narrows. In the limit, as $N \to \infty$,

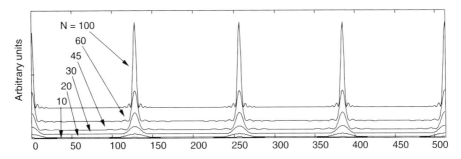

Figure 5.8 Intensity distribution of light in the observation plane produced by diffraction from a discrete coherent line source of finite extent for increasing values of N.

this array of sinc functions reduces to a unit impulse array. In reality we must consider the continuum limit of this result, where not only N approaches ∞ but the source separation a approaches zero, with the provision that the individual source strengths remain finite. Then the result is

$$E_p = E_o \int_{-D/2}^{D/2} \frac{\sin(\omega t - kr)}{r} dy \tag{5.43}$$

where $r = r(y)$ and D is the source extent.

Although the ideal line source that we have just described is not physically realizable in a strict sense, it is a good mathematical model for calculating elementary diffraction patterns. As we shall see, the position of point P relative to the array of point emitters distinguishes Fresnel from Fraunhofer regions. Consider now several examples of Fraunhofer diffraction.

EXAMPLE 5.1 Fraunhofer Diffraction from a Single Slit Consider the narrow aperture or slit illustrated in Fig. 5.9. The differential strip with dimensions dz by l is a coherent line source. Each segment dy makes a contribution to the far field of

$$dE_p = \frac{E_o}{R} \sin(\omega t - kr) dy$$

where

$$r = R - y \sin\theta + \frac{y^2}{2R} \cos^2\theta + \cdots$$

The Fraunhofer approximation given by $r \sim R - y \sin\theta$ implies that the total field at P is

$$E_p = \frac{E_o}{R} \int_{-l/2}^{l/2} \sin[\omega t - k(R - y\sin\theta)] dy$$

or

$$E_p = \frac{E_o l}{R} \frac{\sin\left(\frac{kl}{2}\sin\theta\right)}{\left(\frac{kl}{2}\sin\theta\right)} \sin(\omega t - kR)$$

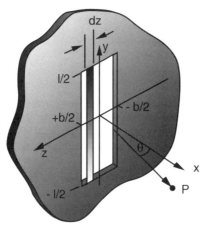

Figure 5.9 Narrow slit geometry for calculation of single slit diffraction pattern.

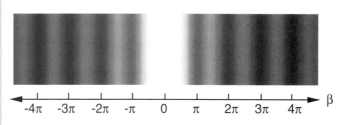

Figure 5.10 Intensity pattern caused light diffraction of light from a single rectangular slit.

assuming $r = R$ in the denominator. The irradiance distribution from such a source is then

$$H(\theta) = c\varepsilon_o \langle E_p^2 \rangle = H(0) \operatorname{sinc}^2 \left(\frac{kl \sin \theta}{2} \right)$$

where a time average is taken to obtain the result shown. When $l \gg \lambda$, $H(\theta)$ drops off very rapidly with θ, so that the slit source emits predominantly in the xz plane, and the E field looks like that from a point source. Thus we can think of the original differential strip (dz by 1) as a point emitter. All such strips from $-b/2$ to $+b/2$ correspond to a linear array of point sources along the z axis. Then

$$H(\theta) = H(0) \operatorname{sinc}^2 \left(\frac{kb \sin \theta}{2} \right)$$

This result is plotted in Fig. 5.10 versus $\beta = (kb \sin \theta)/2$.

EXAMPLE 5.2 Multiple-Slit Diffraction Pattern The diffraction pattern that results from multiple slits is derived in a manner similar to that for a single slit; however, a simple geometric construction yields the appropriate formula for intensity at the observation plane. Figure 5.11 shows one possible source distribution. Assuming all coherent line sources, spaced apart by a, are alike, the resultant amplitude at P is

$$E_p = 2E_0 \sin(N\delta/2)$$

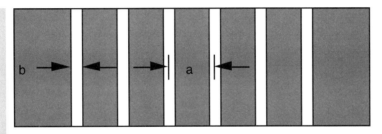

Figure 5.11 Typical multiple slit aperture or source distribution for calculating diffraction, where $a = 4b$.

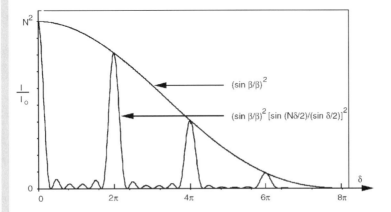

Figure 5.12 Multiple slit diffraction pattern for the case of six slits ($N = 6$) spaced apart by $a = 4b$. The overall envelope (element factor) is indicated.

where $\delta = \pi k a \sin \theta$. The initial amplitude is $E_{pi} = 2E_0 \sin(\delta/2)$. Therefore the relative amplitude given by E_p/E_{pi} and the initial value of the intensity $I_0 = |E_o|^2$ combine to give

$$I = I_0 \left[\frac{\sin \beta}{\beta}\right]^2 \left[\frac{\sin(N\delta/2)}{\sin(\delta/2)}\right]^2$$

where the leading sinc term is the diffraction pattern due to a single slit of width b as derived in Example 5.1, which is the so-called element factor (also termed the interference term), and the second periodic sinc function is due to the multiple slits and is called the array factor (also termed the diffraction term). This pattern is plotted in Fig. 5.12 for $N = 6$ and a slit spacing of $a = 4b$.

EXAMPLE 5.3 Fraunhofer Patterns of Rectangular and Circular Apertures It is instructive to highlight the far-held diffraction patterns of rectangular and circular apertures. The intensity of the rectangular aperture diffraction pattern is

$$I(Y, Z) = I(0) \operatorname{sinc}^2\left(\frac{kaZ}{2R}\right) \operatorname{sinc}^2\left(\frac{kbY}{2R}\right)$$

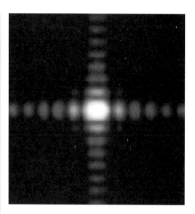

Figure 5.13 Fraunhofer diffraction pattern from a rectangular aperture (with sides $a = 6$ and $b = 8$).

Figure 5.14 Fraunhofer diffraction pattern of a circular aperture (classic Airy pattern).

a result which is plotted in Fig. 5.13, where a and b are the aperture dimensions in x and y coordinates, R is the distance from the aperture to the diffraction plane, and (x, y) are diffraction plane coordinates.

The diffraction pattern of a circular aperture is illustrated in Fig. 5.14. The corresponding expression for intensity is

$$I(\theta) = I(0) \left[\frac{2 J_1(ka \sin \theta)}{ka \sin \theta} \right]^2$$

where $J_1(\ldots)$ is the first-order Bessel function, a is the radius of the circular aperture, and $\sin \theta = q/R$, where q is the radial coordinate in the diffraction pattern plane (plane of observation), and R is the distance between the aperture (source) plane and the diffraction (observation) plane. This result is the classic Airy pattern. The first null is at $ka \sin \theta = 3.83$ or $\theta_1 = 1.22 \, R\lambda/2a$. In fact, the Rayleigh resolution criterion $1.22\lambda/D$ (where $D = 2a$) for resolving point objects with a telescope objective of diameter D is derived from this.

The various examples cited so far serve to illustrate the Fourier transform relationship between the aperture function and its corresponding diffraction pattern. This result occurs at a long distance from the aperture compared to its diameter, consistent with the Fraunhofer (far-field) approximation. In order to enable the Fourier transform to be recovered at short distances from an aperture, it is necessary to use a collimating lens and an intermediate Fourier

5.5 FRESNEL DIFFRACTION

To understand how the Fresnel pattern is calculated from first principles consider the schematic of a circular aperture shown in Fig. 5.16. Here the distance from the source to the point of observation is finite. The contribution of the annular surface element of area dS to the field at P has amplitude given by

$$E_{0n} = g(\theta)\frac{dS}{r} \qquad (5.44)$$

where the subscript n denotes that the annulus in question is the so-called nth zone. The phase of the wave at point P due to dS is

$$\delta = 2\pi r/\lambda \qquad (5.45)$$

We wish to add all the contributions from all rings for $\lambda \ll r_o$ at increments of $\lambda/2$ from P—that

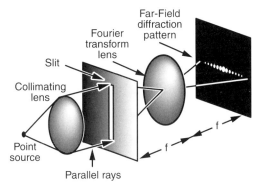

Figure 5.15 Use of intermediate (or collimating) lens and Fourier transform lens to obtain Fraunhofer diffraction pattern at close range with respect to the diffracting aperture.

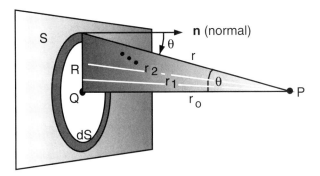

Figure 5.16 Circular aperture and geometry used to calculate corresponding Fresnel diffraction pattern.

is, at $r_1 = r_o + \lambda/2$, $r_2 = r_1 + \lambda/2$, and so on. Each ring of area dS contributes an amplitude and phase given by Equations (5.44) and (5.45), respectively, which constitutes a vector quantity: $E_{0n} \exp(j\delta)$. For the first "ring" (circle) out from point O (center of the aperture), successive vectors from within the ring add to yield an amplitude at P equal to E_{00} for that ring. For the next ring we sum all such contributions again, yielding E_{01}. Because $g(\theta)$ and $1/r$ factors become smaller and smaller, the resulting vector addition $E_{00} + E_{01} + E_{02} + \ldots$ should converge. If incremental distances are separated by $\lambda/2$, then the phase difference between each of the terms (or rings) is π, and convergence will be guaranteed. To see this we can rewrite the sum of the fields

$$E_{\text{total}} = E_{00} + E_{01} + E_{02} + \cdots \tag{5.46}$$

as

$$E_{\text{total}} = \frac{1}{2}E_{00} + \left(\frac{1}{2}E_{00} - E_{01} + \frac{1}{2}E_{00}\right) + \left(\frac{1}{2}E_{02} - E_{03} + \frac{1}{2}E_{04}\right) + \cdots \tag{5.47}$$

so that for amplitudes virtually equal between neighboring zones $E_{\text{total}} = E_{00}/2$ for an infinite plane.

The radii of the Fresnel zones (in the plane of the aperture) are important in constructing Fresnel zone plates. These are given by

$$R_n = (r_{n^2} - r_{o^2})^{1/2} \tag{5.48}$$

but since $r_n = r_o + n\lambda/2$ then $R_n \sim n\lambda r_o$.

Thus all Fresnel zones have virtually the same area. If an obscuring screen with a circular aperture is placed at position Q such that it obscures everything but the first zone and then is increased in diameter unveiling successively more concentric zones (like a camera shutter), the

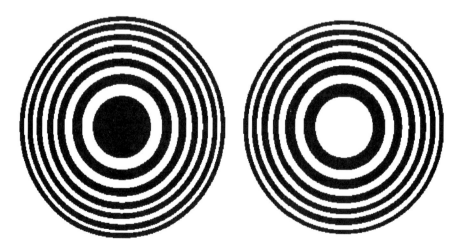

Figure 5.17 Examples of Fresnel zone plates (a) with and (b) without a central obscuration.

intensity at point P will fluctuate. If only odd or even zones are used, this will yield intensities exceeding the original intensity at P without the screen, because we obtain coordinated phase reinforcement at P. This is the basis of Fresnel zone plates that act like lenses, which are illustrated in Figs. 5.17(a) and 5.17(b).

Finally, an important feature of diffraction is worth discussing here: the transition from the near-field to far-field. Fraunhofer diffraction occurs when both the incident and diffracted waves are essentially plane—that is, at ranges with respect to the diffracting plane where wave curvature can be neglected. If the observation point P is close, then the wavefront curvature cannot be neglected. Assuming that the source is at $-\infty$—that is, that light incident on the aperture consists of plane waves—then Fraunhofer diffraction occurs for ranges (beyond the aperture) such that $r \gg D^2/2\lambda$, where D is the aperture diameter and λ is the wavelength of incident light. A more complete description of the transition in two dimensions is given in Fig. 5.18. It shows the diffraction pattern from a rectangular aperture (see Fig. 3.1 for the one-dimensional case) illustrating that the near-field Fresnel diffraction pattern is an approximation (similar to that obtainable by Fourier *synthesis*) to the aperture function, whereas the far-field Fraunhofer diffraction pattern is arbitrarily close to the Fourier transform of the aperture function (similar to that obtainable by Fourier *analysis*). In the intermediate zone the shape of the diffraction pattern is neither the aperture function nor its Fourier transform but a result that, in general, is only obtainable from the full propagation integral between the plane of the aperture and the observation plane.

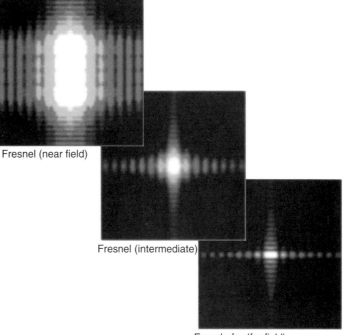

Figure 5.18 Transition from Fresnel (near-field) to Fraunhofer (far-field) diffraction for a rectangular aperture (32 × 16), viewed in two dimensions, at ranges of 64 and 256 and in the far-field with respect to aperture dimensions.

PROBLEM EXERCISES

1. Two plane waves of the same wavelength λ are incident on a screen making an angle θ with respect to each other and perpendicular to the screen. Write an expression for each wave's electric field, add their amplitudes and show that the distance between adjacent interference bands is given by λ/θ.

2. In Young's double-slit experiment the distance between slits is 0.2 mm. The wavelength of incident light is 633 nm. What screen distance is required to achieve a fringe spacing of 0.5 mm?

3. Calculate and plot the interference pattern that results from the use of three equally spaced slits rather than two as in Young's experiment.

4. What is the line width in nanometers and hertz for GaAlAs laser light of 830-nm mean wavelength and 10-cm coherence length?

5. A pinhole source of light is used to create a diffraction pattern by illuminating a 1-mm-diameter aperture at a distance of 5 m. For $\lambda = 633$ nm, at what distance does the transition from Fresnel to Fraunhofer diffraction occur?

6. In the Fraunhofer diffraction pattern of a double slit the fourth secondary maximum is missing. Explain in terms of the element and array factors. What is the ratio of slit width to slit separation?

7. A point source (S) is placed 1 m in front of an aperture consisting of a hole with a circular stop. The radius of the hole is 1 mm, and the radius of the stop is 0.5 mm. At 1 m behind the aperture (point P), what is the intensity compared to that without the stop?

8. An object is located in the front of a positive lens of focal length $f = 50$ cm at a distance $s_o = 100$ cm. If the lens has a diameter of 2.5 cm and the object is illuminated with red light of 630 nm, what is the minimum resolution distance in the focal plane?

9. Consider a linear array of N identical circular apertures, of separation d, normally illuminated by a monochromatic plane wave.
   (a) Accurately plot a diagram illustrating the array using MATLAB.
   (b) Write down an expression that describes this array mathematically, using appropriate symbols—for example, circ(...), δ(...), and so on—and explain each term.
   (c) Determine the far-field irradiance distribution.
   (d) Identify the element factor and array factor, and plot the irradiance with MATLAB and interpret your results.

10. The transmittance distribution $t(x_0)$ for a one-dimensional Fresnel zone plate is given by

$$t(x_o) = \sum_{n=-\infty}^{\infty} \frac{\sin\left(\frac{n\pi}{2}\right)}{n\pi} \exp\left[-jn\left(\frac{\pi}{\lambda f}\right) x_o^2\right] \text{rect}\left(\frac{x_o}{a}\right)$$

where n corresponds to the number of the band (black or white) and f is the principal focal length. Accurately plot this function using MATLAB. Then calculate the amplitude distribution $u(x_i, y_i)$ on a screen at a distance z_i as a function of position off-axis in the Fresnel approximation and show that it is a sinc($ax_i/\lambda z_i$) function. Plot your result using MATLAB.

11. Recall that the diffraction pattern at point P of a circular aperture of radius a is given by

$$I(P) = \frac{2J_1(\pi\delta)}{\pi\delta}$$

where $d = (ka \sin\theta)/2\pi$ (or $(kb \sin\theta)/2\pi$). In terms of this result, derive the diffraction pattern of an annulus as shown in the figure below and plot it with MATLAB.

12. A diffracting screen contains a finite $(n \times m)$ array of circular apertures of radius r_o and spacing a in x and b in y as shown in the figure below. Calculate and plot the corresponding Fraunhofer diffraction pattern using MATLAB tools in Appendix C.

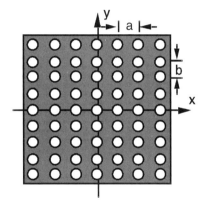

13. A plane wave of wavelength λ is incident on a diffracting screen with transmission given by

$$t(x, y) = \tfrac{1}{2} + m \sin f_o x$$

where f_o is the angular spatial frequency and m is the modulation index.

(a) Calculate the Fresnel diffraction pattern formally and using MATLAB.

(b) For small modulation index ($m \ll 1/2$) determine at what points along the z axis where amplitude modulation and approximate phase modulation occur.

14. Show that the approximation of the obliquity factor: $\cos(\mathbf{n}, \mathbf{r}_{01}) \approx 1$ is comparable to the approximation

$$\frac{1}{r_{01}} \approx \frac{1}{z}$$

Use the definition of the scalar (dot) product, namely,

$$\mathbf{A} \cdot \mathbf{B} = |\mathbf{A}| \, |\mathbf{B}| \, \cos\theta$$

15. If a sinusoidal phase grating is formed, given by

$$t(x, y) = \exp[ja \, \sin(f_o x)] \, \text{rect}\left(\frac{x}{L}\right)$$

where f_o is the grating angular spatial frequency, a is a constant that sets the maximum phase variation, and L is the grating width, calculate the intensity distribution in the focal plane of a lens used to observe the diffraction pattern as suggested in Fig. 5.15. Assume that $Lf_o \gg 1$. Show that the resulting line spectrum is similar to that shown in Fig. 5.12, but where the envelope of the intensity pattern is controlled by the function $J_n^2(a)$, where $J_n(a)$ is the Bessel function of the first kind of order n. Plot your results using MATLAB.

BIBLIOGRAPHY

M. Alonso and E. J. Finn, *Fundamental University Physics, Volume II: Fields and Waves*, Addison-Wesley, Reading, MA (1967).

G. R. Fowles, *Introduction to Modern Optics*, Holt, Rinehart and Winston, New York (1968).

E. J. Hecht and A. Zajac, *Optics*, Addison-Wesley, New York (1974).

L. J. Pinson, *Electro-Optics*, John Wiley & Sons, New York (1985).

Chapter 6
FOURIER TRANSFORM AND IMAGING PROPERTIES OF OPTICAL SYSTEMS

6.1 OVERVIEW

In addition to the basic properties of diffraction that were just reviewed, there are a number of important features of optical systems that can be explained in terms of a formal mathematical analysis of wavefront propagation through apertures. We will review here the important properties of lens systems, where we will derive from first principles not only diffraction through apertures, but phase modulation caused by actual lens structures.

A second treatment directed at explaining imaging systems is then provided to explain lenses from a linear systems perspective. Lastly, a summary of Fourier optics and imaging systems in a simplified signal processing form is provided to draw important analogies between optics and time-domain signal processing, which may be more familiar to the practicing electrical engineer.

6.2 EFFECT OF LENS ON A WAVEFRONT

The most fundamental approach to the analysis of optical (lens) systems is to consider the effect of a lens on a wavefront. Consider what happens when a plane wave is incident on a lens from the left as shown in Fig. 6.1(a). A lens causes an incident wave to focus at a distance s_i with wave curvature expressed as a phasor $\exp(-jkr)$, where $r^2 = s_i^2 + x^2 + y^2$. If x and y are $\ll s_i$, then

$$r \sim s_i + \left(\frac{x^2 + y^2}{2s_i}\right) \tag{6.1}$$

There is an overall average phase shift of ks_i radians and another phase shift versus x and y. Ignoring the average phase shift ks_i, the phase delay right after the lens is

$$\exp[-jk(x^2 + y^2)/2s_i]$$

If there is a diverging wave in front of the lens (to the left) it has a phase factor

$$\exp[-jk(x^2+y^2)/2s_o]$$

This case is illustrated in Fig. 6.1(b).

Now the phase factor right after the lens is

$$\exp[jk(x^2+y^2)/2s_o]p_1 = \exp[-jk(x^2+y^2)/2s_i] \quad (6.2)$$

where p_1 is the phase factor accounting for the actual lens. Therefore,

$$p_1 = \exp\left[-jk(x^2+y^2)\frac{1}{2}\left(\frac{1}{s_o}+\frac{1}{s_i}\right)\right] \quad (6.3)$$

which implies that if a wave coming from point P is to converge to point P', the lens must change the phase of the incident wave by the amount given by p_1.

Now, the focal length of a lens is defined in terms of the effect on a plane wave, in which case we require that s_o approach infinity. Then

$$p_1 = \exp[-jk(x^2+y^2)/2f] \quad (6.4)$$

where we have replaced s_i by f. In other words, for s_o approaching infinity a lens of focal length f is needed to converge the wave to P' at a distance s_i given by

$$\frac{1}{f} = \frac{1}{s_o} + \frac{1}{s_i} \quad (6.5)$$

A similar analysis for a negative lens will show that

$$p_1 = \exp\left[-jk(x^2+y^2)\frac{1}{2}\left(\frac{1}{s_o}-\frac{1}{s_i}\right)\right] \quad (6.6)$$

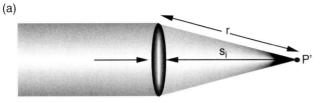

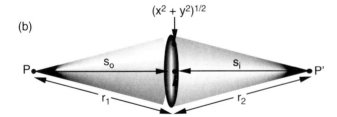

Figure 6.1 (a) Plane wave incident on positive lens and (b) diverging wave incident on the same lens.

A more detailed look at the geometry of a lens can determine the form of p_1 in terms of geometric and material parameters. The result is then given by

$$p_1 = \exp(jknT_0) \exp\left[-jk(n-1)\left(\frac{1}{R_1} - \frac{1}{R_2}\right)\left(\frac{x^2+y^2}{2}\right)\right] \tag{6.7}$$

Since the focal length is given by the lensmaker's formula,

$$\frac{1}{f} = (n-1)\left(\frac{1}{R_1} - \frac{1}{R_2}\right) \tag{6.8}$$

we then obtain

$$p_1 = \exp(jknT_o)\exp[-jk(x^2+y^2)/2f] \tag{6.9}$$

The approximation leading to Equation (6.7) requires that the $f/\#$ of the lens be large. Apart from the first phase factor, which is constant with respect to x and y, a lens creates a quadratic phase factor that scales inversely with focal length.

6.3 FOURIER TRANSFORM PROPERTY OF A SINGLE LENS

Now that the phase transformation properties of a simple thin lens have been described, we can consider the overall effect of a lens on an incident amplitude distribution. Refer to Fig. 6.2 and for the following discussion. We will assume that the incident wavefront at P_1 is created by passing a uniform plane wave through a transparency $t(x_1, y_1)$ such that the resulting complex amplitude distribution right after P_1 is $u_1(x_1, y_1) = t(x_1, y_1)$. In propagating the light from plane P_1 to plane P_2 the diffraction propagation integral gives the complex amplitude distribution just in front of P_2:

$$u_2(x_2, y_2) = \frac{\exp(jks_o)}{j\lambda s_o} \int_{-\infty}^{\infty}\int_{-\infty}^{\infty} u_1(x_1, y_1) \exp\{jk[(x_2 - x_1)^2 \tag{6.10}$$
$$+ (y_2 - y_1)^2]/2s_o\} dx_1 dy_1$$

which can be rewritten (using vector notation) as

$$u_2(\mathbf{x}_2) = \frac{\exp(jks_o)}{j\lambda s_o} h(\mathbf{x}_2; s_o) \int_{-\infty}^{\infty}\int_{-\infty}^{\infty} u_1(\mathbf{x}_1)h(\mathbf{x}_1; s_o) \exp(-jk\mathbf{x}_1 \cdot \mathbf{x}_2/s_o) d\mathbf{x}_1 \tag{6.11}$$

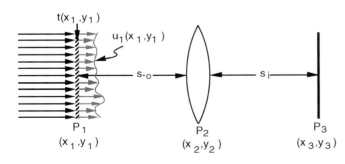

Figure 6.2 Layout of single lens configuration for arbitrary finite conjugate distances s_i and s_o. The special case of $s_i = s_o = f$ yields the exact Fourier transform.

140 FOURIER TRANSFORM AND IMAGING PROPERTIES OF OPTICAL SYSTEMS

where $h(\mathbf{x}; s) = \exp(jk\mathbf{x}\cdot\mathbf{x}/2s)$ and where $d\mathbf{x}_1$ is the vector differential. To get the distribution in the (x_3, y_3) plane u_2 must be multiplied by $p_1(\mathbf{x}_2)$ and $P(\mathbf{x}_2)$, where $p_1(\mathbf{x}_2)$ is the previously discussed phase transformation, and $P(\mathbf{x}_2)$ is the pupil function of the lens. This product is then used in the diffraction integral to yield

$$u_3(\mathbf{x}_3) = \frac{\exp(jks_i)}{j\lambda s_i} h(\mathbf{x}_3; s_i) \int_{-\infty}^{\infty}\int_{-\infty}^{\infty} P(\mathbf{x}_2) u_2(\mathbf{x}_2) h(\mathbf{x}_2; -f) h(\mathbf{x}_2; s_i)$$
$$\times \exp(jk\mathbf{x}_2 \cdot \mathbf{x}_3/s_i)\, d\mathbf{x}_2 \qquad (6.12)$$

Using Equation (6.11) in Equation (6.12), dropping the lens thickness factor $\exp(jknT_o)$ and the total phase delay of the path, $\exp[jk(s_o + s_i)]$, and assuming

$$P(\mathbf{x}_2) = \begin{cases} 1 & \text{(inside lens)} \\ 0 & \text{(outside lens)} \end{cases} \qquad (6.13)$$

we obtain

$$u_3(\mathbf{x}_3) = \frac{-1}{\lambda^2 s_o s_i} h(\mathbf{x}_3; s_i) \int_{-\infty}^{\infty}\int_{-\infty}^{\infty} u_1(\mathbf{x}_1) h(\mathbf{x}_1; s_i) \qquad (6.14)$$
$$\times \int_{-\infty}^{\infty}\int_{-\infty}^{\infty} P(\mathbf{x}_2) P\left[\frac{jk\mathbf{x}_2 \cdot \mathbf{x}_2}{2}\left(\frac{1}{s_o} + \frac{1}{s_i} - \frac{1}{f}\right)\right]$$
$$\exp\left[-j2\pi\left(\frac{\mathbf{x}_1 \cdot \mathbf{x}_2}{\lambda s_o} + \frac{\mathbf{x}_2 \cdot \mathbf{x}_3}{\lambda s_i}\right)\right] d\mathbf{x}_2 d\mathbf{x}_1$$

If we assume that the lens is sufficiently large that $P(\mathbf{x}_2)$ is unity over the range of integration, then

$$u_3(\mathbf{x}_3) = \frac{-1}{\lambda^2 s_o s_i} h(\mathbf{x}_3; s_i) \int_{-\infty}^{\infty}\int_{-\infty}^{\infty} u_1(\mathbf{x}_1) h(\mathbf{x}_1; s_i) \times \int_{-\infty}^{\infty}\int_{-\infty}^{\infty} h(\mathbf{x}_2; w) \qquad (6.15)$$
$$\exp\left\{-j2\pi\left[x_2 \cdot \left(\frac{x_1}{\lambda s_o} + \frac{x_3}{\lambda s_i}\right) + y_2 \cdot \left(\frac{y_1}{\lambda s_o} + \frac{y_3}{\lambda s_i}\right)\right]\right\} d\mathbf{x}_2 d\mathbf{x}_1$$

where

$$\frac{1}{w} = \frac{1}{s_o} + \frac{1}{s_i} - \frac{1}{f} \qquad (6.16)$$

The inner integral describes the Fourier transform of $h(\mathbf{x}_2; w)$ which can be expressed as

$$j\lambda w h\left(\frac{x_1}{\lambda s_o} + \frac{x_3}{\lambda s_i}, \frac{y_1}{\lambda s_o} + \frac{y_3}{\lambda s_i}; \frac{-1}{\lambda^2 w}\right)$$

After regrouping exponentials, Equation (6.15) now becomes

$$u_3(\mathbf{x}_3) = \frac{w}{j\lambda s_o s_i} \exp[-jk\mathbf{x}_3 \cdot \mathbf{x}_3(1 - w/s_i)/2s_i] \int_{-\infty}^{\infty}\int_{-\infty}^{\infty} u_1(\mathbf{x}_1)$$
$$\times \exp[-jk\mathbf{x}_1 \cdot \mathbf{x}_1(1 - w/s_o)/2s_o] \exp[-jk\mathbf{x}_1 \cdot \mathbf{x}_3 w/s_i s_o]\, d\mathbf{x}_1 \qquad (6.17)$$

The integral in Equation (6.17) can be made a Fourier transform if the exponential term with $\mathbf{x}_1 \cdot \mathbf{x}_1$ in the integrand is eliminated. This will occur if $1 - w/s_o = 0$ or $w = s_o$, which using Equation (6.16), is the same as requiring $s_i = f$. Then

$$u_3(\mathbf{x}_3) = \frac{1}{j\lambda f} \exp[-jk\mathbf{x}_3 \cdot \mathbf{x}_3(1 - s_o/f)/2f] \int_{-\infty}^{\infty} \int_{-\infty}^{\infty} u_1(\mathbf{x}_1) \\ \times \exp[-j2\pi \mathbf{x}_1 \cdot \mathbf{x}_3/\lambda f]\, d\mathbf{x}_1 \quad (6.18)$$

Now, if we further demand that $s_o = f$, the field at P_3 is an exact Fourier transform of the field at P_1 because the phase term outside of the integral will vanish.

Note that the resulting Fourier transform has a exponential kernel in the $(x_3/\lambda f, y_3/\lambda f)$ domain, which means that the scale of the amplitude distribution in the transform plane depends on wavelength and lens focal length. We can replace $x_3/\lambda f$ and $y_3/\lambda f$ by u and v, respectively. Then

$$u_3(\mathbf{x}_3) = \frac{1}{j\lambda f} \int_{-\infty}^{\infty} \int_{-\infty}^{\infty} u_1(\mathbf{x}_1) \exp(j2\pi \mathbf{u} \cdot \mathbf{x}_1)\, d\mathbf{x}_1 \quad (6.19)$$

where $\mathbf{u} = (u, v)$.

In practical situations we must take account of the finite aperture of the Fourier transform lens. We can do this by simply employing geometrical optics—that is, by essentially taking account of the projection of the lens pupil back into the object plane as shown in Fig. 6.3. This is permissible under most circumstances in practice provided that the object distance s_o is small enough that the object is within the Fresnel region for the lens aperture. Then the light amplitude at a particular point in the focal plane (x_3, y_3) is given by the sum of all the rays reaching that point with direction cosines given by x_3/f and y_3/f, where f is the focal length of the lens. Since the finite aperture of the lens cannot collect all of these rays, the object defined by $t(x_1, y_1)$ is essentially vignetted by the lens pupil function $P(x_1 + s_o x_3/f, y_1 + s_o y_3/f)$. Thus the value of $u_3(x_3, y_3)$ at point (x_3, y_3) in the focal plane is given by

$$u_3(\mathbf{x}_3) = \frac{1}{j\lambda f} \exp[-jk\mathbf{x}_3 \cdot \mathbf{x}_3(1 - s_o/f)/2f] \int_{-\infty}^{\infty} \int_{-\infty}^{\infty} t(\mathbf{x}_1) P(\mathbf{x}_1 + s_o \mathbf{x}_3/f) u_1(\mathbf{x}_1) \\ \times \exp[-j2\pi \mathbf{x}_1 \cdot \mathbf{x}_3/\lambda f]\, d\mathbf{x}_1 \quad (6.20)$$

The portion of the object subtended by the pupil function P is centered at the coordinates $x_1 = -s_o x_3/f$ and $y_1 = -s_o y_3/f$ as indicated by the argument of P in the integrand of Equation (6.20). As a result of this vignetting effect, it is preferable to place the object close to

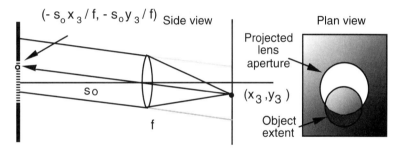

Figure 6.3 Side view of Fourier transform lens configuration for arbitrary object distance $s_o (\neq f)$ and plan view of object plane and vignetting effect of finite lens aperture (looking toward P_1 from P_3).

or actually directly against the lens to minimize vignetting. The Fourier transform, however, will not be exact except at $s_o = f$, as already indicated.

6.4 IMAGING PROPERTIES OF LENSES

By cascading two lenses in series, we can obtain two successive Fourier transforms. We assume that the transparency $t(x_1, y_1)$ at the object plane is illuminated with a uniform plane wave such that $u_1(x_1, y_1) = t(x_1, y_1)$ so that the transform of $u_1(x_1, y_1)$ appears at the intermediate focal plane such that

$$u_2(\mathbf{x}_2) = \frac{1}{j\lambda f} U_1(\mathbf{x}_2/\lambda f_1) \qquad (6.21)$$

where f_1 = focal length of the first lens, and we use an uppercase U_1 to signify a Fourier transform of u_1. The second lens picks up $u_2(\mathbf{x}_2)$ and transforms it again to yield

$$u_3(\mathbf{x}_3) = \frac{1}{j^2 \lambda^2 f_1 f_2} \int_{-\infty}^{\infty}\int_{-\infty}^{\infty} U_1(\mathbf{x}_2/\lambda f_1)\, \exp[-j2\pi \mathbf{x}_2 \cdot \mathbf{x}_3 / \lambda f_2]\, d\mathbf{x}_2 \qquad (6.22)$$

To exactly recover the original function u_1, we would need an inverse transform for the second leg (from P_2 to P_3), rather than a forward transform. We can in essence achieve this by reversing the sign of x_3 and y_3. Substituting for $U_1(\mathbf{x}_2/\lambda f_1)$ yields

$$u_3(\mathbf{x}_3) = \frac{f_1}{f_2} u_1(-f_1 \mathbf{x}_3/f_2) \qquad (6.23)$$

indicating that the two successive Fourier transform lenses cause a change in amplitude (or brightness of the object distribution if $f_1 \neq f_2$), a magnification, and an inverse of the image of u_1.

From geometrical optics we know that a single lens will image an object at finite conjugate distances s_o and s_i. In this configuration we assume that the illuminating waveform incident on the transparency $t(x_1, y_1)$ is not plane but a diverging spherical wavefront. Expressing this wavefront as $e(x_1, y_1)$ then yields $u_1(x_1, y_1) = e(x_1, y_1)\, t(x_1, y_1)$.

Then the diffraction propagation from plane P_1 to plane P_2 and to plane P_3 is

$$u_3(\mathbf{x}_3) = \frac{-1}{\lambda^2 s_o s_i} h(\mathbf{x}_3; s_i) \int_{-\infty}^{\infty}\int_{-\infty}^{\infty} u_1(\mathbf{x}_1) h(\mathbf{x}_1; s_o) \times \int_{-\infty}^{\infty}\int_{-\infty}^{\infty} P(\mathbf{x}_2) h(\mathbf{x}_2; w)$$

$$\exp\left\{-j2\pi\left[x_2\left(\frac{x_1}{\lambda s_o}+\frac{x_3}{\lambda s_i}\right)+y_2\left(\frac{y_1}{\lambda s_o}+\frac{y_3}{\lambda s_i}\right)\right]\right\} d\mathbf{x}_1\, d\mathbf{x}_2 \qquad (6.24)$$

In examining the conditions for image formation, several assumptions have to be made to obtain a tractable derivation. We start by assuming that the lens is perfect and large enough to replace $P(\mathbf{x}_2)$ with unity over the region of integration. Furthermore, u_3 can be written directly in terms of u_1 if the integral over x_2 and y_2 produces a delta function. This delta function would remove u_1 from the integral over x_1 and y_1. If, in turn, $h(\mathbf{x}_2; w) = 1$, then the integral over x_2 and y_2 becomes the Fourier transform of unity, which yields another delta function. Then we let $h(\mathbf{x}_2; w)$ equal 1, which occurs if $1/w = 0$, which turns out to be the Gaussian lens law. Now, if $P(\mathbf{x}_2) = 1$ over the region of integration, then by using the similarity theorem, and recalling that the incident wave is spherically diverging, yields a direct (imaging) relationship between u_3 and the transparency t. Therefore,

$$u_3(\mathbf{x}_3) = \frac{1}{m} u_1(-\mathbf{x}_3/m) \tag{6.25}$$

Now, however, we must go back and consider the effect of the pupil function $P(\mathbf{x}_2)$. First note that the integral in Equation (6.24) over coordinates (x_2, y_2) can be written in terms of the convolution of transforms of P and h—that is,

$$u_3(\mathbf{x}_3) = \frac{1}{\lambda^2 s_o s_i} h(\mathbf{x}_3; s_i) \int_{-\infty}^{\infty} \int_{-\infty}^{\infty} |\text{FT}\{P(\mathbf{x}_2)\} \otimes \text{FT}\{h(\mathbf{x}_2; w)\}|_{\left(\frac{x_1}{\lambda s_o} + \frac{x_3}{\lambda s_i}, \frac{y_1}{\lambda s_o} + \frac{y_3}{\lambda s_i}\right)}$$
$$\times u_1(\mathbf{x}_1) h(\mathbf{x}_1; s_o) d\mathbf{x}_1 \tag{6.26}$$

where the Fourier transform in the integrand of Equation (6.26) is evaluated at $(x_1/\lambda s_o + x_3/\lambda s_i, y_1/\lambda s_o + y_3/\lambda s_i)$, as indicated.

Equation (6.26) is in the form of a correlation between $(\text{FT}\{P\} * \text{FT}\{h\})$ and $u_1 h$. If the $h(\mathbf{x}_2; w)$ term could be removed (from the second term in the integrand), the above equation would represent the correlation of the image distribution with the transform of the pupil function. This will occur if w approaches infinity. Then $h(\mathbf{x}_2; w) = 1$ and the Fourier transform again yields a delta function. The requirement that w approach infinity is again the imaging condition. If this resulting delta function is convolved with $\text{FT}\{P\}$, then the diffraction propagation (6.26) can now be written

$$u_3(x_3) = \frac{s_o}{s_i} h(x_3; s_i) \int\int u_1(x_1) h(x_1; s_o) \mathbf{P}\left(x_1 + \frac{s_o x_3}{s_i}, y_1 + \frac{s_o y_3}{s_i}\right) dx_1 \tag{6.27}$$

where $\mathbf{P} = \text{FT}\{P\}$. The above correlation can be changed so as to look like convolution by reversing coordinates. Then

$$u_3(x_3) = \frac{1}{m} h(x_3; s_i) \mathbf{P}\left(\frac{x_3}{m}\right) * \left[u_1\left(\frac{-x_3}{m}\right) h\left(\frac{x_3}{m}; s_o\right)\right] \tag{6.28}$$

where $*$ denotes convolution. This expression shows that the image is magnified and inverted and that resolution is reduced by convolution with the Fourier transform of the pupil function denoted by bold $\mathbf{P}(x_3/m)$. We will now explain what the Fourier transform of the pupil function actually is in the case of imaging with coherent and incoherent light.

6.5 LINEAR SYSTEM PROPERTIES OF IMAGING SYSTEMS

In considering imaging systems, which in general can have more than one or two lenses, we have to take account of two situations: coherent and incoherent illumination. For a point source emitting a spherical wave we have

$$u(\mathbf{r}, t) = \frac{u_o}{r} \cos[\mathbf{k} \cdot \mathbf{r} + \omega t + \phi(t)] \tag{6.29}$$

where $r = (x_1^2 + y_1^2 + z_1^2)^{1/2} \cdot u(\mathbf{r}, t)$ can also be expressed as

$$u(\mathbf{r}, t) = \text{Re}\left\{\frac{u_o}{r} \exp[j\mathbf{k} \cdot \mathbf{r} + j\omega t + j\phi(t)]\right\} \tag{6.30}$$

We will ignore explicitly writing Re{...} and carrying along the time-varying components, since they are understood. For coherent sources all points have a fixed phase relationship and vary in unison (like a laser). For incoherent sources, all points have a randomly varying phase relationship; hence $\phi(t)$ will be averaged to zero (like a light bulb). For coherent sources, the convolution operation representing the optical system depicted in Fig. 6.4 must be performed in terms of complex amplitudes; that is, the system is linear in complex amplitude. For incoherent sources the convolution operation must be performed on intensities, and the system is therefore linear in intensity.

6.6 POINT SPREAD FUNCTION

If we look specifically at the case of a circular pupil function of diameter d:

$$P(x, y) = \text{rect}\left(\frac{\sqrt{x_1^2 + y_1^2}}{d}\right) \tag{6.31}$$

illuminated with a converging spherical wave from the left (as shown after the lens in Fig. 6.4):

$$u(x_1) = \frac{1}{R}\exp(-jkR) \tag{6.32}$$

the Huygens–Fresnel diffraction integral is

$$u_3(\mathbf{x}_3) = \frac{1}{j\lambda}\int_{-\infty}^{\infty}\int_{-\infty}^{\infty} P(\mathbf{x}_1)\frac{1}{R}\exp(-jkR)\frac{1}{r}\exp(jkr)\cos\theta\, d\mathbf{x}_1 \tag{6.33}$$

where $P(\mathbf{x}_1)$ is the aperture pupil function and where $R = [x_1^2 + y_1^2 + z_1^2]^{1/2}$ and $r = [(x_3 - x_1)^2 + (y_3 - y_1)^2 + s_i^2]^{1/2}$.

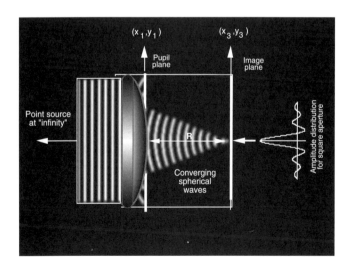

Figure 6.4 A plane wave is transformed into a spherical wave through a lens in an imaging configuration.

6.6 POINT SPREAD FUNCTION

In the above integral we can assume small θ so that $\cos\theta \sim 1$ and that R and r can be replaced by s_i (except in the exponential terms where k may be large). Using the Fresnel approximation

$$R = s_i \left[1 + \frac{x_1^2}{s_i^2} + \frac{y_1^2}{s_i^2}\right]^{1/2} \tag{6.34}$$

and

$$r = s_i \left[1 + \frac{(x_3-x_1)^2}{s_i^2} + \frac{(y_3-y_1)^2}{s_i^2}\right]^{1/2} \tag{6.35}$$

and the binominal expansion for $|x| \ll 1$ such that

$$\sqrt{1+x} = 1 + \frac{1}{2}x - \frac{1}{8}x^2 + \cdots \tag{6.36}$$

then

$$R \approx s_i \left[1 + \frac{1}{2}\frac{x_1^2}{s_i^2} + \frac{1}{2}\frac{y_1^2}{s_i^2}\right] \tag{6.37}$$

and

$$r \approx s_i \left[1 + \frac{1}{2}\frac{(x_3-x_1)^2}{s_i^2} + \frac{1}{2}\frac{(y_3-y_1)^2}{s_i^2}\right] \tag{6.38}$$

Using these approximations the field at the image plane is now

$$u_3(\mathbf{x}_3) = \frac{1}{j\lambda s_i^2} \int_{-\infty}^{\infty}\int_{-\infty}^{\infty} P(\mathbf{x}_1) \frac{1}{R} \exp\left[-jks_i\left(1 + \frac{x_1^2}{s_i^2} + \frac{y_1^2}{s_i^2}\right)\right]$$
$$\times \exp\left\{jks_i\left[1 + \frac{1}{2}\frac{(x_3-x_1)^2}{s_i^2} + \frac{1}{2}\frac{(y_3-y_1)^2}{s_i^2}\right]\right\} d\mathbf{x}_1 \tag{6.39}$$

Expanding exponents and collecting terms yields

$$u_3(\mathbf{x}_3) = \frac{\exp\left[\frac{jk}{s_i^2}(x_1^2+y_1^2)\right]}{j\lambda s_i^2} \int_{-\infty}^{\infty}\int_{-\infty}^{\infty} P(\mathbf{x}_1) \exp\left[-j\frac{2\pi}{\lambda s_i}(x_3 x_1 + y_3 y_1)\right] d\mathbf{x}_1 \tag{6.40}$$

Letting $u = x_1/\lambda s_i$ and $v = y_1/\lambda s_i$, we then have

$$u_3(\mathbf{x}_3) = -j\lambda \exp\left[\frac{jk}{s_i^2}(x_1^2+y_1^2)\right] \int_{-\infty}^{\infty}\int_{-\infty}^{\infty} p(\lambda s_i u, \lambda s_i v)$$
$$\exp[-j2\pi(x_3 u + y_3 v)] \, du\, dv \tag{6.41}$$

This result gives the amplitude distribution in the image plane due to a point source in the object plane. The integral in Equation (6.41) is the Fourier transform of the pupil function (with scaled arguments) and is termed the *point spread function* (PSF) of the optical system.

The additional complex exponential factor only affects the phase in the image plane. The system is shift invariant, because if the source is shifted off-center in the (x_1, y_1) plane, it does not change the above result. Thus it is generally observed that for reasonable off-axis distances in well-designed optical systems the shape of the spot image does not change. In the peripheral field of low-quality optical systems, this does not hold; but in general, shift invariance is violated gradually. Therefore, typically there exist smaller, local regions where shift invariance holds true, called *isoplanatic regions*.

Essentially, what we have derived is the impulse response of the optical system defined by $h(x, y) = \text{FT}[p(\lambda s_i u, \lambda s_i v)]$ such that

$$u_3(\mathbf{x}_3) = \int_{-\infty}^{\infty} \int_{-\infty}^{\infty} h(\mathbf{x}_3 - \mathbf{x}_1) u_o(m\mathbf{x}_1) \, d\mathbf{x}_1 \qquad (6.42)$$

where $u_o(m\mathbf{x}_1)$ is the amplitude distribution of the object after magnification but without degradation. Thus, imaging is a two-step process: projection (with magnification and inversion) followed by convolution in the image plane with the PSF. If another Fourier transform is taken, so that we have two Fourier transforms of the pupil function, we arrive at the coherent transfer function given by

$$\text{FT}\{\text{FT}[p(\lambda s_i u, \lambda s_i v)]\} = P(-\lambda s_i u, -\lambda s_i v) \qquad (6.43)$$

which has a reflection about the origin relative to the original pupil function. For symmetrical apertures, the pupil function (properly scaled) is therefore the coherent transfer function.

6.7 OPTICAL TRANSFER FUNCTION

For incoherent radiation we must observe intensity, in which case we will be concerned with

$$I_3(\mathbf{x}_3) = E\{u_3(\mathbf{x}_3) u_3^*(\mathbf{x}_3)\} \qquad (6.44)$$

Applying Equation (6.42) again we have

$$I_3(\mathbf{x}_3) = E \left\{ \int_{-\infty}^{\infty} \int_{-\infty}^{\infty} h(\mathbf{x}_3 - \mathbf{x}_1) u_1(m\mathbf{x}_1) \, d\mathbf{x}_1 \int_{-\infty}^{\infty} \int_{-\infty}^{\infty} h^*(\mathbf{x}_3 - \mathbf{x}_1') u_1^*(m\mathbf{x}_1') \, d\mathbf{x}_1' \right\} \qquad (6.45)$$

Since $h(x, y)$ is independent of time, then

$$I_3(\mathbf{x}_3) = \int_{-\infty}^{\infty} \int_{-\infty}^{\infty} \int_{-\infty}^{\infty} \int_{-\infty}^{\infty} h(\mathbf{x}_3 - \mathbf{x}_1) h^*(\mathbf{x}_3 - \mathbf{x}_1') E\{u_1(m\mathbf{x}_1) u_1^*(m\mathbf{x}_1')\} \, d\mathbf{x}_1 d\mathbf{x}_1' \qquad (6.46)$$

For incoherent images the expectation operator signifies cross-correlation. For distinct image points, this cross-correlation is zero, signifying that the response is a spatial impulse. If $x_1 = x_1'$, then the value of the expectation operator yields the intensity. Thus

$$E\{u_1(m\mathbf{x}_1) u_1^*(m\mathbf{x}_1')\} = I_1(m\mathbf{x}_1)\delta(\mathbf{x}_1 - \mathbf{x}_1') \qquad (6.47)$$

Therefore,

$$I_3(\mathbf{x}_3) = \int_{-\infty}^{\infty} \int_{-\infty}^{\infty} |h(\mathbf{x}_3 - \mathbf{x}_1)|^2 I_3(m\mathbf{x}_1) \, d\mathbf{x}_1 \qquad (6.48)$$

We can note immediately that the incoherent PSF is the squared modulus of the coherent PSF. Since $h(x, y)$ is the Fourier transform of the pupil function, then the incoherent PSF is the power spectrum of the pupil function. The PSF as just derived is valid in an isoplanatic region. For perfect optical systems it is limited only by diffraction, but for practical systems it will also include the effects of aberrations.

We can also define the optical transfer function (OTF) of the optical system with incoherent illumination. It is the normalized autocorrelation of the pupil function, given by

$$\text{OTF}(u, v) = \frac{\int_{-\infty}^{\infty} \int_{-\infty}^{\infty} p(\lambda s_i \mathbf{u}') p[\lambda s_i (\mathbf{u}' - \mathbf{u})] \, d\mathbf{u}'}{\int_{-\infty}^{\infty} \int_{-\infty}^{\infty} p^2(\lambda s_i \mathbf{u}') \, d\mathbf{u}'} \qquad (6.49)$$

Since the OTF(u, v) is, in general, a complex number, we know that we can define the modulus of the OTF as the modulation transfer function (MTF) and the phase of the OTF as the phase transfer function (PTF) such that OTF$(u, v) = \text{MTF}(u, v) \exp[j\text{PTF}(u, v)]$. It is clear that the OTF is simply the Fourier transform of the PSF and hence entirely analogous to the frequency response of a time-domain filter.

Control of the OTF can be effected by selection of the aperture shape—that is, pupil function modification, where $p(x, y) = T(x, y) \exp[jkw(x, y)]$, where $T(x, y)$ is the pupil transmittance, and $w(x, y)$ accounts for aberrations and corrections. $w(x, y)$ is essentially the path length difference in wavelengths between the actual and ideal (spherical wave) propagation from a point (x, y) in the pupil aperture to the image plane.

EXAMPLE 6.1 PSF and OTF of a Rectangular Aperture By exploiting the form of the OTF expression given in Equation (6.49) we can calculate the OTF of a rectangular aperture using the geometric construction shown in Fig. 6.5(a). The OTF is thus seen to be

$$H(u, v) = \frac{\text{Area of overlap}}{\text{Total area}}$$

Thus

$$H(u, v) = \text{tri}(\lambda s_i u / w) \, \text{tri}(\lambda s_i v / l)$$

where

$$\text{tri}(\lambda s_i u / x) = \begin{cases} x(1 - \lambda s_i u / x), & u \leq x / \lambda s_i \\ 0, & \text{otherwise} \end{cases}$$

This is plotted in Fig. 6.5(b) along with the corresponding PSF given in Fig. 6.5(c) which has the form

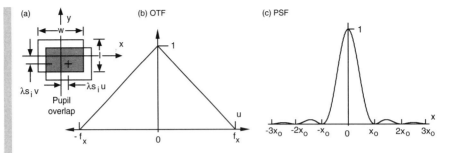

Figure 6.5 (a) Geometric construction for calculating OTF and PSF of a rectangular aperture. (b) Corresponding OTF and (c) corresponding PSF.

$$h(x, y) = \frac{\sin^2(\pi x/x_o)}{(x/x_o)^2} \frac{\sin^2(\pi y/y_o)}{(y/y_o)^2}$$

where $x_o = \lambda s_i/w$ and $y_o = \lambda s_i/l$.

EXAMPLE 6.2 PSF and OTF of a Circular Aperture Using a similar analysis, we can calculate the OTF of a circular pupil function. Using the geometric construction shown in Fig. 6.6(a), we can note from inspection of one quadrant shown in the inset that the area of overlap is four times the area of the half-chord such that

$$\text{Area of half-chord} = \text{Area of circular sector} - \text{Area of triangle}$$

$$= \frac{\theta}{2\pi} \pi \left(\frac{l}{2}\right)^2 - \frac{1}{2}\frac{\lambda s_i u}{2} \sqrt{\left(\frac{l}{2}\right)^2 - \left(\frac{\lambda s_i u}{2}\right)^2}$$

$$= \frac{\cos^{-1}(\lambda s_i u/l)}{2\pi} \pi \left(\frac{l}{2}\right)^2 - \frac{1}{2}\frac{\lambda s_i u}{2} \sqrt{\left(\frac{\ell}{2}\right)^2 - \left(\frac{\lambda s_i u}{2}\right)^2}$$

where ℓ is the diameter of the circular aperture. This result was derived earlier in Chapter 2 as an example of convolution. Because of circular symmetry of the OTF, it is only necessary to calculate the OTF along the positive u axis. Therefore

$$H(u, v) = H(u, 0) = \frac{4(\text{Area of circular sector} - \text{Area of triangle})}{\pi(l/2)^2}$$

which yields

$$H(\rho) = \begin{cases} \frac{2}{\pi}\{\cos^{-1}(\rho/2\rho_o) - (\rho/2\rho_o)[1 - (\rho/2\rho_o)^2]^{1/2}, & \rho \leq 2\rho_o \\ 0, & \text{otherwise} \end{cases}$$

where ρ = radial spatial frequency and ρ_o = radial cutoff = $1/2\lambda s_i$. The corresponding MTF is shown in Fig. 6.6(b). A plot of the PSF is shown in Fig. 6.6(c). The PSF is of the form

$$h(r) = \left[\frac{2J_1(\pi r/r_o)}{(\pi r/r_o)}\right]^2$$

where $r = [x^2 = y^2]^{1/2}$ and $r_o = \lambda s_i/d$.

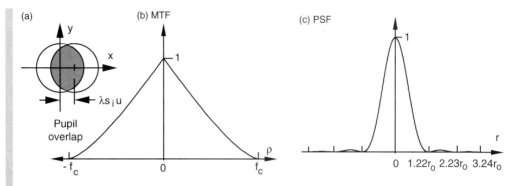

Figure 6.6 (a) Geometric construction for calculation of OTF and PSF of a circular aperture. (b) Corresponding OTF and (c) corresponding PSF.

Let us now consider a more practical definition of the modulation transfer function (MTF). As a quantifiable measure of image quality, it describes the ability of a lens or optical system to transfer object contrast to an image. Using a transparency as an object with maximum and minimum transmittances of $T_{\max}$ and $T_{\min}$, respectively, the object contrast is

$$M_O = \frac{T_{\max} - T_{\min}}{T_{\max} + T_{\min}} \tag{6.50}$$

The corresponding the image contrast is

$$M_I = \frac{I_{\max} - I_{\min}}{I_{\max} + I_{\min}} \tag{6.51}$$

where $I_{\max}$ and $I_{\min}$ are image irradiances. Then the MTF is $H(u) = M_I/M_O$, where u is the spatial frequency. For a perfect circular lens the cutoff spatial frequency is $u_c = 1.22/r$, where r is the radius of the Airy disk [see Fig. 1.13(b)]. Thus, given a lens of focal length F configured as in Fig. 6.7 and a test object with (linear) spatial frequency f'_o measured in cycles per millimeter [cy/mm], the (angular) spatial frequency f_o in cycles per milliradian [cy/mrad], which appears in the frequency response of the lens $H(u)$, is given by f_o[cy/mrad] $= 10^{-3}$ [rad/mrad] f'_x[cy/mm]s_o[mm], where s_o = object distance. This frequency is the output spatial frequency of the lens. The corresponding linear spatial frequency at the image plane in cycles per millimeter is

$$f'_i = f'_o \frac{s_o}{s_i} = \frac{f'_o}{m} = \frac{10^3}{s_i} f_o \tag{6.52}$$

Units of (angular) spatial frequency given for f_o (i.e., cycles/milliradian) are intrinsic to the lens and independent of configuration. Linear spatial frequency as measured by f'_i (cy/mm) in the focal plane, however, is of important practical value in particular instances. Hence we are interested in their relationship. The intrinsic frequency cutoff of a telescope objective can be derived from the previous equation simply by inverting it and assuming that s_i approaches f for viewing objects at "infinity." Then $f_o = F f'_i / 10^3$. Hence, if f'_i is measured and F is known by design, we can determine the intrinsic spatial frequency limit of the lens.

A complete description of the MTF would not be complete without citing some of the restrictions or caveats on its proper use. Of particular interest in complex optical systems is the

150 FOURIER TRANSFORM AND IMAGING PROPERTIES OF OPTICAL SYSTEMS

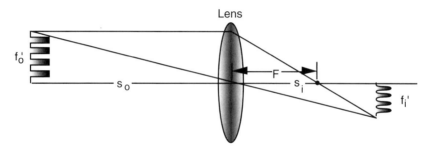

Figure 6.7 Configuration of lens for defining a practical measure of MTF.

cascading of individual MTFs to yield the total system MTF. Ordinarily if we chained a series of electrical filters together and properly buffered them, the total frequency response would be the product of the individual frequency transfer functions. In order to legitimately multiply the MTFs of cascaded components, however, the following must be true:

1. The first system must be isoplanatic and diffraction-limited throughout the shared field.
2. The PSF of the first component (or system) must be smaller than the PSF of the second component (or system) in the object space of the second.

When the above criteria are satisfied, it is necessary prior to multiplication to mathematically transform MTFs to a common image surface (otherwise it is wrong to multiply incoherent MTFs of, say, a telescope objective and eyepiece). Since the objective PSF is larger, the intermediate image formed by the objective appears as a partially coherent virtual object to the eyepiece. Thus multiplication will not be totally accurate. In other words, in cascaded systems, magnification factors allow projection of PSFs into the image plane. Electronic devices such as amplifiers have equivalent MTFs if the magnification factor reflects a change from time to space.

Furthermore, there are additional factors that affect the spatial resolution of an image created by an optical system. After the lens there is often a detector array. Each detector samples the image spatially and then is sampled temporally. In addition, noise is added in the detection process. Furthermore, in practical optical systems the PSF varies across the field of view. These conditions mean that the four requirements for strict MTF applicability—incoherent light, linearity, stationarity, and deterministic signals—will not be met exactly. Nonstationarity in this context means that the impulse response varies across aperture, and/or the scan may be nonlinear, and that the detector array, multiplexer, and A/D converter are sampled data devices. Signals are nondeterministic because of noise in the detectors and amplifiers. Linearity will be violated (perhaps deliberately) because the post-detection amplifier, video processing, and display may contain nonlinear transfer characteristics. We will talk more about this in the next chapter.

Finally a few comments about the phase transfer function (PTF) are worthy of note at this point. In the absence of aberrations, the Fourier transform of the pupil function is Hermitian. This means that the PTF must be even regardless of the symmetry of the pupil. Aberrations can create an odd PTF under some conditions. The phase of an image transform is often far more important than the modulus. An image in which the modulus is drastically reduced is still recognizable, but the corresponding phase changes will render the image unrecognizable. Hence the PTF can be very important.

6.8 SIGNAL PROCESSING ANALOGIES FOR OPTICS

We will examine several analogies between optics and signal processing described by Johnson as a way of summarizing some of the concepts introduced in this chapter. This will also serve to address the connection between optics and signal processing. These analogies are very clarifying and elegant, especially when expressed in one dimension (rather than two).

Consider the general linear shift-invariant optical system represented by

$$g(x) = \int_{-\infty}^{\infty} \int_{-\infty}^{\infty} f(x')h(x-x')\,dx' \tag{6.53}$$

In compact notation the output is related to the input by $g(\mathbf{x}) = f(\mathbf{x}) * h(\mathbf{x})$ and by the convolution theorem: $G(\mathbf{u}) = F(\mathbf{u})H(\mathbf{u})$.

The first analogy is between temporal (electrical) frequency and spatial (optical) frequency. Electrical frequency is given by

$$f = \frac{d\phi(t)}{dt} \tag{6.54}$$

whereas spatial frequency is given by

$$u = \frac{d\phi(x)}{dx} \tag{6.55}$$

Thus the impulse response relation $h(t) = h(0) \exp[j(\omega t - \theta)]$ for a single frequency, along with the more general expression

$$h(t) = \frac{1}{2\pi} \int_{-\infty}^{\infty} H(\omega) \exp\{j[\omega t - \theta(\omega)]\}\,d\omega \tag{6.56}$$

for a band of frequencies, holds true for spatial frequencies as well. This is also consistent with the fact that in certain systems, spatial frequency in cycles per unit length or cycles per unit-angle are convertible to temporal frequency (in Hz) by a temporal scanning (or spatial sampling) of the signal in an interval of time. If the scan velocity is v (in units of unit length/sec or unit angle/sec), then electrical frequency $= f = uv$ (where u is spatial frequency).

The second analog is that Fresnel diffraction is like an all-pass filter. Consider an optical disturbance in an aperture given by $f(x')$. Then, as we have already seen by Huygen's principle, the optical field in the (x, y) plane (at point z along the optical axis) given by $g(x, z)$ in one-dimension is

$$g(x, z) = \frac{-j \exp(jkz)}{\lambda z} \int_{-\infty}^{\infty} f(x') \exp\left[\frac{jk(x-x')^2}{2z}\right] dx' \tag{6.57}$$

which can be written as $g(x) = f(x) * \exp(j\alpha x^2)$, where $\alpha = k/2z$ (ignoring the proportionality factor outside the above integral). Thus the impulse response of the system is $h(x) = \exp(j\alpha x^2)$. The corresponding transfer function is

$$H(u) = \sqrt{\lambda z} \, \exp(-j\lambda z u^2/4\pi) \tag{6.58}$$

Replacing spatial with temporal frequency yields $g(t) = f(t) * \exp(-j\alpha t^2)$, where $\exp(j\alpha t^2)$ is $h(t)$. The corresponding temporal frequency transfer function is

$$H(\omega) = \sqrt{\pi/\alpha} \, \exp(-j\omega^2/4\alpha) \tag{6.59}$$

the magnitude of which is constant versus ω. Thus $h(t)$ can be considered an all-pass filter with instantaneous frequency $\omega = 2\alpha t$ and group delay $d\theta(\omega)/d\omega = \omega/2\alpha$.

The third analogy is that the Fraunhofer diffraction pattern is the Fourier transform, a fact that has been abundantly obvious, namely,

$$g(x, z) = \int_{-\infty}^{\infty} f(x') \exp\left(-j\frac{k}{z} x' \cdot x\right) dx' \tag{6.60}$$

where we assume the plane of observation (at z) is at

$$z \gg \left|\frac{k}{2}(x'^2 + y'^2)\right|_{\max} \tag{6.61}$$

and that $z \gg (x^2 + y^2)^{1/2}$ to ensure that

$$\exp\left[j\frac{k}{2z}(x^2 + x'^2)\right] \approx 1 \tag{6.62}$$

The fourth analogy is that a lens acts like a linear frequency modulated signal mixed with the input signal (incident illumination). The lens is equivalent to the transparency $\tau(x) = \exp(-j\beta x^2)$ as shown in Fig. 6.8 where $\beta = k/2f$ and we ignore the finite extent of the lens.

Thus we can characterize the lens as a linear spatial-FM generator with instantaneous spatial modulation frequency $= \nu(x) = 2\beta x$. If we assume the input field is $f(x) = \exp(j\nu_o x)$, then the field exiting the lens with finite aperture is

$$\alpha(x) = P(x) \exp(-j\beta x^2 - \nu_o x) \tag{6.63}$$

where

$$P(x) = \begin{cases} 1, & |x| \leq d \\ 0, & |x| > d \end{cases} \tag{6.64}$$

Equation (6.63) is analogous to the form of the impulse response $h(t)$ of a bandpass filter of the form

$$h(t) = \exp[j(\alpha t^2 + \omega_o t)] \tag{6.65}$$

The instantaneous spatial modulation frequency decreases linearly from $\nu_o + 2\beta d$ to $\nu_o - 2\beta d$. By introducing an input field given by $\exp(j\nu_o x)$ (i.e., a plane wave), we can consider this

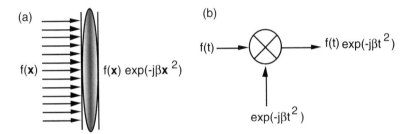

Figure 6.8 (a) Effect of a simple lens on a wavefront, and (b) its equivalent electrical model.

analogous to introducing an offset frequency (heterodyning to a carrier frequency) into the local oscillator of a linear-FM radar receiver to (partially) compensate for any mismatch between receiver and transmitter frequency (e.g., from Doppler shift).

The fifth analogy that Johnson introduced is that between the field in the focal plane and real-time spectral analysis. An expression for this process, as described optically in Fig. 6.9(a), is written

$$g(x) = [f(x)\exp(j\beta x^2)] * \exp(j\gamma x^2) \tag{6.66}$$

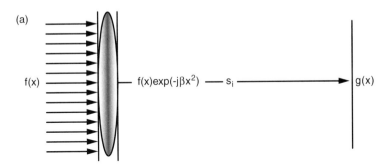

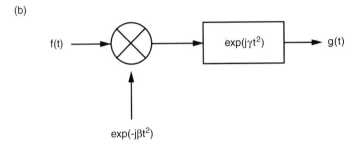

Figure 6.9 (a) Schematic of optical Fourier transform layout and (b) corresponding electrical model.

where $\gamma = k/2s_i$. The corresponding signal-flow block diagram (for temporal signals) is shown in Fig. 6.9(b). If we choose $s_i = f$, then $\beta = \gamma$ and Equation (6.68) becomes

$$g(x) = \int_{-\infty}^{\infty} f(x') \exp(-j\beta x'^2) \exp[j\beta(x-x')^2] dx' \qquad (6.67)$$

This result is a Fourier transform with a quadratic phase factor. As we have seen, $g*(x)g(x)$ is proportional to the power spectrum.

The sixth analogy is that between imaging and pulse compression. For a finite lens, Equation (6.66) becomes

$$g(x) = [P(x) f(x) \exp(-j\beta x^2)] * \exp(j\gamma x^2) \qquad (6.68)$$

At the focal plane we have $s_i = f$, and the field is determined by taking the Fourier transform of the aperture field distribution. For instance, if $f(x) = 1$ for $|x| \le d$ as shown in Fig. 6.10(a), then

$$g(x) = \exp(j\beta x^2) 2d \frac{\sin(2d\beta x)}{2d\beta x} \qquad (6.69)$$

as illustrated in Fig. 6.10(b), where at the focal plane we have $4d^2\beta \gg \pi^2$ such that $\exp(j\beta x^2) \sim 1$.

This result is clearly analogous to pulse compression radar where the temporal duration of a pulse of width τ is compressed into a duration $2/B$, where B is the modulation bandwidth. The instantaneous peak power increases by $B\tau$, analogous to the $2d^2\beta/\pi$ factor in optics. The optical and electric analogs are shown in Figs. 6.11(a) and 6.11(b).

The seventh analogy relates magnification to time scaling. In electrical systems we know that changing $f(t)$ to $f(ct)$ creates a new time-scaled function $g(t) = f(ct)$. This can be realized analytically using two all-pass filters with quadratic phase and an FM signal generator $\exp(j\beta t^2)$ as shown in Fig. 6.12(b). There the system output is

$$g(t) = \{[f(t) * \exp(j\alpha t^2)] \exp(-j\beta t^2)\} * \exp(j\gamma t^2) \qquad (6.70)$$

If $\alpha + \gamma = \beta$ it can be shown that

$$g(t) = \frac{\pi}{\alpha} \exp(j\gamma\beta t^2/\alpha) f(-\gamma t/\alpha) \qquad (6.71)$$

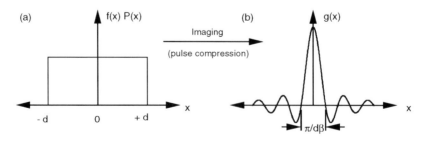

Figure 6.10 (a) Input rectangular aperture distribution (in one dimension) and (b) corresponding Fourier transform plane distribution (sinc) illustrating concept of pulse compression.

6.8 SIGNAL PROCESSING ANALOGIES FOR OPTICS   155

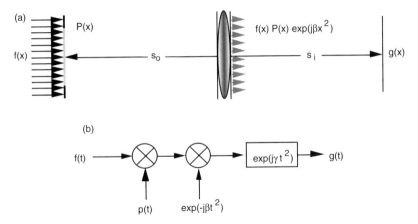

Figure 6.11 (a) Optical configuration for imaging and (b) corresponding electrical model.

If $f(t) = 0$ for $|t| < \tau/2$ and $|\gamma t/\alpha| \ll 1$, then

$$g(t) = \frac{\pi}{\alpha} f(ct) \quad (6.72)$$

where $c = -\gamma/\alpha$ and $|\gamma t/\alpha| < \tau/2$. To see the optical interpretation of this, examine Fig. 6.12(a). The field propagated from the object and incident on the lens is $f(x) * \exp(j\alpha x^2)$, where $\alpha = k/2s_o$, which is analogous to passing the signal through the first filter of Fig. 6.12(a). The field exiting the lens is $[f(x) * \exp(j\alpha x^2)] \exp(-j\beta x^2)$ where $\beta = k/2f$. Propagation of the field to the image plane (not necessarily the focal plane) yields

$$g(x) = \{[f(x) * \exp(j\alpha x^2)] \exp(-j\beta x^2)\} * \exp(j\gamma x^2) \quad (6.73)$$

where $\gamma = k/2s_i$. Now it is clear that $\alpha + \gamma = \beta$ is analogous to

$$\frac{1}{f} = \frac{1}{s_i} + \frac{1}{s_o} \quad (6.74)$$

since $\alpha = k/2s_o$, $\beta = k/2f$ and $\gamma = k/2s_i$. Therefore, $g(x) = \lambda s_o f(x/m)$, where $m =$ magnification $= -s_i/s_o$.

The eighth analogy relates image enhancement to spectral domain filtering. Consider the optical lens diagram of Fig. 6.13(a). Propagation from the object plane to the lens yields $a(x) = f(x) * \exp(j\alpha x^2)$. Passage through the lens yields $a(x) \exp(-j\alpha x^2)$. Propagation from the lens to the focal plane then yields

$$g(x) = [a(x) \exp(-j\alpha x^2)] * \exp(j\alpha x^2) \quad (6.75)$$

Thus $g(x) = \exp(j\alpha x^2) A(2\alpha x)$, where $A(2\alpha x)$ is the Fourier transform of $a(x)$. Then

$$g(x) = \sqrt{\frac{\pi}{\alpha}} \exp(j\pi/4) F(2\alpha x) \quad (6.76)$$

156 FOURIER TRANSFORM AND IMAGING PROPERTIES OF OPTICAL SYSTEMS

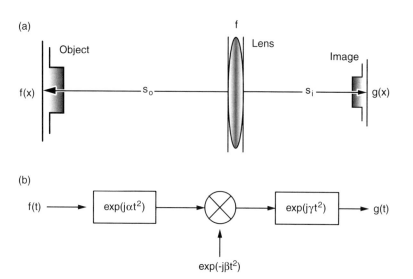

Figure 6.12 (a) Optical diagram for image magnification and (b) corresponding electrical analog.

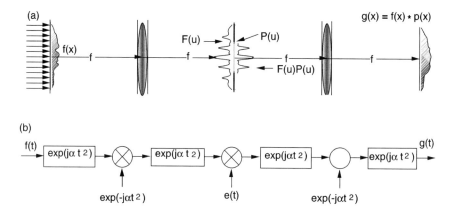

Figure 6.13 (a) Optical diagram for image enhancement and (b) corresponding time domain filter model.

since

$$A(2\alpha x) = \sqrt{\frac{\pi}{\alpha}} \exp(j\pi/4) \exp(-j\alpha x^2) F(2\alpha x) \qquad (6.77)$$

Therefore the distribution over the focal plane is an exact Fourier transform of the object distribution (neglecting the constant phase delay and finite extent of the lens).

This last analogy introduces the idea of image enhancement. As discussed previously, a second lens applies another Fourier transform to the output (focal plane) of the first, which is the same as an inverse Fourier transform except for a sign change. This operation is diagrammed in Fig. 6.13(b). Thus, for instance, to remove periodic noise at a spatial frequency v_o and bandwidth Δv_o, we place an opaque mask at position $\lambda f v_o/2\pi$ in the common focal plane with width $\lambda f \Delta \nu_o / 2\pi$. This is the direct analog to stopband filtering, as suggested by the electrical analog in Fig. 6.13(b). Putting a mask at the center of the common focal plane will thus result in so-called DC-background suppression.

The ninth and last analogy is that between optical apodization and pulse shaping. We already know that the field at the focal plane of a lens is the Fourier transform of the incident field distribution, whereas the field at the exit pupil is directly related to the incident field, since entrance and exit pupil are conjugate images. If the field is unity over the entrance pupil (corresponding to a plane wave), then the field at the focal plane is the Fourier transform of the exit pupil complex transmittance (neglecting scale factors). This field is, in fact, the lens impulse response $h(x)$. The image spectrum is then $G(\nu) = F(\nu)H(\nu)$, where $F(\nu)$ is the object spectrum and $H(\nu) = \text{FT}[h(x)]$ is the lens OTF. For a point source (object) $G(\nu) = H(\nu)$. Thus to modify the shape of the signal pulse we must adjust the frequency response (or transfer function). Analogously, in optics we must adjust the OTF, which is realized by adjusting the complex transmittance of the pupil function.

PROBLEM EXERCISES

1. If two positive lenses are cascaded and have focal lengths f_1 and f_2 and separation d, and they are illuminated with a monochromatic plane wave, what is their total phase transformation?

2. Consider the situation where we account for vignetting of an object by a Fourier transform lens. The object is a circular aperture of diameter d and the lens has a diameter D, where $D > d$.

   (a) What is the maximum spatial frequency for which the intensity in the focal plane is an accurate Fourier transform of the circular aperture in the object plane?

   (b) What is the spatial frequency above which the measured spectrum is zero in the focal plane?

   (c) For the case where the object distribution is partially vignetted, write an expression for the light distribution in the focal plane.

3. (a) Derive an expression for the complex light amplitude distribution and corresponding irradiance at the focal plane of a lens shown below, when a transparency is placed at plane P_2 (behind plane P_1) with amplitude transmittance $t_1(x', y')$. Assume the incident illumination is a monochromatic plane wave.

   (b) If a second screen with transmission $t_2(x', y')$ is placed a distance F in front of the lens P_1, determine the expression for the irradiance at plane P_3. Assume that the expression for $t_1(x', y')$ corresponds to a circular aperture of diameter $2r$. What is the effect of varying d?

4. Using two thin lenses, one convex-spherical (L_1) with focal length $f_1 = 4$ cm and the other convex-cylindrical (L_2) with focal length $f_{2y} = 6$ cm, how can they be arranged to give an image of an object in the y axis and the magnitude of the Fourier transform of the same object in the x axis, at a common output plane? What are their corresponding object (s_o) and image (s_i) distances? Sketch the arrangement to clarify your result.

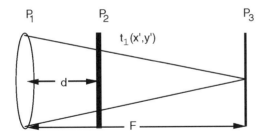

5. An object with a square-wave amplitude transmittance with fundamental frequency of 10^3 cycles/cm is imaged by a lens of focal length of 10 cm, where the object distance is 20 cm and the wavelength is 10^{-4} cm. What is the minimum lens diameter that will yield any variations of intensity across the image plane for (a) coherent and (b) incoherent object illumination?

6. In the two-lens architecture shown below,

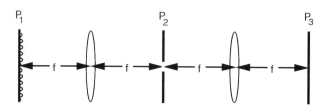

a stop is placed in plane P_2, with diameter $\Delta f \ll f_o$, where f_o is the spatial frequency of the input light distribution at P_1 described by

$$t(x_1, y_1) = 0.5 + 0.5\cos(2\pi f_o x_1)$$

What is the output light amplitude distrbution at P_3? If the output light distribution is square-law detected, what signal processing function has this architecture performed on the input function?

7. Given the aperture described by the transmittance function

$$t(x_1, y_1) = \left\{\left[\text{rect}\left(\frac{x_1}{X}\right)\text{rect}\left(\frac{y_1}{Y}\right)\right] * \left[\frac{1}{\Delta}\text{comb}\left(\frac{y_1}{\Delta}\right)\delta(x_1)\right]\right\}\text{rect}\left(\frac{y_1}{N\Delta}\right)$$

where N is odd and $\Delta > Y$, do the following:

(a) Plot the aperture function using MATLAB and identify the element and array factors.

(b) Determine an expression for the intensity distribution in the focal plane of a Fourier transform architecture, assuming that the incident illumination is a monochromatic plane wave and $N \gg 1$.

(c) Plot the result of part (b) using MATLAB, label the axes and key abscissa and ordinate values, and identify the element and array factors.

(d) If the aperture function is the input pattern to the Fourier transform architecture of Fig. 6.2, where the Fourier transform lens has focal length $f = 1$ m, and the wavelength of light used is $\lambda = 633$ nm, what is the size of the resulting Fourier transform pattern for $\Delta = 150\,\mu\text{m}$, $X = Y/2 = 50\,\mu\text{m}$, and $N = 101$?

8. A diffracting aperture has a circularly symmetric amplitude distribution defined by

$$t(r) = \left[\frac{1}{2} + \frac{1}{2}\cos(\alpha r^2)\right]\text{circ}\left(\frac{r}{r_o}\right)$$

(a) Show that the transparency acts like a combination of three lenses: flat, convex, and concave.

(b) What are their effective focal lengths?

(c) For a normally incident monochromatic plane wave, what is the resulting light distribution in the back focal plane?

(d) If the incident light is polychromatic, what factors limit the transparency's usefulness for imaging?

9. Consider the circular pupil function with a central stop shown below.

(a) What is the OTF with and without the central stop?

(b) Plot the limiting form of the OTF when the size of the stop approaches the pupil using MATLAB.

(c) What is the PSF in the limit of part (b)?

10. A transparency with sinusoidal amplitude transmission defined by

$$0.5 + 0.5 \cos(2\pi f_o x_o)$$

is placed in front of a lens of focal length F as shown below and illuminated by a monochromatic plane wave a oblique incidence (θ) in the $x_o z$ plane.

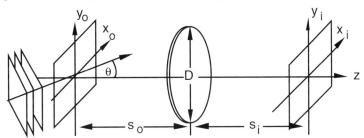

(a) What is the Fourier transform of the input light distribution seen at the output plane?

(b) If the architecture is arranged in an imaging configuration so that $s_o = s_i = 2f$, what is the maximum value of θ for which any variations of intensity occur in the image plane?

(c) Using the maximum value of θ, what is the intensity distribution in the image plane and how does it compare to the case where $\theta = 0$?

(d) Compare the spectra for the θ_{max} and $\theta = 0$ cases. What is the maximum sinusoidal pattern frequency that will allow variations of intensity in the image plane? How does this affect the image distribution versus the case for $\theta = 0$?

11. Assume that the spatial impulse response of a one-dimensional optical system is given by

$$h(x, y) = \text{rect}\left(\frac{x}{\Delta x}\right) - \text{rect}\left(\frac{x - \Delta x}{\Delta x}\right)$$

(a) If the optical system is illuminated by an incoherent wavefront defined by

$$f(x, y) = \text{rect}\left(\frac{x}{\Delta x}\right)$$

calculate the irradiance distribution at the output plane and plot it with MATLAB.

(b) If the optical system is illuminated by a coherent wavefront also described by $f(x, y)$, calculate the corresponding output amplitude distribution.

12. Consider the case of Problem 3 where the object distribution $t(x', y')$ is placed behind the Fourier transform lens at plane P_1 of focal length F. In order to get an exact Fourier transform at the output plane, another convex lens can be used. Where should it be placed and what should its focal length be? In this configuration, if we take account of the effective pupil function in the object plane, we must incorporate an extra scale factor F/d in the argument of the pupil function in the expression for the Fourier transform at the output plane. This distance (d) of separation between the object and the lens is also present in the Fourier kernel. What advantage does this afford us in obtaining the Fourier transform in the output plane?

13. A Fresnel zone plate diffracting aperture is described by the expression

$$t(r) = \left[\frac{1}{2} + \frac{1}{2}\text{sgn}(\cos ar^2)\right]\text{circ}\left(\frac{r}{r_o}\right)$$

and it is illuminated at normal incidence with a monochromatic plane wave. Show that the zone plate acts like a lens with multiple focal lengths and indicate their magnitudes. Determine the intensity at each focal plane.

14. In the architecture shown below, two transparencies, $t_1(x_1, y_1)$ and $t_2(x_2, y_2)$, are placed each in front of the two lenses shown. The first lens L_1 has focal length $f_1 = 2d$, and the second lens has focal length $f_2 = d$. They are spaced apart by $2d$. The input light is a plane wave. Determine the expression for the output plane light amplitude distribution and plot it with MATLAB. In what applications is this architecture useful?

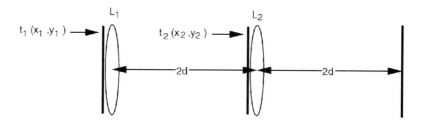

15. An object is located in front of a positive lens of focal length $f = 50$ cm at a distance $s_o = 100$ cm. If the lens of diameter is 2.5 cm and the object is illuminated with red light at 630 nm, what is the minimum resolution distance in the focal plane?

16. If two stars are separated by an angle of 1 μrad, at a wavelength of 550 nm, what is the diameter of a telescope objective required to just resolve the images?

17. An example of a space-variant (astigmatic) optical processing system is an architecture that Fourier transforms in one axis (the x axis) and images in the orthogonal axis (y axis). The mathematical operation is defined as

$$F(u, y) = \int_{-\infty}^{\infty} f(x, y) \exp(-j2\pi u x) \, dx$$

(a) Construct an architecture using three lenses that accomplishes this operation. Specify the type of lenses and their spacing between input and output planes. Assume that the light is coherent and sketch the architecture in top and side views.

(b) Construct an architecture using two lenses, specifying lens type, spacing and sketch top and side views, as before. Can this architecture preserve the phase?

18. Using Equation (6.19) and the fact that both u and v depend on λ through the relations $u = x_3/\lambda f$ and $v = y_3/\lambda f$, consider what effect a finite spectral bandwidth $\Delta\lambda$ has on spatial frequency resolution in the Fourier plane. Assume that the input light distribution is described by a rectangular truncated cosine function and look at the magnitude of the Fourier transform for a finite spectral brand average over λ_1 to λ_2.

19. For the case of a finite aperture Fourier transform lens, Equation (6.20) determines its actual vignetting effect in the Fourier plane. Assuming that $s_o = f$, that the lens pupil function is circular, that the object plane aperture is rectangular, and that the incident illumination is a plane wave, calculate the result in the Fourier plane. Plot your result using Matlab and interpret the effects of the rectangular aperture—that is, the effect of aperture size on the exact Fourier transform in the two orthogonal directions with respect to the lens pupil size.

20. Using the analysis of a single-lens MTF—that is, Equation (6.52) and the discussion on pages 149–150 and the diagram of a simple telescope given in Fig. 4.20—show explicitly that the MTF of a simple telescope is dominated by the diffraction MTF of its objective lens aperture.

21. By following the discussion of the seventh analogy on pages 154–155, explicitly derive the result of Equation (6.73) and hence Equation (6.74). Model this process after Fig. 6.12 using MATLAB and verify the results for a rectangular input function.

22. Consider a lens constructed in the form of a portion of a cone as shown below. Construct a detailed description of the lens geometry in plan and side view. Assume that the thin-lens approximation holds, so that only phase changes need be considered for determining the effect on an incident wavefront. Assume that a plane wave is incident on this lens. The phase changes are in proportion to the lens thickness such that the phase factor $\phi(x, y)$ is given by

$$\phi(x, y) = k\Delta_o + k(n-1)\Delta(x, y)$$

where k is the wavenumber, n is the index of refraction, Δ_o is the maximum thickness, and $\Delta(x, y)$ is thickness variation. Show that the phase change introduced by this lens is

$$k\left[n\Delta_o + \frac{(n-1)Ry}{h} - \frac{x^2}{2f(y)}\right]$$

where

$$f(y) + \frac{R(1+y/h)}{n-1}$$

is the line of focus.

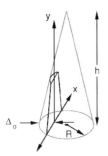

BIBLIOGRAPHY

J. D. Gaskill, *Linear Systems, Fourier Transforms, and Optics*, John Wiley & Sons, New York (1978).

J. W. Goodman, *Introduction to Fourier Optics*, McGraw-Hill, San Francisco (1968).

K. Iizuka, *Engineering Optics*, Springer-Verlag, Series in Optical Sciences, New York (1987).

R. B. Johnson, "Radar Analogies for Optics," SPIE Vol. 128, *Proceedings of Effective Utilization of Optics in Radar Systems* (1977), pp. 75–83.

H. Lavin and M. Quick, "The OTF of Electro-Optical Imaging Systems," SPIE Vol. 46, *Proceedings of Image Assessment and Specification*, May 20–22, 1974, Rochester, NY, pp 279–286.

E. L. O'Neill, "Transfer Function for an Annular Aperture," JOSA **46**, 285–288 (1956); and errata, *JOSA* **46**, 1096 (1956).

R. V. Shack, "On the Significance of the Phase Transfer Function," SPIE Vol. 46, *Proceedings of Image Assessment and Specification*, May 20–22, 1974, Rochester, NY, pp. 39–43.

F. T. S. Yu, *Optical Information Processing*, John Wiley & Sons, New York (1983).

Chapter 7
LIGHT SOURCES AND DETECTORS

7.1 OVERVIEW

Optical signal processing would not be practical without electro-optical devices that interface between the optical architecture and the outside world. Ultimately, electro-optical devices are necessary for converting optical signals into *electrical* signals or vice versa. The "outside world" consists of (a) incoming optical radiation detected by appropriate sensors, (b) outgoing electrical or optical signals to enable command, control, communication, computation, or display, and (c) intermediate (optical or electrical) input signals for modulating the light distributed within the optical architecture. In the block diagram for the basic optical processor shown in Fig. 7.1, we are looking at the input and output devices, as highlighted below.

The input light sources we will consider for optical signal processing include the light emitting diode (LED) and laser diodes (both single elements, linear and matrix arrays). The output light detectors we will cover include single photodiodes, linear self-scanned arrays, and two-dimensional focal plane arrays (FPAs). Their principles of operation and their configurations will be given in some detail, and some general requirements on these devices for optical signal processing will also be summarized.

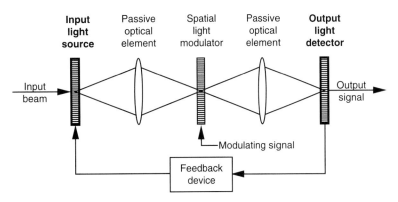

Figure 7.1 Block diagram of a generic optical processing architecture.

7.2 LASER PRINCIPLES OF OPERATION

Lasers need an active medium in order to amplify light. Amplification is achieved inside a suitable Fabry–Perot cavity via feedback, in which case the laser amplifier is also an oscillator in much the same way that an electronic amplifier achieves oscillation by feedback. Under these conditions a laser beam is produced in a so-called Fabry–Perot resonator as described below.

A Fabry–Perot resonator consists of an active medium between two mirrors in which the gain and loss mechanisms are distributed. Thus, an electric field $E(0)$ at point $x = 0$ evolves to $E(x, t)$ at point x and time t and is given by

$$E(x, t) = E(0) \exp(\gamma x) \exp(-\alpha x) \exp[j(\omega t - kx)] \tag{7.1}$$

where γ is the gain constant, α is the attenuation constant, ω is the radian frequency, and $k = 2\pi/\lambda$ is the propagation constant in the absence of gain or loss.

Writing Equation (7.1) as $E(x, t) = E(0) \exp[j(\omega t - k'x)]$, we define $k' = k + \Delta k + j(\gamma - \alpha)$, which is the propagation constant accounting for dispersion (i.e., causal wave propagation), which is consistent with a Hilbert transform relationship between the real and imaginary parts of k'. The real part of k' determines wavelength, and the imaginary part depends on the gain and/or loss in the medium.

Applying the properties of a Fabry–Perot cavity described earlier in Chapter 4, we obtain

$$\frac{E_{out}}{E_{in}} = \frac{t_1 t_2 \exp(-jk'L)}{1 - r_1 r_2 \exp(-jk'2L)} \tag{7.2}$$

where E_{out} is the electric field of the output transmitted beam, E_{in} is the electric field of the input incident beam, and L is the cavity length between mirrors 1 and 2 with respective amplitude transmittances t_1 and t_2 and amplitude reflectances r_1 and r_2.

As a condition for oscillation we require the denominator to be zero, so that the transfer function E_{out}/E_{in} becomes infinite (similar to an amplifier with feedback). At a certain threshold gain constant γ_{th} given by

$$\gamma_{th} = \alpha - \frac{1}{2L} \ln(r_1 r_2) \tag{7.3}$$

the laser starts oscillating. γ_{th} is an equilibrium value above which saturation occurs and below which oscillations die out. In solving for the condition of oscillation in which the denominator of Equation (7.2) is zero, we obtain a relation $r_1 r_2 \exp(-jk'2L) = 1$. If we equate real and imaginary parts of this equation, we can determine the discrete allowable oscillation frequencies of the laser cavity, which are $f_m = mv/2L$, where v is the velocity of light in the medium, and m is an integer. That is, the separation between resonant wavelengths is $2L/m$, which corresponds to standing wave modes that enter and leave the cavity all in-phase and constructively interfering.

The transmission function of a Fabry–Perot laser cavity in which the medium is thermally excited (whether a gas or diode laser) is given by the well-known Lorentzian line-shape:

$$\frac{T(f)}{T_{max}} = \frac{1}{1 + \pi^2 F(f/f_o - m)^2} \tag{7.4}$$

where $F = [2r/(1 - r)^2]^2$ is the contrast and $f_o = v/2L$ is the cavity resonant frequency. This result determines the shape of the spectral components of oscillation shown in Fig. 7.2.

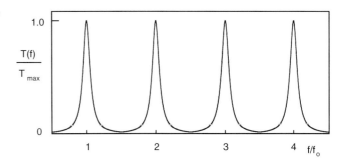

Figure 7.2 Typical laser line spectrum versus frequency determined by Fabry–Perot cavity.

A more sophisticated analysis of the electric field produced by a laser, which incorporates the theory of Fresnel diffraction, starts with an assumed Gaussian distribution in x and y coordinates in the laser given by

$$E(x, y) = \sqrt{\frac{2}{\pi}} \frac{1}{w(0)} \exp\left[\frac{-(x^2 + y^2)}{w(0)^2}\right] \quad (7.5)$$

and applies the diffraction integral:

$$E(x, y, z) = \frac{\exp[j(\omega t - kz)]}{j\lambda z} \int_{-\infty}^{\infty}\int_{-\infty}^{\infty} E(x', y', 0)$$
$$\exp\left\{-j\frac{k}{2z}[(x-x')^2 + (y-y')^2]\right\} dx'dy' \quad (7.6)$$

to yield the field at point z outside the laser:

$$E(x, y, z) = \frac{1}{j\lambda z}\frac{1}{w(z)} \exp\left[-j\frac{k}{2z}(x^2 + y^2)\right] \exp[j(\omega t - kz - \phi(z))] \quad (7.7)$$

where $w(z) = w(0)[1 + (z/z_R)^2]^{1/2}$ is the laser beam waist, z_R is the Rayleigh distance $= \pi w^2(0)/\lambda$ (transition to Fraunhofer region), and $\phi(z) = \tan^{-1}(z/z_R)$. As a result the Gaussian beam shape is preserved. Up to the Rayleigh distance the Gaussian waist, which determines spot size, is essentially constant. For higher-order modes of oscillation the laser beam shape is more complicated and described by Hermite polynomials.

If we treat the laser in terms of the particle picture of light, it emits photons of energy E given by $E = h\nu$. The simplest laser system has two energy levels E_1 and $E_2 (> E_1)$, where the emitted light has a center frequency given by $f_o = (E_2 - E_1)/h$ and spread of $\Delta f \approx 1/\tau_{\text{spont}}$, where τ_{spont} is the spontaneous emission rate. That is, the laser has a decaying electric field

$$\mathbf{E} = \mathbf{E}_o \exp(j\omega t) \exp(-t/\tau_{\text{spont}}) \quad (7.8)$$

with Fourier transform (or amplitude spectrum)

$$\text{FT}\{E(t)\} = \frac{E_o}{1 + (\omega - \omega_o)^2 \tau_{\text{spont}}^2} \quad (7.9)$$

which has a center frequency of f_o and is plotted in Fig. 7.3.

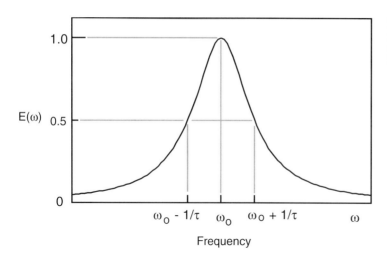

Figure 7.3 Fourier transform of laser temporal modulation envelope. Note the resemblance to a single Fabry–Perot cavity bandshape shown in Fig. 7.2.

For a laser material in thermal equilibrium, energy is absorbed and emitted at the same rate normally spontaneously and randomly so that the emitted energy is diffused and widely dispersed in wavelength. Stimulated emission which occurs when the medium "lases," however, has energy levels E_1 and E_2 determined by specific energy states in the material, which result in a high degree of spatial coherence and directionality. At the atomic level the differences between spontaneous and stimulated emission are illustrated in Fig. 7.4. We can consider a population of such atoms in which the number of photons generated by the laser is the difference between the photons emitted and absorbed. Provided that we pump the medium, the electron density can be inverted; that is, we can make $N_2 > N_1$, where N_2 and N_1 are the number of electrons in levels E_2 and E_1 respectively. Pumping can be accomplished in a number of ways, including (a) flash lamps for gas lasers like the HeNe and (b) high-current density injection as in laser diodes. To actually achieve gain in a laser medium with more than one state per energy level (degeneracy), we require $N_2 > g_1 N_1 / g_2$, where g_1 and g_2 are the degeneracies for each level. Helium–neon (HeNe) lasers are often used to provide light for "breadboarding" optical processors because they can produce up to 100 mW of continuous wave (CW) power output in a well-collimated, stable, single-mode coherent beam. However, because of their compact size and high bandwidths, we will concentrate on laser diodes, especially for applications.

Before describing LEDs and laser diodes in engineering terms, an important technique will be covered that is used to correct the lack of lateral beam coherence. This type of laser beam degradation is characteristic of most lasers that are used in practice—for example, HeNe and especially the laser diode, where unwanted intensity fluctuations are created by interference effects on an otherwise uniform Gaussian beam. These effects are caused by light scattering off dust on lenses and in the air, as well as scattering from window and lens defects. What is seen in practice can be roughly modeled as a Gaussian intensity profile with high-frequency fluctuations superimposed on it as shown in Fig. 7.5. The ideal beam profile can be described by

$$I(r) = I_o \exp(-2r^2/a^2) \qquad (7.10)$$

Figure 7.4 Mechanisms for photon absorption and emission for (a) spontaneous and (b) stimulated emission.

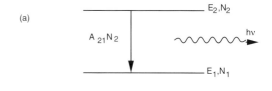

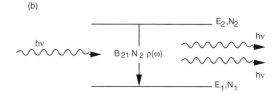

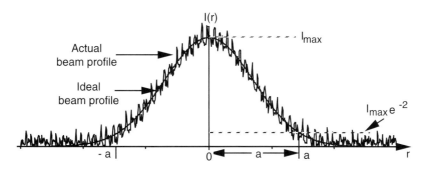

Figure 7.5 Gaussian laser beam profile with higher frequency noise superimposed on it.

where $I_o = 2P_T/\pi a^2$, P_T is the total laser beam power, and a is the laser beam radius at $I(r) = I_o \exp(-2)$.

In addition to the ideal beam profile, there is a noise on the beam given by $I_N(r)$. This noise usually varies rapidly and randomly over distances less than a. When we focus the actual light distribution $I(r) + I_N(r)$ onto a focal plane, a positive lens forms the optical power spectrum of light distribution given by

$$I(r) = I_o \exp(-2r^2/a^2) + I_N(r) \tag{7.11}$$

where r is the radius within optical power spectrum plane (measured in units of distance), $a(= \lambda F/\pi a)$ is the radius of optical power spectrum at which Gaussian is down by e^{-2}, F is the objective lens focal length, and λ is the laser wavelength.

The power spectrum is illustrated by Fig. 7.6. Since the higher-frequency content of the noise component, $I_N(r)$ is usually separated from the focused Gaussian spot, $I(r)$, this unwanted noise can be blocked by a spatial filter (pinhole), which passes a fraction of the power $P(r)/P_T$ given by

$$P(r)/P_T = 1 - \exp[-2(r/a)^2] \tag{7.12}$$

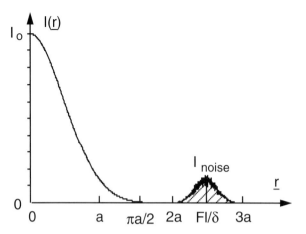

Figure 7.6 Power spectrum of noisy Gaussian laser beam. Note that ideal Gaussian is located at low spatial frequencies and that the noise is isolated at higher frequencies from the desired ideal beam.

The optimum pinhole diameter (D) can be selected to pass most (e.g., 99.3%) of the total beam power and block almost all components of the noise spectrum, in which case $D = 2\lambda F/a$. Then the fraction of total power passed is

$$\frac{P(D)}{P_T} = 1 - \exp\left[-\frac{1}{2}(\pi a D/\lambda F)^2\right] \tag{7.13}$$

and the maximum frequency blocked is $\rho_{\max} = D/2F\lambda$.

7.3 LIGHT EMITTING DIODES AND LASER DIODES

In an LED, electrons and holes are injected into the p–n junction by forward biasing. After the electrons and holes recombine (radiatively and thermally), light is emitted of wavelength $\lambda = hc/E_g$ at a typical rate of 10–20 mW/A and linearly with respect to injected current. E_g is the energy gap between valence and conduction bands in the semiconducting medium, as we will discuss further. The modulation bandwidth is limited to several hundred megahertz by the recombination time. Because the electrons are nonuniformly distributed in energy, the frequency content of photon emission is spread and is typically ≥ 20 nm. Radiation from the LED is primarily from spontaneous emission, which takes place regardless of external illumination.

The laser diode, on the other hand, emits primarily by stimulated emission when light impinges on the p–n junction at a specific wavelength λ corresponding to a frequency $\nu = (E_2 - E_1)/h$, where the energy difference $E_2 - E_1$ may correspond to E_g. A laser diode also differs in that the stimulated emission is amplified by feedback in the laser cavity. Both ends of the laser diode cavity are polished on cleaved facets, resulting in multiple reflections and multiple passes through the medium, which lead to further stimulated emission. Finally the laser diode differs from the LED because photons created by stimulated emission have the same phase as the initial excitation photon.

Length of the cavity is a few hundred microns, and the shorter it is, the wider the spacing of resonant lines in a multiline emission spectrum. The index of refraction of the active region is higher than the cladding regions so that a light guide is formed. If the thickness of the active region is sufficiently thin ($\sim$0.4 μm), only one mode is excited in the transverse direction. The electrode may also be shaped to confine action to a narrow width ($\sim$10 μm).

The most efficient and useful lasers for optical processing now and in the future are laser diodes and laser diode arrays. They are preferred for their compact size and high modulation bandwidth, but are also beginning to be favored for their high power and improved coherence. A comparison of light emitting diodes (LEDs) and laser diodes in Table 7.1 reveals the differences between these most common optoelectronic semiconductor emitters.

The principles of operation of LEDs and laser diodes are very similar. The LED essentially acts like the amplifier for a laser diode. They are both constructed as a *p–n* junction and have an energy level diagram as shown in Fig. 7.7. Unlike HeNe lasers, where the emitted photon is generated by electron transitions in a He–Ne gaseous mixture, in a semiconductor the transitions are between the mostly filled valence band and partially populated conduction band. Thus, electron–hole pairs recombine across the band gap to generate a photon of energy, E_g, and wavelength λ given by $\lambda(\mu m) = 1.2394/E_g(\text{eV})$.

When an LED is current-modulated, the carrier density, the active layer, and optical output are directly modulated, provided that the modulation frequency ω_m is much less than $1/\tau$, where τ is the recombination time of electrons in the active region. The rate equation for electron concentration n in the active region is

$$\frac{dn}{dt} = \frac{1}{ed}j - \frac{1}{\tau}n \qquad (7.14)$$

where j is the current density, and d the thickness of the active region. In the small-signal limit we may assume $n \sim n_o + \delta n \exp(j\omega t)$ and $j \approx j_o + \delta j \exp(j\omega t)$, where n_o and j_o are

Table 7.1
Comparison of Semiconductor Diode Emitters

	LEDs	Laser Diodes
Angular radiation pattern	Not very directional	Directional
Spectral width	$\sim$40 nm	$\sim$0.01 nm (single mode)
Response time	Few nsec	$\sim$0.01 nsec
Modulation bandwidth	$\sim$200 MHz	Several GHz
Light output vs. current input	Linear w/o I_{th}	Nonlinear $< I_{th}$/linear $> I_{th}$

Figure 7.7 Energy level diagram of a simple pn junction, which is the basis of LED and laser diode operation.

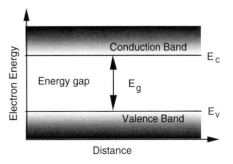

steady-state values, δn and δj are the small-signal terms, and ω is the modulation frequency. The resulting frequency response is

$$H(\omega) = \frac{\delta ned}{\delta j \tau} = \frac{1}{1 + j\omega\tau} \tag{7.15}$$

A typical value of $\tau = 3$ nsec yields a 3-dB bandwidth of 53 MHz.

Semiconductor lasers typically consist of two layers of p- and n-type materials with two sides cleaved or polished perpendicular to the p–n junction plane to form a Fabry–Perot cavity and the remaining sides left roughened to eliminate lasing in that direction. When the current reaches a threshold, stimulated emission occurs and a highly directional, small spectral bandwidth, divergent beam of light is emitted from the active region near the p–n junction. This layer is approximately 1 μm thick or less.

Laser diode structures come in several categories. Of particular interest are the so-called double heterostructure lasers, which consist of n- and p-doped solid solutions separated by a second p-doped solution. Figure 7.8 illustrates the light intensity distribution, index of refraction, and band gap (or confinement) for the double heterostructure compared to the overall energy-level diagram. The power (P) is confined to a region where, by virtue of the changing material characteristics, the index of refraction (n) and the energy gap (E_g) vary rapidly. Laser diodes are the most useful for optical signal processing applications because they are so compact and efficient and can be constructed as arrays or pulsed to achieve sufficient power output. The charge carriers are confined to the active region by heterojunction potential barriers on both sides, and the light field is confined within the active region by the abrupt change in the refractive index outside the active region. These confinements enhance stimulated emission and substantially reduce the required threshold current density (J_{TH}).

As a result of bringing the p- and n-type materials together in a junction, an electric field is created and a depletion layer is formed. The corresponding energy level diagram is shown in Fig. 7.8(d). By applying a forward bias voltage, a large number of electrons and holes (i.e., minority carriers) are injected into the depletion region where they diffuse and recombine, spontaneously generating photons. Lasing action occurs in the active region centered on the depletion layer if a population inversion is caused.

Emissions are spontaneous at low current and have wide spectral distributions (100–500 nm at half-power). As current exceeds the threshold, the spectral distribution narrows. The spectral linewidths of emissions from p–n junctions (depending on the type: LED, multimode laser diode, or single-mode laser diode) are shown in Figs. 7.9(a)–(c).

For most laser diodes, the linewidth is less than 1 nm. Some lasers have many lines or modes separated by $\Delta\lambda$ at lower currents and collapse into a single mode at higher currents. The separation between modes is

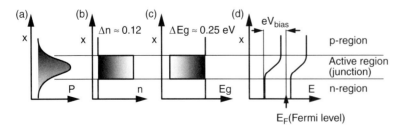

Figure 7.8 (a) Power distribution in laser diode cavity, (b) index of refraction variation, and (c) energy gap variation with spatial distance normal to the depletion region. The energy-level diagram under a bias voltage is shown in (d) for comparison.

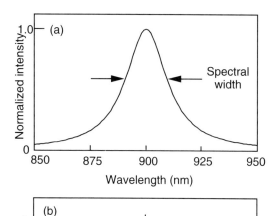

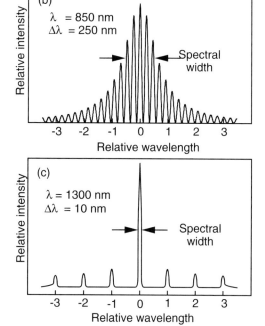

Figure 7.9 Spectral linewidths for (a) typical LED, (b) multi-mode laser diode and (c) single mode laser diode.

$$\Delta\lambda = \frac{\lambda^2 \Delta m}{2nL[1 - \lambda/(n\, dn/d\lambda)]} \quad (7.16)$$

where λ is the wavelength of the mode, n is the refractive index of the medium, L is the length of the diode, and Δm is the integral number of half-wavelengths between the reflection planes. Note the dispersive term in the denominator. Clearly the separation is inversely proportional to the diode length.

Although the laser diode beam is coherent and monochromatic, it is more divergent than a laser beam from a gas laser. (Coherent emission is necessary especially for supporting interferometric optical processing architectures.) The far-field beam is elliptical in shape, having a beamwidth (θ) parallel to the plane of active region of about 10 degrees, and perpendicular (θ') of approximately 35–65 degrees (depending on thickness of the active layer and its composition). The electric field dependence across the diode can be substituted into the following equation for intensity at arbitrary θ relative to intensity at $\theta = 0$:

$$\frac{I(\theta)}{I(0)} = \frac{\cos^2\theta \left| \int_{-\infty}^{\infty} E_y(x,0) \exp(jkx \sin\theta)\, dx \right|^2}{\left| \int_{-\infty}^{\infty} E_y(x,0)\, dx \right|^2} \qquad (7.17)$$

where $E_y(x,0) = E_y(0,0) \cos(kx) \exp(j\omega t)$.

Another characteristic of importance for optical signal processing applications of the laser diode is turn-on delay and modulation frequency. Delay time t_d is derived from the continuity equation for electrons in a p-type semiconductor—that is, $dn/dt = I/eAd - N/\tau_e$, where n is injected electron concentration, e is electron charge, A is area, and τ_e is carrier lifetime. The first term is the generation rate and the second term is the recombination rate. By solving for time we obtain

$$t = \tau_e \ln\left[\frac{I}{I - \frac{gn(t)Ad}{\tau_e}} \right] \qquad (7.18)$$

When $n(t)$ reaches the threshold value $en_{TH}Ad/\tau_e$, then $t = t_d = \tau_e \ln[I/(I - I_{TH})]$. Typical delay time is ~ 10 psec, which allows the diode to be modulated at frequencies in the gigahertz range, but output power drops off as frequency increases.

The output power of a laser diode is proportional to the injection current (I) [above a threshold current $I_{th}(T)$] as shown in Fig. 7.10(a). The current dependence on voltage given by the I–V characteristic is shown in Fig. 7.10(b). [Note that $I_{TH}(T)$ is a function of junction temperature.] The instantaneous optical output power (P) is given by

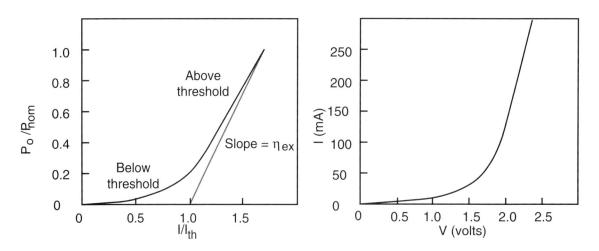

Figure 7.10: (a) Output power versus bias current, and (b) current-voltage characteristic of a typical laser diode.

$$P = [I(V) - I_{\text{th}}(T)]\frac{\hbar\omega}{2\pi e}\eta_{ex}\eta_{\text{mod}}(\omega_{\text{RF}})\cos^2\omega t \qquad (7.19)$$

where T is the ambient laser temperature, ω is the spectral frequency, ω_{RF} is the applied RF modulation frequency, η_{ex} is the external quantum efficiency, $\eta_{\text{mod}}(\omega_{\text{RF}})$ is the modulation efficiency, and V is the applied voltage. The modulation efficiency $\eta_{\text{mod}}(\omega_{\text{RF}})$ is given by

$$\eta_{\text{mod}}(\omega_{\text{RF}}) = \frac{\eta_{\text{mod}}(0)}{[(\omega_0^2 - \omega_{\text{RF}}^2)^2 + \gamma^2\omega_{\text{RF}}^2]^{1/2}} \qquad (7.20)$$

and remains flat up to a resonance frequency (ω_0) of several GHz, where γ = damping constant $\sim$10 MHz and $\eta_{\text{mod}}(0)$ is the rate of electron injection at DC.

In the most common type of laser diode, aluminum-doped Ga and As are brought in contact to form a pn junction, thus forming a sandwich-type structure of GaAlAs as shown in Fig. 7.11.

As can be seen in Fig. 7.11, the angular spread of the beam is greatest perpendicular to the plane of the junction because the light emitted is confined more ($\sim$0.5 μm) than in the direction of the junction ($\sim$10 μm), a result clearly understood as diffraction from finite apertures. The degree of lateral confinement of the injected current also depends on the contact geometry, and the degree of confinement perpendicular to the junction depends on dopant concentration levels. (Spatial variation of injected current yields "gain-guided" operation, whereas spatial variation of material composition yields "index-guided" operation.)

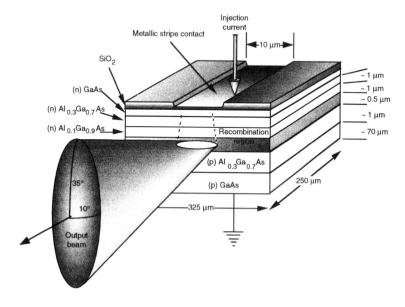

Figure 7.11: Sandwich-like structure of Ga-Al-As which serves as the physical basis of LED's and laser diode construction.

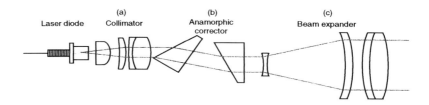

Figure 7.12 Beamforming optics used to shape a typical laser diode beam. A short focal length singlet (a) is used to collect the rather divergent laser beam, followed by a doublet to collimate the beam. An anamorphic prism pair (b) makes the beam cross section approximately circular. A second larger diameter, longer focal length lens is used in conjunction with a negative lens to (c) expand the beam after the anamorphic corrector.

EXAMPLE 7.1 Typical Laser Diode Output Power   Using Equation (7.19) we can calculate a typical laser diode output power assuming the following characteristics:

I_{th} = 50 mA

$I(V)$ = 250 mA (bias) (from the I–V curve for a particular DC voltage)

λ = 830 nm

η_{ex} = 0.1

η_{mod} = 1

Then

$$P_o = (250 \text{ mA} - 50 \text{ mA}) \frac{(1.055 \times 10^{-34} \text{ J} \cdot \text{sec})(2.27 \times 10^{15} Hz)}{(1.6 \times 10^{-19} \text{C})} (0.1)(1)$$

$$P_o = 3.0 \times 10^{-2} \text{W} = 30 mW$$

In this example we have assumed a DC bias condition so that $\eta_{mod} \approx 1$.

In practice, single laser diodes can be used to illuminate an optical architecture using the beam forming optics illustrated in Fig. 7.12. To achieve sufficient instantaneous power the laser may be pulsed at a relatively low duty cycle as suggested in Fig. 7.13. Typical parameters for a pulsed laser diode are given in Table 7.2.

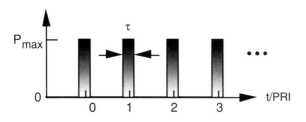

Figure 7.13 Typically a low-duty cycle waveform is used to modulate a laser diode, where the duty cycle D $\sim 10^{-4}$.

7.3 LIGHT EMITTING DIODES AND LASER DIODES

Table 7.2
Parameters for Single Pulsed Laser Diode

Total peak radiant flux:	3 W (Min) – 9 W (Max)
Threshold current:	6 A (Min) – 17.5 A (Max)
Peak forward voltage:	8 V
Wavelength:	800–900 nm
Spectral bandwidth:	5 nm
Half-angle 3-dB beam width:	7.5° (||) × 15° (perpendicular)
Risetime:	<1 nsec (10–90%)
Emitting region size:	∼ 76 × 2 μm
Typical PRF:	1 KHz
Nominal pulse length:	100 nsec

EXAMPLE 7.2 Duty Cycle and Coherence Length Tradeoffs The above diode characteristics are typical of single laser diodes, which can also be pulsed to achieve useful peak output powers, provided that duty cycle limits are met. The actual duty cycle must be less than the maximum (damage-limited) duty cycle. The maximum duty cycle that can be achieved without damage or degradation to the laser diode output is given by

$$D_{\max} = \text{PRF} \tau_{\max} \frac{P_{\text{op}}}{P_{\max}}$$

where PRF is the pulse repetition frequency, $\tau_{\max}$ is the maximum pulse length, P_{op} is the operating peak pulse radiant flux, and $P_{\max}$ is the maximum allowed peak radiant flux.

If we use a waveform compatible with synchronous operation of the laser diode with a TV raster-line compatible optical processor, we would employ the waveform shown in Fig. 7.13. In this case the PRF exceeds the typical manufacturer recommended PRF of 1 kHz and is determined by PRF = 1/PRI = 1/63.5 μsec = 15.748 kHz, where PRI is the pulse repetition interval (corresponding to one video raster line). To maintain a limiting duty cycle of $D_{\max} = 10^{-4}$ we require that either $\tau_{\max}$ or P_{op} be limited to less than the maximum recommended rating. To maximize useful power, set $P_{\text{op}} = P_{\max}$. Then $\tau_{\max}$ is $D_{\max}/\text{PRF} = 10^{-4}/15.748$ kHz $= 6.35$ nsec, which is achievable since the intrinsic diode response time is often ≤ 1 nsec.

The problem with short pulse modulation of laser diodes is that during the risetime the laser diode cavity is not thermally stable. This will result in multimode emission and loss of phase coherence. We can always reduce the peak operating flux (P_{op}) and extend the pulse length ($\tau_{\max}$) by a corresponding amount. For instance, if we still assume a TV raster line compatible processor, we can require that each video line processed in the optical processor be illuminated by the laser diode for a period of one pixel dwell-time (assuming that a freeze-frame action is desired with the pulsed laser—that is, a strobe action that avoids more than one-pixel blur). Typically, one pixel dwell time is $\tau_{\text{pixel}} = \text{PRI}/512 = 124$ nsec. Therefore the maximum operating flux is $P_{\text{op}} = (D_{\max}/\text{PRF}\tau_{\text{pixel}})P_{\max} = 5.12 \times 10^{-2} P_{\max}$. For a $P_{\max} = 9$ W, P_{op} is 461 mW.

Now consider the phase coherence of the laser diode. Given the nominal spectral bandwidth of 5 nm under conditions prescribed to be 1-kHz PRF and 100-nsec pulse length, the estimated coherence length is

$$l_c = \frac{\lambda^2}{\Delta\lambda} = \frac{(850 \text{ nm})^2}{5 \text{ nm}} = 0.1445 \text{ nm}$$

The choice of $\tau = \tau_{\text{pixel}}$ is consistent with the pulse length condition (124 nsec > 100 nsec) for this degree of phase coherence. This, however, will not be sufficient to support coherent optical processing in typical optical processing components or architectures. A coherence length at least several orders of magnitude higher is required. Thus high power and adequate phase stability cannot be simultaneously achieved with single pulsed laser diodes in most instances. To obviate this tradeoff between peak power and coherence length, we must turn to multiple emitters—that is, laser diode arrays.

7.4 LASER DIODE ARRAYS

Single laser diodes can be arranged in linear monolithic arrays to achieve higher power and narrower beamwidth. Adjacent laser elements in the array can emit as limited by diffraction in the end portions near the laser cavity mirrors. Figure 7.14 shows an array of laser diodes on a single chip which are coupled by diffraction. Such diffraction coupling results in phase locking and leads to narrowing of the output beam. Diode laser arrays can be operated to powers in excess of 1 W continuously.

The shape of the far-field diffraction pattern depends on the phase of each individual laser element and the degree of field coupling from each element to its neighbors. In general, the far-field intensity distribution for an N-element coupled array of identical laser elements is given by $I(\theta) = |E(\theta)|^2 A(\theta)$, where θ is measured with respect to the facet normal, $E(\theta)$ is the element factor, and $A(\theta)$ is the array factor (representing interference effects between coupled emitters). The array factor $A(\theta)$ is a complicated expression, which in its simplest form is

$$A(\theta) = \frac{\sin^2[(N+1)\alpha/2 + n\pi/2]}{\sin^2(\alpha/2) - \sin^2[n\pi/2(N+1)]} \tag{7.21}$$

where $\alpha = 2\pi d \sin\theta/\lambda$ and d is the spacing between elements. $A(\theta)$ yields the solid curves (after modulation by the element factor $|E(\theta)|^2$) in Figs. 7.15(a)–(c). In most cases the element factor $|E(\theta)|^2$ can be modeled as a Gaussian, given by

$$|E(x)|^2 = E_0^2 \exp(-2x^2/w_0^2) \tag{7.22}$$

where $2w_0$ is the Gaussian beamwidth and $x = r \sin\theta$. This yields the dashed lines in Figs. 7.15(a)–(c). When all lasers operate in phase the output pattern has a single sharp lobe as shown in Fig. 7.15(a). If one or more laser elements differ in phase from the others, multiple sharp lobes and beam broadening occur as shown in Figs. 7.15(b) and (c). Nonuniformities

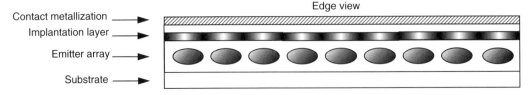

Figure 7.14 Linear laser diode array using simple epitaxial edge emitting diodes coupled by optical difffraction.

Figure 7.15 Far field angular distribution of diffracted light from a laser diode array when (a) all emitters are in-phase, (b) some emitters are out-of-phase, and (c) all emitters are out of phase, where $N = 8$.

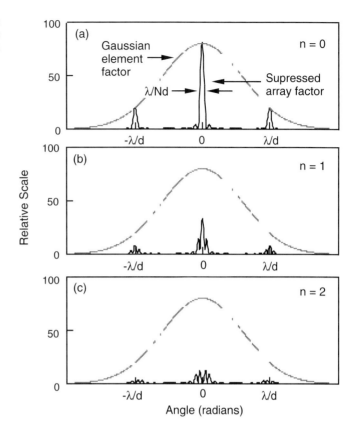

will widen the beam from ideal. Bias current is fed to all laser elements in parallel, and all elements operate at the same level above threshold. Typical currents are 200–400 mA, and output powers are ≥ 1 W. Quantum efficiencies are typically 40% or more.

In addition to traditional edge emitting laser diodes, surface emitting laser diodes have been developed. Surface emitting laser diodes emit radiation perpendicular to the surface of the array. This will enable two-dimensional (2-D) arrays to be fabricated and will allow optical interconnects to be implemented. Two-dimensional arrays should provide considerably less beam divergence than conventional single-element edge emitting laser diodes by virtue of emitting over larger areas.

Two basic configurations of the optical cavity are used for surface emitting lasers: vertical and horizontal cavity. In the vertical cavity shown in Fig. 7.16(a) the optical cavity

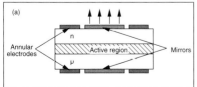
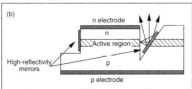

Figure 7.16 Basic configurations for surface emitting laser diode arrays: (a) vertical cavity configuration and (b) horizontal cavity configuration.

is formed from mirrors on the top and bottom of the wafer. Mirrors are typically 5–20 μm in diameter and are encircled with annular electrode rings. Because the length of the region is short perpendicular to the active layer, the gain is small. In the horizontal cavity configuration shown in Fig. 7.16(b) the optical cavity lies in the plane of the active layer, which is the same as conventional laser diodes. In this case the gain is high since active region dimensions are 100–400 μm in length. Surface emission is achieved by deflecting part of the light perpendicular to the axis of the cavity using angled mirrors (or appropriate gratings). Large, overall efficiencies of at least 50% can be obtained. Typical flux densities of 50 W/cm^2 have been achieved. Overall phase-locking of these arrays are more difficult than 1-D arrays. An example of a surface emitting array is shown in Fig. 7.17.

7.5 OUTPUT LIGHT DETECTORS

In this section of the chapter we focus on the last important component in the optical processing architecture, the detector. Sometimes it may limit the speed, dynamic range, and spatial resolution achievable with a given architecture or approach. Again we will start with a single element and describe its fundamental principles of operation. Then we will consider 1-D (linear self-scanned arrays) and 2-D focal plane arrays—for example, charge coupled devices (CCDs) and related charge injection devices (CIDs). In addition to these devices, which ultimately multiplex signals after detection for readout, there are parallel readout schemes that increase the available rate of post-detection signal processing and provide for direct-coupling to parallel

Figure 7.17 Photograph of a surface emitting laser diode array.

digital signal processors. [In the interest of addressing state-of-the-art photoelectric detectors, which appear to be the detectors of choice for real-time optical processors now and in the future, we will not treat photographic (or photochromic) film as a detector.]

The simplest form of image sensor is a single-element detector serially scanned in a raster format, which covers the entire 2-D image scene. Pixel-to-pixel nonuniformities (fixed pattern noise) are nonexistent, but complex and less reliable mechanical scanning is required. Staring arrays eliminate the necessity of mechanical scanning at the expense of fixed pattern noise. As we shall see, there are methods of focal plane and post-detection signal processing that ameliorate the fixed pattern noise (and temporal detector noise), namely time delay and integration (TDI), as well as more sophisticated nonuniformity compensation methods.

7.6 SINGLE DETECTORS

There are two general types of solid-state detectors of optical radiation: thermal and photon detectors. We will focus on photon detectors since they are faster and generally more sensitive than thermal detectors in the optical region. They include photoemissive, photoconductive, and photovoltaic types. In photoemissive detectors, charge carriers are created in the detector material in proportion to the incident radiation. Typical examples are photomultiplier tubes of various types. In photoconductive materials the resistivity changes in proportion to the incident radiation, and in photovoltaic detectors it is the voltage that changes. We will confine our attention primarily to photodiodes used in the photoconductive or photovoltaic modes in conjunction with MOSFET addressed arrays for readout. Consider first, however, the basic physics of detection in a typical silicon photoelectric device.

In the p–i–n photodiode the incident photons create free charge carriers by elevating carriers from the valence band to the conductance band, which constitute the output current. By virtue of the photoelectric effect, an absorbed photon can excite an electron from the valence band (or an intermediate state) to the conduction band across an energy gap as shown in Fig. 7.18(a). Because of the presence of an energy gap (E_g) there is a minimum wavelength for this to occur given by $\lambda(\mu m) = 1.24/E_g$ (eV), a relation seen earlier for laser diodes. [This means that for silicon detectors with a band gap of 1.1 eV they respond only to wavelengths shorter than 1.13 μm. This is not surprising since laser diodes and photodiodes are similar in structure, just used in reverse. The ideal wavelength dependence is shown in Fig. 7.18(b).] At finite temperatures electrons and holes exist, and any excess carriers will be created by photoexcitation. These photoinduced charges carry a current given by

$$I_s = \frac{E_s \eta w \tau e \mu V_b}{l} \qquad (7.23)$$

where

η = detector quantum efficiency (charge carriers per photon),

μ = charge carrier mobility (m^2/V · sec),

e = electron charge (C),

τ = intrinsic response time of detector (sec),

E_s = signal photon flux per unit time (photons/s · m^2),

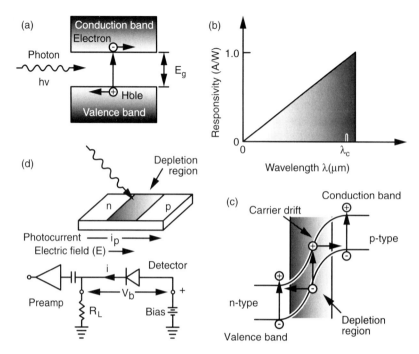

Figure 7.18 Single photodiode characteristics. (a) Energy band picture of photon-induced electron-hole production, (b) ideal wavelength dependence of response, (c) distortion of band structure across depletion region for photovoltaic operation, and (d) typical electrical interface circuit.

V_b = bias voltage (volts),
w = width of detector sensitive area (m), and
l = length of detector-sensitive area (m).

Analysis of the current–voltage characteristics of a photodiode indicates that output current from a photodiode depends linearly on the incident light intensity (H_s) with proportionality constant given by responsivity (R) such that $I_s = RH_s$. Thus as the incident light level increases here, the voltage (V_s) across a load resistance (R_L) increases proportionately; that is, $V_s = I_s R_L$.

The two most common types of quantum detectors that will be of interest in optical signal processing are photoconductive and photovoltaic. In the photoconductive detector the conductance is increased as absorbed photons produce free charge carriers. The change in conductance is measured by applying an external bias voltage across the detector and measuring the change in current through the detector. In the photovoltaic detector, photons absorbed at a p–n junction [as shown in Fig. 7.18(c)] produce electron–hole pairs that change the junction barrier potential. The consequent change in the open-circuit output voltage is a measure of the incident optical radiation. This mode does not require an external bias voltage. In the photoconductive mode a bias is applied to the photodiode in series with a load resistance to ground as shown in Fig. 7.18(d). The resulting voltage is AC-coupled through a capacitor to feed a preamplifier that has some electronic gain. The spectral sensitivity for both modes is governed by the energy levels in the semiconductor.

There are also two *modes* of detection: video and heterodyne. Often video detection is used in imaging and optical signal processing. In video detection the incident optical radiation is modulated at some low frequency ω_m before it is collected at the detector surface, although this modulation is more commonly employed in communication than imaging. As a result of square-law detection the induced current is

$$I_s(t) = \frac{2\pi P_s e \eta}{h \omega_c}[1 + m \cos(\omega_m t)]^2 \cos^2(\omega_c t) \qquad (7.24)$$

where P_c and ω_c are the power and frequency of the carrier (lightwave). The square-law nature of the detector derives from considering the square of the *amplitude*. In heterodyne detection the incident lightwave amplitude $E_c \cos \omega_c t$ is combined with a local oscillator $E_L \cos[(\omega_c + \omega_L)t]$, where $\omega_L \ll \omega_c$. Then the resulting signal current is square-law detected to yield

$$I_s(t) = \frac{2\pi P_L e \eta}{h \omega_L}[E_L^2 + E_c^2 + 2 E_L E_c \cos^2(\omega_L t)] \qquad (7.25)$$

where the resulting current is seen to be dependent on the local oscillator power, the frequency difference ω_L with respect to the carrier, and the carrier power.

In evaluating photodetectors in any application, including imaging and signal processing, several important performance parameters must be considered. In addition to responsivity (R), there is noise sensitivity, which is generally measured by noise equivalent power (NEP), measured in $W \cdot Hz^{-1/2}$, and/or specific detectivity (D^*), measured in $cm\, Hz^{1/2} W^{-1}$. Frequency response—that is, bandwidth (Δf) and response time (τ_D)—are also important related parameters.

Responsivity (R_λ) at a specific wavelength (λ), called *spectral responsivity*, is defined as the ratio of detected electrical output signal to the incident optical flux in a differential wavelength band $d\lambda$ centered at λ; that is,

$$R_\lambda = \frac{S_\lambda}{P_\lambda d\lambda} \qquad \text{(V/W) or (A/W)} \qquad (7.26)$$

where the incident flux (P_λ) is measured in watts (W) (per unit wavelength) but can also be measured in photons $\cdot sec^{-1}$ (per unit wavelength) and the output signal (S_λ) can be measured in volts (amperes). The responsivity is usually normalized by dividing by the total radiant flux incident on the detector; such that

$$R = \frac{\int_0^\infty R_\lambda P_\lambda\, d\lambda}{\int_0^\infty P_\lambda\, d\lambda} \qquad (7.27)$$

Thus the total responsivity depends not only on the spectral response of the detector (given by R_λ) but also on the spectral properties of the source (given by P_λ). Except for broadband incoherent optical processing these considerations will not normally be important. For optical signal processing we will often work at a specific (laser) wavelength. Responsivity also depends on the temporal modulation frequency, which typically has a low pass characteristic, namely,

$$R_\lambda(\omega) = \frac{R_\lambda(0)}{(1 + \omega^2 \tau_D^2)} \qquad (7.28)$$

where τ_D is the detector time constant. This is plotted in Fig. 7.19. Factors that limit the frequency response include charge-carrier lifetime (i.e., the mean lifetime of photogenerated charge carriers) and charge-transport times (i.e., time for charge carriers to move over the length of the detector, which is essentially a function of detector size and charge-carrier mobilities). Parasitic capacitance, preamp bandwidth, and, if required, light modulation frequency may also affect the overall frequency response.

The other key factor in assessing the performance of a photodetector is noise sensitivity, which is often given in terms of noise equivalent power per unit wavelength (NEP$_\lambda$). The NEP$_\lambda$ is the level of incident flux that produces a detected output SNR of unity in an electrical modulation bandwidth Δf (Hz)—that is, that meets the following condition $V_{\lambda\text{RMS}} = R_\lambda P_\lambda$, where $V_{\lambda\text{RMS}}$ is the root-mean-square (RMS) noise voltage. P_λ is, in fact, equal to the NEP$_\lambda$. The lower the NEP$_\lambda$, the better the detector. The factors contributing to the noise variance that enter into a determination of the NEP for typical CCDs are summarized in Table 7.6 at the end of the next section.

If the modulation frequency is chosen to be above the so-called $1/f$ noise, then the usual source of noise is thermal noise in the detector/preamp combination, or fluctuations in the number of charge carriers in the detector due to dark current or background photon flux (i.e., shot noise).

Another parameter that measures noise sensitivity is spectral detectivity D_λ, which is the reciprocal of NEP$_\lambda$. Hence good detectors have high values of D_λ. Both NEP$_\lambda$ and D_λ depend on noise bandwidth, as shaped by the detector electrical bandwidth $R_\lambda(\omega)/R_\lambda(0)$,

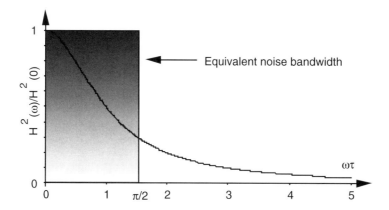

Figure 7.19 Low-pass (power spectrum) frequency response characteristic of photodiode detection and idealized low-pass characteristic (shaded region) often associated with equivalent noise bandwidth (ENB). The ENB region shown has an area equal to the area under the H(ω)/H(0) curve.

which is ideally shaped like a rectangular function. For shot noise (charge carrier fluctuation) the rms noise equals the square root of the number of charge carriers. At a given wavelength the incident power on the detector is $P_\lambda\, d\lambda = A_D H_\lambda\, d\lambda$, where H_λ is the incident light intensity (in W/m^2) per unit wavelength, and A_D is the detector area (in m^2).

In terms of photons, the incident radiant flux is

$$P = \frac{H_\lambda\, d\lambda\, A_D}{\left(\dfrac{hc}{\lambda}\right)} \qquad (7.29)$$

where hc/λ is the energy per photon at wavelength λ. The number of photoinduced charge carriers is then

$$n = P\eta\tau = \frac{H_\lambda \lambda\, d\lambda\, A_D}{hc}\eta\tau \qquad (7.30)$$

where η is the quantum efficiency (number of charge carriers produced per incident photon), and τ is the detector response time. Typical values of η and τ range from 0.15 to 0.6 and 1 nsec to 1 μsec, respectively. If we then integrate over the noise bandwidth (Δf), the mean noise power is

$$V_n^2 = \frac{H_\lambda \lambda\, d\lambda\, A_D}{hc}\eta\tau\Delta f \qquad (7.31)$$

over a 1-ohm resistive load. Then the total rms noise voltage is

$$V_{\lambda\text{RMS}} = \left(\frac{H_\lambda \lambda\, d\lambda\, \eta\tau}{hc}\right)^{1/2} (A_D \Delta f)^{1/2} \qquad (7.32)$$

Hence NEP$_\lambda$ is proportional to the square root of detector area and electronics bandwidth. The other parameters in the above equation are either wavelength or material dependent (quantum efficiency and carrier lifetime). Rather than specifying bandwidth and detector area, we should normalize by these parameters to get the *specific* spectral detectivity D_λ^* defined as

$$D_\lambda^* = (A_D \Delta f)^{1/2} D_\lambda = \frac{(A_D \Delta f)^{1/2}}{\text{NEP}_\lambda} \quad (\text{cm}\cdot\text{Hz}^{1/2}\cdot\text{W}) \qquad (7.33)$$

or

$$D_\lambda^* = \frac{(A_D \Delta f)^{1/2} R_\lambda(\omega)}{V_{\lambda\text{RMS}}} \qquad (7.34)$$

For quantum detectors operating in an ideal background-noise-limited mode, D_λ^* becomes

$$D_\lambda^* = \frac{\lambda}{2hc}\left(\frac{\eta}{H_B}\right)^{1/2} \qquad (7.35)$$

where H_B is the incident background flux density on the detector from all sources. Notice that D_λ^* and $R_\lambda(\omega)$ depend linearly on wavelength as depicted in Fig. 7.18(b). Typical values of D_λ^* are 10^{10} to 10^{12} cm $\cdot$ Hz$^{1/2}$/W.

7.7 LINEAR AND MATRIX ARRAYS

The photogenerated current collected at each detector site can be rather small, typically a few picoamperes. Thus charge must be integrated and stored, typically for an entire TV frame time. These charges are stored in the photosensor array for subsequent transfer to a CCD array arranged to be adjacent to it with corresponding elements at each photodetector site as suggested by Fig. 7.20(a). After the conversion and collection process has been completed, the array of collection sites is interrogated in a scanning pattern so that the signal corresponding to the charge at each photosite is developed and delivered to the output terminal of the device. This scanning process also erases the stored charge pattern after scanning and resets the CCD so that the sensor is ready for the next sample of incident light.

This interrogation process is enabled by the use of MOSFET switches which link each photodiode to an actual CCD shift register, as shown in Fig. 7.20(b). The bus is connected through a load resistor to a reset voltage supply. The switch is closed to back-bias the photodiode to the supply voltage, and then it is opened to isolate the diode that collects the photogenerated charge carriers during the next integration time. When the address switch is closed again, current flows through R_L and recharges the diode so that the peak signal voltage equals the collected charge divided by the output bus capacitance C_s. (Although other schemes are possible.) Individual sensor elements are then combined into linear arrays as shown in Fig. 7.20(a) or into matrix (2-D) arrays as shown in Fig. 7.21. The linear array consists of six key elements:

- Image sensor elements
- Transfer gate
- Analog shift registers

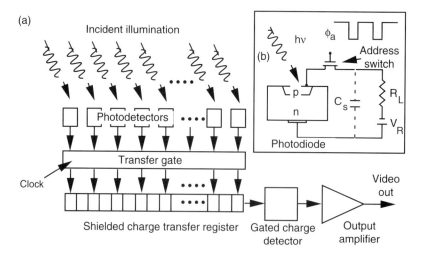

Figure 7.20 (a) Linear photodetector array and parallel CCD charge transfer register, and (b) electrical interface circuit between each photodetector and its corresponding CCD charge well.

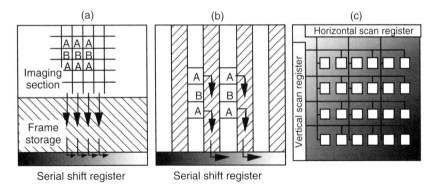

Figure 7.21 Basic architectures for CCD (or CID) imagers: (a) frame transfer, (b) line transfer, and (c) charge injection (CID) devices.

- Clock driver circuitry
- Gated charge detector/amplifier
- Dark and white reference circuitry

The 2-D arrays are simple extensions of the 1-D arrays, but with differing architectural arrangements. Charge transfer devices (CCD or CID) have three basic architectures for use as 2-D imagers: frame transfer, line transfer, and charge injection, which are illustrated in Figs. 7.21(a)–(c). The frame transfer architecture works in the imaging mode by holding the pixel potential high during exposure. Then the accumulated photogenerated charge is clocked out after each exposure. The line transfer architecture transfers photogenerated charge to an adjacent CCD through a single gate. We will concentrate on the line transfer architecture since it is essentially embodied in the linear self-scanned array, and it is frequently used in current commercial 2-D CCD imagers. The 2-D CID imager is a unique architecture, so it will be treated separately from the 2-D CCD imager.

In the linear array, each address switch is turned on and off in sequence by a pulse derived from an on-chip shift register running at a prescribed clock rate. A start pulse derived from the shift register initiates each scan of the array for continuous operation. The address switch MOSFET array is positioned next to the photosites but is covered by a light shield. Two shift registers can be used with two sets of address switches and output buses, one set addressing even and the other odd sensor elements. This allows for narrower bandwidth amplifiers for lower noise. Typical characteristics for a linear array are given in Table 7.3.

Charge transfer efficiency (CTE) is a measure (or percentage when multiplied by 100) of charges successfully transferred to the next cell in a CCD transfer register under the action of a clock drive signal. It is related to the charge transfer inefficiency, ε, which is a measure of the charge left behind, so that CTE $= 1 - \varepsilon$. ε is typically 10^{-4} or less. Since charge is generated by incident light, charge transfer inefficiency results in some degradation of the image sharpness (or MTF) of a CCD camera. (See Problem 12.)

Area arrays are more complicated since each address component must be placed in close proximity to its sensing element, thus putting it into the photosensitive area. Two address switches are needed for each photodiode unless a separate control lead is provided to each

Table 7.3
Typical Linear Self-Scanned Array Characteristics

Characteristics	Value(s)
Dynamic range	1500:1
Saturated exposure	0.07 $\mu J \cdot cm^{-2}$
RMS noise equivalent exposure	9×10^{-5} $\mu J \cdot cm^{-2}$
Charge transfer efficiency (CTE)[a]	9.9999×10^{-1}
Peak-to-peak temporal noise	1.0 mV
Photoresponse nonuniformity	60 mV
Saturation output voltage	1.5 V
Responsivity	3.0 V/($\mu J \cdot cm^{-2}$)

[a]CTE = $1 - \varepsilon$.

pixel. Each row can be enabled as a unit and each column subsequently addressed by using two switches per element. Only the element where both row and column switches are closed is readout. All others have at least one switch open and are isolated and actively collecting photogenerated carriers. A vertical shift register provides a pulse to turn on the MOSFET switch gates of an entire row at a time through a single control bus per row. The horizontal shift register is operated at higher speed but synchronously with the first. It is used to bias a column of enable switch-drains at one time and turns on a single address switch at the selected intersection. Therefore, every pixel in an $m \times n$ array can be addressed by $m+n$ control buses but requires two control elements at each photodiode location. Some performance characteristics for a typical CCD imager are given in Table 7.4.

The typical CCD imager shown in Fig. 7.22 has several major components on the chip, including:

- Image sensor elements
- Vertical analog transport registers

Table 7.4
Typical CCD Performance Characteristics[a]

Characteristics	Value(s)
Frame rate	30 Hz
Field rate	(2:1 Interface) 60 Hz
Line rate	15.75 KHz
Pixel rate	7.16 MHz
Saturation output voltage	0.7 Vpp
Saturation exposure	0.28 $\mu J/cm^2$
Responsivity	2.5 V/($\mu J \cdot cm^{-2}$)
Horizontal pixel (resolution)	380
Vertical pixel (resolution)	488
Aspect ratio	4:3
Dynamic range	≥ 30 dB
Array size	8.8×11.4 mm

[a]For Fairchild CCD.

7.7 LINEAR AND MATRIX ARRAYS   187

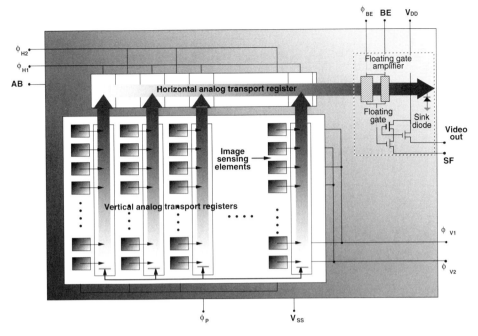

Figure 7.22 2-D (matrix) array of photodetectors and analog shift registers making up typical CCD focal plane.

- Horizontal analog transport registers
- Floating gate amplifier

The heart of the shift registers is an array of metal oxide semiconductor (MOS) capacitors. Each element looks like that pictured in Fig. 7.23 (for a two-phase device). A charge packet of minority carriers is formed under the metal electrode when a voltage is applied. Charge transfer is achieved with a series of gates (MOS capacitors) sequentially biased by a clock-driven voltage waveform. The basic CCD consists of an input section which provides minority carriers, a transfer section which transports the charge packets, and an output section which recovers the analog signal. The transfer section consists of metal electrodes, which control potential wells at the semiconductor–oxide interface. When proper voltages are applied sequentially to the electrodes, the potential wells are essentially moved along the structure, and the charge packets follow.

As mentioned earlier, time delay and integration (TDI) is used to enhance signal-to-noise after signal processing. For monolithic detector arrays (in which both detector and CCD shift-register are made of silicon) the TDI mode operates as follows. A charge packet is generated at the detector site from exposure to an image point scanned across it. The CCD is clocked so that the transfer of these charge packets down the CCD to the output amplifier is synchronous with the scan rate of the image point. This causes the signal charge of the first detector to be delayed so that its charge reaches the output gate associated with the second detector at the same instant that direct photoinduced charge is generated at that second detector, and so on. The signal charge from both detectors (and all subsequent detectors) is combined accordingly.

188 LIGHT SOURCES AND DETECTORS

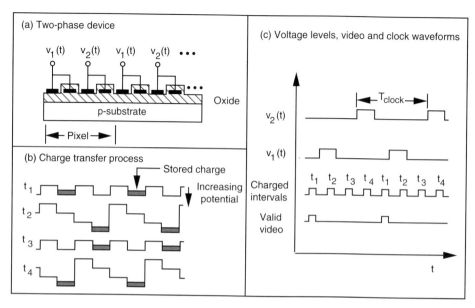

Figure 7.23 (a) Typical layout of MOS charge transfer cell for a two-phase device, (b) corresponding charge transfer process, and (c) clock drive, charging intervals, and video waveforms for a CCD.

This accumulation of charge acts as a photogenerated charge integrator as a function of space location in the scene or image. Thus TDI can be considered an optoelectronic correlator or charge integrator. The result for imaging is to enhance the signal-to-noise ratio by the square root of the number of detector elements involved in TDI, assuming that the noise from each detector is uncorrelated.

In charge injection device (CID) imaging the operation is different. The CID uses *intra*cell charge transfer and charge injection to achieve image sensing. Again, photon-generated charge signals are collected and stored in an array of MOS charge storage capacitors. The level of signal charge is detected *in situ* so that an excess charge transfer structure is unnecessary. Charge injection into the underlying semiconductor is used to clear the sensing region of accumulated signal charge and sometimes to provide a mechanism for readout. Either linear addressing (line imager) or 2-D coincident addressing (area imager) can be achieved.

A MOS capacitor operated in a CID mode is illustrated in Fig. 7.24. A pulsed drive voltage induces a displacement current. This displacement current is integrated over the injection interval. The difference between levels before and after injection is proportional to the net injected charge. Charge is removed from the sensor site by injecting minority carriers from the surface inversion region into the underlying semiconductor.

A simplified (4 × 4) CID imaging array is illustrated in Fig. 7.25. In operation, photons penetrate the sensor surface and generate carriers, which are collected by charge storage areas for each pixel. Each sensing site is covered by two gate electrodes side-by-side as suggested in the figure. The right-hand gate of each pair is connected to the row bus at each location, and the left-hand gate is connected to the column bus. The column storage area of each pixel and the interconnection of each for pixels in that column is ordered by a single strip of high conductivity polysilicon, which vertically crosses the thin oxide regions of all pixels in a column of the CID matrix. Column and row storage areas are dielectrically separated. Charge collected under a

Figure 7.24 Charge injection device mode of operation.

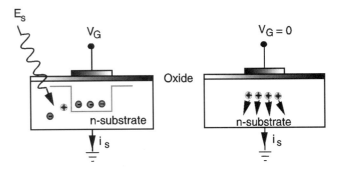

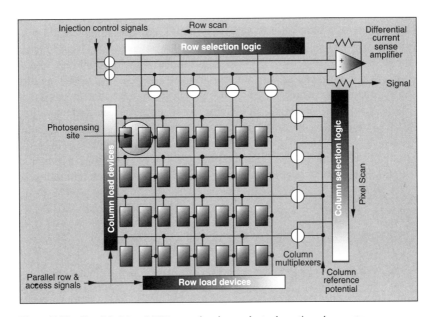

Figure 7.25 Simplified 4 × 4 CID array showing readout schematic and geometry.

specific pixel's electrode area can be transferred or injected between column and row of the pixel storage area. This results in no spacing between pixels and thus minimum obscuration of incident illumination. Rows and columns form a differential row-read CID imager. Specific columns and rows are addressed through column and row multiplexers by a logical "1" output from the column and row selection logic. Simultaneous selection of a column and row causes signal charge collected under the column electrode to be transferred to the row electrode. Then it is sensed and amplified by a high-gain, low-noise preamplifier. On-chip noise, especially fixed pattern noise, is removed at the imager output. Typical CID performance characteristics are summarized in Table 7.5.

The basic noise sources in CCDs (and CIDs) are the photodetectors and readout electronics. In the detector there is shot noise from incident (signal or background) illumination as well as shot noise due to leakage or dark current. The expression for shot noise variance

Table 7.5
Typical CID Performance Characteristics[a]

Characteristics	Value(s)
Number of pixels	248 (V) × 388 (H)
Pixel size	27 μm × 23 μm
Image size	11 mm (diagonal)
SNR	50 dB (at saturation)
MTF	> 80% at 250 lines
Spectral range	350–1100 nm
Sensitivity	1 V at 0.8 foot-candles

[a]GE CID

due to incident illumination shown in Table 7.6 indicates that it varies with incident irradiance H and photo responsivity R, whereas that due to dark current is fixed at a given operating temperature. Typically the product τR is ~ 2.5 mJ/cm^2, and dark current density (I_D/A_d) is ~ 10–100 nA/cm^2, where τ is response time, I_D is dark current, and A_d is detector area. Readout noise includes contributions from the on-chip MOSFET amplifier as well as any off-chip preamp, which vary with design but might contribute ~ 10–100 rms electrons to the total charge variance. Δf_n is the noise bandwidth, C_s is the gate capacitance, I_n is the noise current, and g_m is the transconductance gain. There are other contributions to CCD noise, such as generation-recombination noise, trapping noise, and reset noise. The key differences between CCDs and CIDs are that, although CIDs have higher random noise, they have lower dark current, lower nonuniformity, and lower susceptibility to blooming.

7.8 OPTICAL SIGNAL PROCESSING REQUIREMENTS

The optical signal processing requirements placed upon electro-optical interfaces differ from those requirements that were originally used to develop most such devices because the original

Table 7.6
Basic Contributions to Noise Variance of a CCD (or CID) Imager

Noise Type	Charge Variance
Signal shot noise	$\sigma_1^2 = HR\tau/e$
Dark shot noise	$\sigma_2^2 = I_D\tau/e$
On-chip MOSFET	$\sigma_3^2 = V_{n^2}\Delta f_n C_{s^2}/e^2$
Off-chip preamp	$\sigma_4^2 = I_{n^2}\Delta f_n C_{s^2}/g_{m^2}e^2$

where the total noise (charge) variance is the root-mean-square (rms) value:

$$\sigma_{\text{tot}}^2 = \frac{1}{4}\sum_{i=1}^{4}\sigma_i^2 \tag{7.36}$$

assuming all the noise terms are statistically independent.

requirements were driven by communications or sensing applications. For instance, laser diodes were created, in many instances, to support fiber-optic communications, and focal plane arrays were usually developed to support TV-compatible imaging. Sometimes the optical processor designer can live within the constraints of existing devices by employing clever architectures or selecting the less demanding application, but, in general, this is not possible.

The primary optical signal processing requirement for increased signal-to-noise ratio drives-up the necessary power and saturation levels on laser diodes as well as the dynamic range on spatial light modulators and focal plane arrays. For coherent optical processing, the demand for increased coherence (narrow-band signal-mode) of the laser source is difficult to meet with a single high-power laser diode. To circumvent this limitation, we must resort to pulsed sources and multiple emitters, both of which must maintain excellent coherence. Coherence lengths of 1–10 cm and output powers of 1–10 W are desirable depending on the application. To be competitive with or better than digital parallel processors, spatial light modulators generally need dynamic ranges of at least 30 dB, although some processing techniques have been considered where only one to four discrete levels of dynamic range are necessary (phase-only matched filtering, binary correlation, low-bit level Fourier transforms, etc.). Acousto-optic modulators, however, may provide more dynamic range (e.g., 50 dB or more) when addressed by signals containing only a few frequencies. Focal plane arrays provide approximately 25–30 dB of dynamic range for imaging, but much more is needed for optical processing (as much as 50 dB or more for spectral analysis). The noise sensitivity (e.g., noise equivalent power) required of focal plane arrays is also a key factor, especially if a single source (e.g., laser diode) is used to illuminate a relatively large aperture 2-D optical processor. Reduction in intensity, as seen in the focal plane, occurs as a result of beam spreading and optical propagation losses within the architecture.

Another key parameter for optical processing is space-bandwidth product. This parameter is simultaneously a requirement on the number of elements (or pixels) and spatial resolution (pixel size) when translated into device geometrical parameters. In matched filtering (correlation), the time–bandwidth product is also a measure of the gain or peak-to-side lobe ratio (i.e., matched filter output signal-to-noise ratio). (Time-bandwidth and space-bandwidth are essentially the same parameter since temporal electrical signals often get converted to spatial optical modulation in optical processing architectures.) Space-bandwidth product is equal to the number of pixels or resolution cells available to partition up the signal, which may be measured along one axis of an image, correlator output, or power spectrum. Typically focal plane arrays provide 400^2–500^2 pixels in two dimensions, and linear self-scanned arrays have 1000–2000 pixels. Some focal plane arrays (for astronomical imaging) have 1000^2 pixels.

In many optical processing applications the photodetector array is the limiting factor. For spectrum analysis, for instance, both high dynamic range and high frame rates are required. For correlation these requirements are not as great. In fact, increased dynamic range and sensitivity result from using heterodyne AO spectrum analyzers over direct detection (power) spectrum analyzers. This results from the fact that in heterodyne detection the photodetector current is proportional to the square root of the product of the diffracted signal and reference signal, whereas in direct detection it is proportional to just the signal power. Theoretically, then the dynamic range of the heterodyne detector is the square of the dynamic range of the direct detector. In practice, the margin of gain is only 15–20 dB because of limited reference signal power. In addition to enhanced dynamic range, the heterodyne detection mode maintains phase coherence, which is good for spectrum analysis as well as frequency excision (since complex Fourier transforms can be obtained).

To handle large dynamic range, compression of the signal at the detector is required. This is frequently done using a logarithmic transfer characteristic. If an 8-bit A/D converter is subsequently used, the least significant bit is set to the noise level and the compression is set to yield a maximum voltage corresponding to level 255.

In spectrum analysis, Johnson (thermal) noise is the primary source of noise in the detector and amplifier. In an optimum receiver it is desired to make the amplifier dominant. To do so requires a large load resistance for the detector but a small input amplifier capacitance so that bandwidth is not compromised.

It is desirable that many computational operations implemented with optical processing components be performed in the shortest possible time. For example, state-of-the-art acousto-optic processors can easily perform 500-point Fourier transforms in less than 1 μsec. A linear photodetector array matched to this performance must readout (serially) these 500 points in the same time (2 nsec), which is far in excess of the standard video bandwidth of 5–10 MHz. Typical detector integration times are only ~100 nsec per pixel. Thus, new architectures for parallel readout must be developed that essentially readout at least 50–100 channels per frame time. Other types of optical processing require just the opposite: long integration times, such as for time integrating correlators, that is, 1–100 msec. Thus, the speed of readout is a function of the application, the associated data acquisition time, and required integration time to achieve sufficient signal-to-noise.

PROBLEM EXERCISES

1. What is the linewidth in nm and Hz for a GaAlAs laser of 830-nm mean wavelength and 10-cm coherence length?

2. Light that undergoes spontaneous emission in an LED is incoherent and decays exponentially in time. Hence the electric field for a single frequency carrier obeys the relation

$$E(t) = E(0) \exp(j\omega_0 t) \exp(-t/\tau_{\text{spont}}) u(t)$$

where $u(t)$ is the (single-sided) unit step function. Derive the normalized power spectrum of this light and show that its spread in frequency Δf about the center frequency f_0 is $\frac{1}{2}\pi \tau_{\text{spont}}$ using the criterion that at $\Delta f/2$ away from f_0 the power is down by 3 dB.

3. In a laser diode cavity of length L and attenuation constant α the index of refraction is $n = 3.5$. For each round-trip in the cavity, write an expression for the amplitude decay. If the amplitude gain per unit distance is g, what is the expression for the gain on each round trip? Assuming that these two processes dominate, combine them to specify the threshold condition of oscillation and thus show that the threshold gain is

$$g = \alpha - \frac{1}{L} \ln r$$

For $L = 400$ μm and $\alpha = 20$ cm^{-1}, find g.

4. In a laser diode the incremental efficiency in converting electrons into output photons is the external quantum efficiency η_{ex} given by

$$\frac{1}{\eta_{\text{ex}}} = \frac{1}{\eta_i}\left(1 - \frac{\alpha L}{\ln r}\right)$$

where $\eta_i (\sim 0.75)$ is the probability that an injected carrier will recombine radiatively in the active region. Using the expression for η_{ex} and wavelength of 830 nm, calculate the power output per applied current above threshold, using the parameters of Problem 3.

5. In a laser diode the resonant frequency as indicated in Fig. 7.3 determines approximately the limit of the modulation bandwidth. It is given by the expression

$$\omega_o = \sqrt{\frac{AP_o}{\tau}}$$

where A is a constant related to laser operation and is typically 2×10^{-6} cm^3/sec, P_o is the output power and τ is the photon lifetime. Given that

$$\tau = \frac{n}{c}\left(\alpha - \frac{1}{L}\ln r\right)^{-1}$$

where n, α, L, and r are given in Problem 3, calculate f_o given $P_o = 10^{15}$ photons/cm^3.

6. The modes in a GaAlAs laser diode are defined by those values on integer p that satisfy the resonance condition $\beta L = p\pi$, where β is the propagation constant, which for a sufficiently thin active region is given by $\beta = 2\pi n/\lambda$, corresponding to the lowest excited mode. Then $p = 2nL/\lambda$. To obtain the wavelength spacing between modes, differentiate this expression with respect to λ and show that for no dispersion we have

$$\Delta\lambda = \frac{\lambda^2}{2nL}$$

between adjacent modes. What is the value of $\Delta\lambda$ for $L = 400$ μm and $\lambda = 830$ nm? Compare this to the 10-dB bandwidth of the Fabry–Perot cavity response [Equation (7.4)] for a contrast (F) corresponding to $n = 3.5$.

7. For the pulsed GaAlAs laser diode described in Problem 6 the Fabry–Perot cavity modes $\lambda = 2nL/p$ are perturbed by transient temperature (T) changes. Differentiate λ with respect to T to obtain $d\lambda/dT$. Assuming that $dn/dT = 6 \times 10^{-8}$ K^{-1}, that the dL/dT term can be neglected, and noting that the intensity of interfering beams is proportional to their phase difference $\Delta\phi$ given by

$$\Delta\phi = \frac{2\pi}{\lambda}\text{OPD}$$

where OPD is the optical path difference (cavity round trip times the index of refraction), determine $d(\Delta\phi)/dT$, assuming a cavity length of 400 μm and an emission wavelength of 830 nm. For $\Delta T = 1$ K, what is $\Delta\lambda$ and $d(\Delta\phi)$? Interpret your results.

8. In calculating the far-field diffraction pattern of a laser diode linear array shown below, the total optical intensity in the far-field is given by

$$I(\theta) \propto \cos^2(\theta)|E(\theta)|^2 A(\theta)$$

where $E(\theta)$ and $A(\theta)$ are defined at point P, and $\cos^2\theta \approx 1$ for small θ. If a single emitter is selected with an electric field at the diode facet given by

$$E(y) = E(0)\exp[-\gamma y^2/2]$$

derive the element factor defined by

$$E(\theta) \propto \cos\theta \int_{-D/2}^{D/2} E(y) \exp(-j2\pi y/\lambda)\, dy$$

and consequently $|E(\theta)|^2$, where D is the element spacing. For identical multiple emitters (e.g., $N = 10$), show that the array factor is

$$A(\theta) = \frac{1 - \cos(2\pi ND \tan\theta/\lambda)}{1 - \cos(2\pi D \tan\theta/\lambda)}$$

For $\lambda = 850$ nm, $N = 10$, and $D = 10$ μm, where are the diffraction peaks versus θ? Plot the array factor.

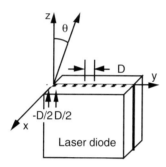

9. Calculate the current induced in a single silicon photodetector element in the focal plane of an optical processor using Equation (7.23), assuming that a laser diode illuminates the optical processor input plane with continuous power of 5 mW. The wavelength is 830 nm, and each detector has a quantum efficiency of 0.2 and a carrier mobility of 1.3×10^3 cm^2/Vsec and is biased to 1 volt. The focal plane is 1 cm^2 in area, contains $(512)^2$ photodiodes with no deadspace, and operates as standard video frame rates. Neglect optical train losses and assume that the laser light illuminates the focal plane evenly, and ignore spectral dependencies.

10. In the previous problem determine the mean noise power for shot-noise-limited operation using Equation (7.31) for a response time of $\tau = 10$ μsec and electrical bandwidth of $\Delta f = 1$ Hz. For a load resistance of 1 Ω, what is the power signal-to-noise ratio?

11. A simple model for the noise equivalent circuit of a CCD is shown below. The voltage source $\langle V_{n^2} \rangle$ in this model is assumed to be associated with Johnson (thermal) noise. Hence

$$\langle V_n^2 \rangle = 4k_B TRB$$

where R is the equivalent CCD channel resistance and B is the bandwidth. Since the output circuit acts as a low-pass filter on $\langle V_n^2 \rangle$ with frequency response $H(f)$, the output mean-squared noise voltage for such a bandwidth is given by $\langle V_n^2 \rangle H(f)\,df$. Derive $H(f)$ explicitly and calculate the mean-squared noise voltage $(\langle V_n^2 \rangle)$ over all frequencies and show that it is independent of resistance. For operation at room temperature with a capacitance of 1 pF, show that $\langle N_n \rangle = 400$ rms electrons using the relation

$$\langle N_n \rangle = C\langle V_n \rangle / e$$

In CCDs an additional common noise term comes from the MOS switch capacitor with $C \approx 10$ pF typically. Calculate the corresponding noise voltage at room temperature given by $V_n = (K_B T/C)^{1/2}$.

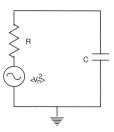

12. When used as an imager a CCD reduces resolution as expressed by its MTF. Its MTF is dominated by two terms due to geometry and charge transfer inefficiency. The MTF due to charge transfer is given by:

$$\text{MTF}_{\text{trans}} = \exp\{-n\varepsilon[1 - \cos(2\pi u)]\}$$

where $n =$ pixel (sample) number and $\varepsilon =$ charge transfer inefficiency ($\ll 1$). The MTF due to geometry is

$$\text{MTF}_{\text{geom}} = \frac{\sin\left(\frac{\pi u \Delta x}{x_o}\right) \sin\left(\frac{\pi v \Delta y}{y_o}\right)}{\frac{\pi u \Delta x}{x_o} \frac{\pi v \Delta y}{y_o}}$$

where Δx and Δy are detector element dimensions, x_o and y_o are spacings, and u and v are spatial frequencies. Derive (a) the MTF for the CCD geometry and (b) the MTF for charge transfer assuming the following model for charge transfer

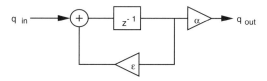

such that $\text{MTF}_{\text{trans}} = (q_{\text{out}}/q_{\text{in}})^k$ corresponding to k such transfers and assuming $fT = u$, where f is temporal frequency and T is the sampling time, and ε is the charge transfer efficiency per CCD cell.

13. Verify the result of Fig. 7.6 for a simulated noisy Gaussian signal using MATLAB. Vary the pinhole diameter D to find the optimum. How much light is lost versus residual noise variance?

14. For a silicon photovoltaic detector used as an element in an optical architecture, as pictured below, the induced current from its output terminals is given by $I_s = P_s \eta e$, where the quantum efficiency η is 0.2, and the incident light flux is $P_s = H_s A_D$. The detector area A_D is 3×10^{-4} cm^2, and the incident irradiance is nominally that of sunlight in the 0.4- to 0.8-μm (visible) band at the earth's surface—that is, $H_s = 6.23 \times 10^{-2}$ W/cm^2. Calculate (a) the corresponding incident photon arrival

rate, $Q_s = \lambda H_s/hc$, assuming $\lambda = 0.633$ mm; (b) the flux P_s; and (c) the induced current I_s. Calculate (d) the carrier generation rate $N = Q_s A_D n$, and compare this to the number of electrons available for conduction (nominally given by Avagadros' number) for a measurement interval (detector dwell time) of $\tau = 10$ μsec. The corresponding current from a photoconductive detector is given by $I_s = P_s \eta e G$, where $G = \mu \tau V_B$ is the photoconductive gain, $\mu = 1.3 \times 10^3$ cm^2/V · sec is the mobility for silicon, and $V_B = 1$ V is the bias voltage. Calculate (e) the gain G and compare the resulting induced currents and (f) the equivalent responsivity $R = I_s/P_s$ for the photoconductive detector in A/W. What is the fundamental expression for R in terms of η, λ, G, and the fundamental constants? (g) Calculate the resulting signal voltage across an $R_L = 50$ Ω load resistance. The resulting RC time constant is given by $\tau_{RC} \approx$ RLC, where C is the diode capacitance given by $C = \varepsilon A_D/w$, where $\varepsilon \sim 16$, $A_D = wl$, and $w = 3\mu$m is the depletion region width. (h) Calculate C, τ_{RC}, and the corresponding maximum effective bandwith B.

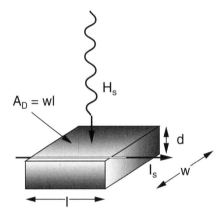

15. Assuming that the detector is background-noise-limited by the incident flux (assumed to be the nominal solar flux at the earth's surface in the previous problem), the shot noise is given by $i_s = e\eta P_s/h\nu$ and the corresponding rms noise current is $\langle i_N^2 \rangle_s^{1/2} = (2eBi_s)$. Calculate (a) the shot noise i_s, (b) rms noise current $\langle i_N^2 \rangle^{1/2}$, (c) the signal-to-noise ratio (SNR), and (d) the NEP $= 2h\nu B/\eta$ assuming B is the Nyquist bandwidth for $\tau = 10$ μsec. (e) Compare this to the thermal (Johnson) noise in the load resistor, given by $\langle i_N^2 \rangle_T^{1/2} = (4k_B T B/R)^{1/2}$ where $T = 300$ K. (f) If both noise sources are counted, what is the effective SNR? (g) What is noise voltage corresponding to each noise source and both combined? (h) Calculate the effective D^*.

BIBLIOGRAPHY

G. W. Anderson, F. J. Kub, and G. M. Borsuk, "Photodetector Arrays and Architectures for Acousto-Optical Signal Processing," *Opt. Eng.* **29**, 58 (1990).

D. F. Barbe, "Imaging Devices Using the Charge-Coupled Concept," *Proc. IEEE* **63**, 38–67 (1975).

D. Botez and D. E. Ackley, "Phase-Locked Arrays of Semiconductor Diode Lasers," *IEEE Circuits and Devices Magazine*, pp. 8–16 (January 1986).

G. M. Borsuk, "Photodetectors for Acousto-Optic Signal Processors," *Proc. IEEE* **69**, 100–118 (1981).

P. S. Cross, R. R. Jacobs, and D. R. Scifres, "Dynamite Diodes," *Photonics Spectra*, pp. 79–86 (September 1984).

P. Das, *Lasers and Optical Engineering*, Springer-Verlag, New York (1991).

H. Haken, *Laser Theory*, Springer-Verlag, Berlin (1970).

J. A. Hall, "Arrays and Charge-Coupled Devices," in *Applied Optics and Optical Engineering*, Vol. VIII, pp. 349–400. Academic Press, New York (1980).

M. Haney and D. Psaltis, "Coherence Properties of Pulsed Laser Diodes," pp. 24–30, Appendix IIIa in "Acousto-Optic Processing of 2-D Signals Using Temporal and Spatial Integration," Report for AFOSR Contract 82-0128, May 31, 1983.

T. V. Higgins, "The Smaller, Cheaper, Faster World of the Laser Diode," *Laser Focus World*, pp. 65–76 (April 1995).

G. C. Holst, *CCD Arrays, Cameras and Displays*, SPIE Optical Engineering Press and JCD Publishing, Bellingham, WA and Winter Park, FL (1996).

J. L. Jewell, A. Schere, S. L. McCall, Y. H. Lee, S. J. Walker, J. P. Harbison, and L. T. Florez, "Low Threshold Electrically-Pumped Vertical-Cavity Surface-Emitting Micro-Lasers," *Optics News*, Vol. 15, No. 12, pp. 25–28, 1989.

G. R. Olbright, "VCSELs Could Revolutionize Optical Communications," *Photonics Spectra*, pp. 98–101 (February 1995).

E. G. Paek, "Microlaser Arrays for Optical Information Processing," *Optics and Photonics News*, pp. 16–23 (May 1993).

L. J. Pinson, *Electro-Optics*, John Wiley & Sons, New York (1985).

G. H. B. Thompson, *Physics of Semiconductor Devices*, John Wiley & Sons, New York (1980).

D. R. Scifres, W. Streifer, and R. D. Burnham, "Experimental and Analytical Studies of Coupled Multiple Stripe Diode Lasers," *IEEE J. Quantum Electr.* **QE-15**, 917 (1979).

C. H. Sequin and M. F. Tompsett, *Charge Transfer Devices*, Academic Press, New York (1975).

J. N. Walpole, "R&D on Surface-Emitting Diode Lasers," *Laser Focus/Electro-Optics*, pp. 66–74 (September 1987).

Chapter 8
SPATIAL LIGHT MODULATORS

8.1 OVERVIEW

Spatial light modulators (SLMs) are devices that produce an output light distribution that results from modulating an input light distribution either optically or by electrical addressing. SLMs may be single-pixel, one-dimensional (1-D), two-dimensional (2-D), or three-dimensional (3-D). We will focus primarily on 1-D and 2-D SLMs, but 3-D SLMs (volume holographic elements) may play an increasingly important role in the future. Figure 8.1 illustrates in very simple terms the typical generic structure of 2-D SLMs. Note that in general, SLMs can be signal multiplying, self-emissive, signal amplifying, or self-modulating. Volume holographic devices consist of photorefractive or nonlinear optical materials. Self-modulating devices include bistable optical devices and passive self-pumped phase-conjugate mirrors. Self-emissive SLMs consist of cathode ray tubes (CRTs) and image intensifiers, both of which employ phosphor screens for image display.

Signal multiplying SLMs are generally three-port devices where the amplitude at each point in the output image is multiplied by an effective transmittance or reflectance of the modulating element or pixel. The amount of modulation may be determined by the write image intensity or by an electrical write signal at the corresponding pixel on the SLM. Electrical addressing requires a means of reading-in a 2-D spatial image usually from a 1-D serial signal.

Some examples of the more common 2-D SLMs include:

- *L*iquid *C*rystal *L*ight *V*alue (LCLV) and *L*iquid *C*rystal *T*ele*v*ision (LCTV)
- *M*agneto-*O*ptic *SLM* (MOSLM)
- *M*icrochannel *SLM* (MSLM)
- *D*eformable *M*irror *D*evice (DMD)
- *P*ockels *R*ead-*O*ut (Optical) *M*odulator (PROM) and *P*reobrasovatel *Iz*obrazheniy (PRIZ)
- *M*ultiple *Q*uantum *W*ell (MQW) SLM

Various parameters can be exploited to spatially modulate light, including intensity (or amplitude), phase, polarization, and spatial frequency. Although intensity and phase are

Figure 8.1 Generic 2-D SLM structure showing basic features of most reflective-type devices. Polarizers are especially useful when the modulating material rotates the plane of polarization as a function of bias volatge as in electrically addressed liquid crystal SLMs.

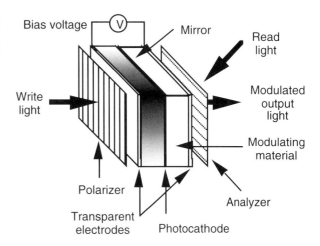

most often used, polarization and spatial frequency may play an intermediate role in effecting intensity or phase modulation.

Finally, in a special class by itself, is the acousto-optic Bragg cell. Although arrays of Bragg cells have been studied and developed recently, the 1-D Bragg cell is the most readily used and implemented in the domestic (U.S.) industry. Furthermore, many novel optical processing architectures have employed 1-D Bragg cells, and we will examine some of these architectures in detail later. Thus the Bragg cell is given special treatment. Then we will focus on some of the planar (2-D) nonholographic devices listed above.

Many fundamental operations performed by SLMs are driven by optical processing applications. SLMs often act as transducers, where they convert incoherent light to coherent light, provide input image amplification, and convert wavelength and reverse contrast. Arithmetic operations can be performed, including addition, subtraction, multiplication, and division. SLMs also have memory, which enables short-term storage, information latching, and low-level signal integration. As we will see, many SLMs possess nonlinear transfer characteristics (between input and output light), which are useful for thresholding, optical limiting (hard clipping with hysteresis), supporting decision processing (e.g., neural networks), and Boolean operations on multiple beams. Of course, linear transfer characteristics are also available and required for arithmetic operations. Finally, device-specific functions can be employed, such as edge detection (using the MSLM or PRIZ), intensity-to-spatial frequency conversion (using a variable grating liquid crystal modulator), and synchronous detection (using a MSLM).

In evaluating the performance of SLMs, there are several important parameters to consider. These are: write-light exposure, framing speed, spatial resolution, switching energy per pixel, contrast ratio or useful dynamic range, storage time, and readout power limitations. Switching energy per pixel is often the limitation on operating speed, especially for very large arrrays. Of course, lowering the switching energy also reduces the total power consumption. The key factors for real-time optical signal processing are spatial resolution, often expressed by the MTF, and dynamic range (or contrast ratio) expressed as $(I_{max} - I_{min})/(I_{RMS})$. A comparison of some of the more common SLMs is given in Table 8.1 for some of the aforementioned parameters.

SPATIAL LIGHT MODULATORS

Table 8.1
Comparison of Commercially Available SLM Characteristics

	Parameter					
	Frame Rate (Hz)	Number of Pixels	Resolution (lp/mm)	Peak Contrast	Addressing Scheme	Vendor
MSLM	7	256^2	10	3000 : 1	Optical	Hamamatsu
DMD	1200	128^2	10	30 : 1	Electrical	TI
PROM	10	350^2	25	—	—	—
MOSLM	200	256^2	13.2	2×10^4 : 1	Electrical	Semetex
	350	128^2	6	—	Electrical	Semetex
LCLV	10	$\sim 400^2$	16	50 : 1	Optical	Hughes
F-LCLV	165	128^2	3	150 : 1	Optical	STC Ltd
	4.6×10^3	400^2	28	20 : 1	Optical	a-Si
LCTV	30	$\sim 380 \times 230$	6×5	30–80 : 1	Electrical	Sharp

Before describing in detail the design and principles of operation of some of these 2-D SLMs, we will consider a unique 1-D SLM mentioned earlier: the acousto-optic Bragg cell.

8.2 ACOUSTO-OPTIC BRAGG CELLS

Sound waves can amplitude- and phase-modulate light, deflect and focus it, and shift its frequency. Diffraction of light by sound in a liquid or solid medium can take on several forms, depending on the relative wavelengths of light and sound as well as the dimensions of the region of interaction.

The two most important forms of sound–light interaction are the Debye–Sears (or Raman–Nath) effect and Bragg reflection. The Debye–Sears effect occurs when light is incident perpendicular to the propagation of sound waves in a slab of crystalline material as shown in Fig. 8.2. The sound wave in the medium varies the index of refraction n by a small amount Δn at a frequency ω_s so that

$$\Delta n(z, t) = \Delta \sin(\omega_s t - k_s z) \tag{8.1}$$

as shown.

For light propagating perpendicular to the sound wave, the medium looks like an optical delay line with phase shift varying with Δn as

$$\Delta \phi = 2\pi \frac{L}{\lambda_0} \Delta n \tag{8.2}$$

where L is interaction distance and λ_0 is the vacuum wavelength of incident light. Thus the acoustically excited crystal acts like a phase grating and essentially phase-modulates an incident plane wave in a manner similar to FM. As a result, sidebands are created in addition to the carrier (incident light wave). These sidebands are diffracted as output wave fronts at increasing angles from the original direction of the incident wave as shown in

Figure 8.2 Interaction between light and sound in an acousto-optic crystal in the Debye–Sears mode, showing sidebands (or multiple images) created in the output light beam.

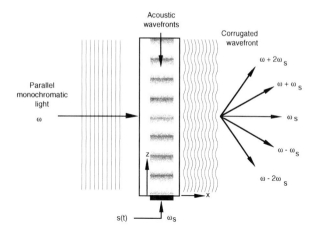

Fig. 8.2. As a result, multiple images of any spatial amplitude modulation superimposed on the incident plane wave front would be seen at the sidebands ($\omega \pm N\omega_s$) located at corresponding diffraction angles.

Further analysis shows that if the interaction length (L) is increased, higher-order sidebands are suppressed. This is unavoidable for practical incident beams of light where diffraction effects account for some minimal degree of beam spreading within the interaction length, leading to destructive interference. To get around this limit and obtain constructive interference, we must rotate the direction of the incident light until it makes an angle $\alpha \approx \lambda/2\lambda_s$ with respect to normal incidence, where λ_s is the acoustic wavelength in the medium. The upper first-order sideband, which makes an angle λ/λ_s with respect to the incident light now reflects from the acoustic wavefronts as if they were mirrors. Now, no matter how large L is, incident light and upper first-order sideband light remain in-phase along each acoustic wavefront, where the higher orders still suffer destructive interference.

This situation is known as the *Bragg condition* and is illustrated in Fig. 8.3. It is more useful for optical signal processing than the Debye–Sears condition. A parameter that distinguishes the two regimes is defined by

$$Q = \frac{2\pi}{n} \frac{\lambda L}{\lambda_s^2} \qquad (8.3)$$

where L is the interaction length. Q is roughly equivalent to the concept of quality factor associated with spectral sharpness of resonators.

For small L such that $Q < 1$ we have the Debye–Sears (or Raman–Nath) regime, but for L sufficiently large that $Q \gg 1$, the Bragg regime ensues. Then the interaction length is sufficient to suppress the higher-order diffraction terms. In this case we can visualize the sound waves as a series of partially reflecting mirrors separated by λ_s and moving at velocity v_s. The diffraction angle is θ_r. The necessary condition for diffraction to occur in a given direction is that all points on a given mirror contribute in-phase along the θ_r direction. This translates into

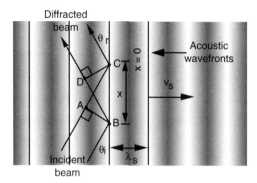

Figure 8.3 Bragg diffraction geometry showing incident and reflected beams in an acousto-optic crystal. The variation of the index of refraction is also shown.

requiring that the optical path difference (OPD) be an integral number of wavelengths along θ_r, namely

$$\text{OPD} = \overline{AC} - \overline{BD} = m\frac{\lambda_0}{n} \tag{8.4}$$

which is equivalent to

$$x(\cos\theta_i - \cos\theta_r) = m\frac{\lambda_0}{n} \tag{8.5}$$

where n is the index of refraction of the acoustic medium. The only way that this condition [Equation (8.5)] can be satisfied for all x is for $m = 0$. Then $\theta_i = \theta_r$. We also require that diffraction from any two acoustic phase fronts add-up in-phase along the direction of the reflected beam. As illustrated in Fig. 8.4, this is equivalent to requiring that the path difference between beams 1 and 2 given by AO + OB be λ_0/n. By inspection of the geometry, we have

$$2\lambda_s \sin\theta_B = \frac{\lambda_0}{n} = \lambda \tag{8.6}$$

which is precisely the Bragg condition.

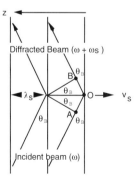

Figure 8.4 Bragg diffraction geometry showing reflection and reinforcement of two portions of an incident beam from two acoustic wavefronts, leading to the Bragg condition. θ is the Bragg angle θ_B.

EXAMPLE 8.1 Bragg Angle To get a feel for the magnitude of the Bragg diffraction angle, rewrite Equation (8.6) as

$$\theta_B = \sin^{-1}\left(\frac{1}{2}\frac{\lambda_o}{n}\frac{f_s}{v_s}\right)$$

For $\lambda_0 = 633$ nm (HeNe line), $n = 2.2$, $v_s = 3600$ m/sec (typical of LiNbO$_3$ materials as shown in Table 8.2), and $f_s = 2$ GHz, θ_B is

$$\theta_B = \sin^{-1}\left[\frac{6.33 \times 10^{-7} m}{2(2.2)} \frac{2.0 \times 10^9 \text{ Hz}}{3.6 \times 10^3 \text{ m/sec}}\right] = 4.58°$$

a rather small deflection angle.

The particle (or quantum mechanical) picture of Bragg diffraction is also a useful way of understanding the effects of Bragg cells. For incident light of wave vector (or momentum vector) $\mathbf{k}_i$ making an angle θ with respect to the normal to the sound propagation direction, as shown in Fig. 8.5(a), the diffracted beam $\mathbf{k}_d$ makes an angle $+\theta'$ with respect to the normal. The condition for Bragg diffraction requires that the triangle ABC in Fig. 8.5(b) be closed using an acoustic wave vector (or momentum wave vector) $\mathbf{k}_s$. We simply apply the conservation of energy and momentum to the Bragg condition embodied in the vector diagram of Fig. 8.5(a), which resembles the vector diagram for a two-particle (photon/phonon) collision as suggested by Fig. 8.5(b).

Conservation of momentum requires $\mathbf{k}_d = \mathbf{k}_i \pm \mathbf{k}_s$ and conservation of energy requires $\omega_d = \omega_i \pm \omega_s$, where + corresponds to an approaching light beam and − corresponds to a receding light beam (relative to the acoustic beam). The frequency relation says that diffracted light is Doppler-shifted up or down when light meets or follows sound, respectively. The wavenumber equation is simply equivalent to the Bragg condition provided that $\mathbf{k}_s \ll \mathbf{k}_i$, which is generally true. Thus in that case we obtain $\mathbf{k}_d \approx \mathbf{k}_i$. Furthermore, by virtue of the frequencies involved ($f_s \leq 10^{10}$ Hz and $f_i \geq 10^{13}$ Hz), we obtain $\omega_d \approx \omega_i$; that is, $\omega_s \ll \omega_d$ or ω_i. Now the Bragg condition is written

$$k_s = 2k_i \sin\theta \tag{8.7}$$

and is equivalent to Equation (8.6) because $\mathbf{k}_s = 2\pi/\lambda_s$ and $\mathbf{k} = 2\pi/\lambda$. Finally, using the

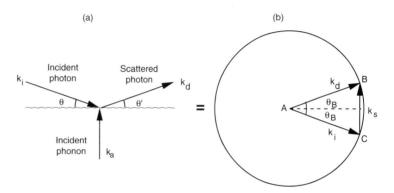

Figure 8.5 (a) Photon–phonon interaction diagram and (b) the corresponding momentum conservation triangle diagram for the Bragg condition.

relations $v_s = \lambda_s \omega_s / 2\pi$ and $c = \lambda \omega / 2\pi$ and Equation (8.7), the amount of Doppler shift can be determined to be

$$\Delta \omega = \frac{2\pi v_s}{\lambda_s} = \omega_s \tag{8.8}$$

which is the induced Doppler shift of light from sound waves moving with velocity v_s in a medium of index of refraction n, a result consistent with the relation $\omega_d = \omega_i \pm \omega_s$. This Doppler shift is generally not important compared to the frequency of light, as can be seen in the following example.

EXAMPLE 8.2 Doppler Shift in Bragg Cell   For light of HeNe wavelength $\lambda_i = 633$ nm, the frequency is $f_i = c/\lambda_i = 4.74 \times 10^{14}$ Hz ≈ 500 THz, whereas the Doppler shift for a modulation frequency of $f_s = 50$ MHz is $\Delta f = f_s = 50$ MHz. Therefore the relative Doppler shift is

$$\frac{\Delta f}{f_i} \approx \frac{50 \text{ MHz}}{500 \text{ Thz}} = 10^{-7}$$

an extremely small number. Even the highest-frequency state-of-the-art devices have relatively small Doppler shifts; for example, for $f_s = 2$ GHz, $\Delta f/f_i \approx 4 \times 10^{-6}$ at $\lambda_i = 633$ nm.

In practice, a Bragg cell is operated as suggested in Fig. 8.6. Sound waves are excited in the acoustic medium by a piezoelectric transducer that is driven by the product of an amplitude modulating input signal $s(t)$ and a carrier signal $\cos(\omega_c t)$. This product can be obtained using a mixer (e.g., single sideband or double sideband) as shown. For a Bragg cell of length w, an acoustic wave of velocity v_s, and a piezoelectric transducer located at position $x = 0$, the diffracted light amplitude at position x and time t immediately after the Bragg cell is

$$A(x, t) = A(x) s(t - T - x/v_s) \exp[\pm j\omega_c (t - T - x/v_s)] \tag{8.9}$$

The factor $A(x)$ is the variation in amplitude due to nonuniform illumination, acoustic attenuation, and other spatial nonuniformities. T is the aperture time of the Bragg cell, given by $T = w/v_s$. The sign of the exponential argument indicates the sign of the diffracted order (± 1).

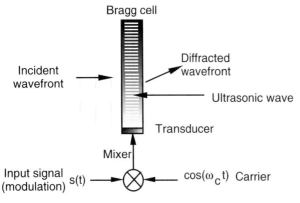

Figure 8.6 Schematic of Bragg cell operation showing electrical, acoustic and optical components. Deflection angle of diffracted beam is somewhat exaggerated for purposes of illustration.

8.2 ACOUSTO-OPTIC BRAGG CELLS

We can derive a more specific result for a practical Bragg cell modulation from Equation (8.9) by the following analysis. If an amplitude-modulated waveform is applied to the transducer as described previously, then a complex wave amplitude transmittance function that is created in the Bragg cell is given by

$$t(x,t) = \exp[js(x - v_s t)] \operatorname{rect}(x/w) \tag{8.10}$$

where we have temporarily ignored the spatial-dependent factor $A(x)$ here, except for the truncating effects of the Bragg cell window expressed by $\operatorname{rect}(x/w)$. $s(x)$ is the transducer-induced elastic strain field in the acousto-optic medium at time $t = 0$ due to the amplitude modulation $v(t)$ at the mixer, v_s is the velocity of the strain field (i.e., acoustic velocity), and w is the width of Bragg cell.

The strain field at the transducer end of the Bragg cell is directly proportional to the input signal voltage $v(t)$; that is, $s[(-w/2) - v_s t] = \varepsilon v(t)$, where ε is the conversion efficiency of the piezoelectric transducer. The input to the cell is a radio-frequency waveform modulated in magnitude and phase, which induces a strain wave (at $t = 0$) given by

$$s(x) = |a(x)| \cos[\omega_c x + \alpha(x)] \tag{8.11}$$

where $\omega_c = 2\pi f_c$ and f_c is the RF carrier spatial frequency. We will use this result in Equation (8.10). First note that

$$t(x,t) = [1 + js(x - v_s t) + \text{higher-order terms}] \operatorname{rect}(x/w) \tag{8.12}$$

by expanding the exponential in Equation (8.10) as a series. If the modulation amplitude $a(x)$ is small, then the higher-order terms can be neglected. Then

$$t(x,t) = \{1 + j|a(x - v_s t)| \cos[\omega_c (x - v_s t) + \alpha(x - v_s t)]\} \tag{8.13}$$

using Equation (8.11). Then expanding the cosine in exponentials yields

$$\begin{aligned} t(x,t) = \{ &1 + j|a(x - v_s t)| \exp[j\omega_c(x - v_s t)] \exp[j\alpha(x - v_s t)] \\ &+ j|a(x - v_s t)| \exp[-j\omega_c(x - v_s t)] \exp[-j\alpha(x - v_s t)] \} \operatorname{rect}(x/w) \end{aligned} \tag{8.14}$$

Defining $s'(x)$ as $s'(x) = a(x) \exp(j\omega_0 x)$, where $s(x) = \operatorname{Re}[s'(x)]$, we can write Equation (8.14) as

$$t(x,t) = \left[1 + j\frac{1}{2} s'(x - v_s t) + j\frac{1}{2} s'^*(x - v_s t) \right] \operatorname{rect}(x/w) \tag{8.15}$$

provided that we identify $a(x)$ as complex—that is, $a(x) = |a(x)| \exp[j\alpha(x)]$. The above result (8.15) is appropriate for a thin acoustic beam as seen in the Debye–Sears (or Raman–Nath) mode. In the Bragg regime, only one or the other (s or s^*) terms survive. The downshifted (-1) component at $\omega_i - \omega_s$ is

$$t(x,t) = \left[1 + j\frac{1}{2} s(x - v_s t) \right] \operatorname{rect}(x/w) \tag{8.16}$$

whereas the upshifted component (+1) at $\omega_i + \omega_s$ is

$$t(x,t) = \left[1 + j\frac{1}{2}s^*(x - v_s t)\right] \text{rect}(x/w) \quad (8.17)$$

[assuming that we revert to the real part of $s'(x)$]. We will use these results later in analyzing acousto-optic signal processors.

For optical signal processing applications there are several important performance parameters for acousto-optic (AO) cells. These are diffraction efficiency (η), bandwidth (Δf_s), aperture time (T), and dynamic range (DR).

When operating AO cells for signal processing, they are often used essentially as light deflectors, but they can also be used as just beam modulators. These two modes of operation are shown in Figs. 8.7(a) and 8.7(b).

When used as modulators the acoustic velocity (v_s) should be high, and beam waist (w_B) should be small. The beam risetime (τ) is $\tau = aw_B/v_s$, where a is a parameter chosen to be in the range of 1.3–3.0, where a larger value of a implies a sharper focus. This equation assumes a Gaussian beam waist and a risetime defined over 10–90%. When used as deflectors we desire to get as many beam positions (or resolvable spots) as possible in the output plane. For small angles we obtain $\theta_B = \pm \lambda/2\lambda_s$, so that for $\lambda_s f_s = v_s$ we have

$$\theta_B = \pm \frac{\lambda}{2v_s} f_s \quad (8.18)$$

Therefore, over the bandwidth (BW) = Δf_s the change in beam angle is

$$\Delta \theta = 2\theta = \frac{\lambda}{v_s} f_s \quad (8.19)$$

For an input laser beam of diameter D the acoustic velocity is $v_s = D/T$. Then

$$\Delta \theta = \frac{\lambda}{D} T \Delta f_s \quad (8.20)$$

But we already know that the diffracted beam divergence is $\sim \lambda/D$. Therefore the total number of beam positions (or resolvable spots) is

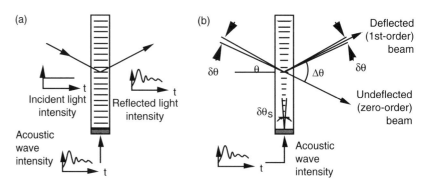

Figure 8.7 Acousto-optic Bragg cell used as (a) a light modulator and (b) a beam deflector.

$$N \approx \frac{\Delta\theta}{(\lambda/D)} = T\Delta f_s \qquad (8.21)$$

which is the time–bandwidth product. Thus the time–bandwidth product will be large if T is large, which requires a low value of v_s (just the opposite of the modulator requirement).

EXAMPLE 8.3 Time–Bandwidth Product of Bragg Cell in Deflector Mode Typical values of aperture time T and bandwidth Δf_s are 40–80 μsec and 24–50 MHz, respectively, for commercial Bragg cells, using the slow-shear wave TeO_2 material (see Table 8.2, p. 211). Thus the time–bandwidth product of such a Bragg cell is $T\Delta f_s = 960$–4000. Recall that we derived a relationship between time–bandwidth product and illuminating laser beam spread $\Delta\theta$ and diffraction-limited resolution (λ/D), namely,

$$T\Delta f_s = \frac{\Delta\theta}{(\lambda/D)}$$

where we used the relation $v_s = D/T$. For a slow-shear TeO_2 Bragg cell considered here we have $v_s = 620$ m/sec. Thus the required Bragg window size (and laser beam waist) is $D = v_s T \approx 2.5$–5 cm. This is the minimum required size of the TeO_2 crystal. The matching laser beam waist corresponds to a fairly wide collimated fan beam.

We have addressed two of the performance parameters (bandwidth and aperture time) for Bragg cells. For signal processing, both should be large (but can be traded-off for a fixed time–bandwidth product). The larger the time–bandwidth product, the higher the frequency of operation, the better the resolution, and the longer the signal integration period can be on any given application, whether it is imaging, spectrum analysis, or correlation.

The next important performance measure is diffraction efficiency—that is, how much light gets diffracted into the first order (at $\pm\theta_B$). The diffraction efficiency is determined by solving the optical wave equation for propagation through an acousto-optic medium. Without going into detail on this, we will simply summarize the method. The optical field is expanded in a Fourier series, which leads (in the Bragg limit) to a pair of coupled first-order differential equations. The solutions of these equations yield intensities of diffracted and undiffracted optical beams relative to the incident beam. The ideal efficiency of light diffracted into the first order is

$$\frac{I_1}{I_o} = \eta \operatorname{sinc}^2[\eta + (\Delta k L/2)^2]^{1/2} \qquad (8.22)$$

where I_1 is the first-order diffracted light intensity, I_o is the incident light intensity, Δk is momentum mismatch between incident light and acoustic propagation vectors (as described in Fig. 8.5), L is the interaction length, and

$$\eta = \frac{\pi^2}{2\lambda^2} M_2 \frac{L P_s}{H} \qquad (8.23)$$

is the diffraction efficiency, where H is the piezoelectric transducer height, P_s is the acoustic power (W), and M_2 is the AO figure of merit.

The geometry for this situation is depicted in Fig. 8.8. As can be seen from inspecting Equation (8.23), η depends not only upon geometry but upon acoustic power and the AO figure

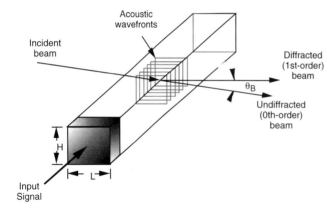

Figure 8.8 Acousto-optic crystal and geometry for Bragg diffraction.

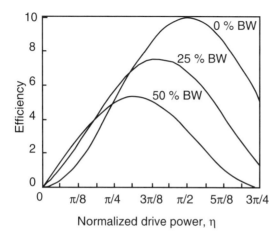

Figure 8.9 Diffraction efficiency versus normalized acoustic drive power for various percentage bandwidths ($100 \times \Delta f/f_c$). The best operating point is at approximately 0.5 efficiency.

of merit M_2. At low acoustic power the efficiency is linearly proportional to the square root of power and thus linearly proportional to the amplitude of the input electrical signal. As it turns out, diffraction efficiency is reduced by acoustic attenuation and diffraction, and it depends on electrical (modulation) bandwidth through a tradeoff discussed later. With these provisos in mind, Fig. 8.9 shows the approximate dependence of η on P_s.

EXAMPLE 8.4 Diffraction Efficiency of Bragg Cell In the small signal limit the diffraction efficiency of a Bragg cell simplifies to

$$\eta = \frac{\pi^2}{2\lambda^2} M_2 \frac{L}{H} \eta_d \eta_a \eta_o \eta_c P_{\text{RF}}$$

where $\eta_d (\sim 0.8)$ is the acoustic diffraction efficiency, $\eta_a (\sim 0.6)$ is the acoustic attenuation efficiency, $\eta_o (\sim 0.9)$ is the optical absorption efficiency, and $\eta_c (\leq 1.0)$ is the RF to acoustic conversion efficiency. η_c relates P_{RF} to P_s [as used in Equation (8.23)]; that is, $P_s = \eta_c P_{\text{RF}}$. We will assume $\eta_c = 1$. Using $L/H = 4$, where L and H are defined in Fig. 8.8, and assuming

$\lambda = 633$ nm (working at HeNe wavelengths) and $M_2 = 34.5 \times 10^{-15}$ s³/kg (for slow-shear Te O$_2$ as given in Table 8.2, p. 211), we obtain

$$\eta = \frac{\pi^2}{2(633 \text{ nm})^2}(34.5 \times 10^{-15} \text{s}^3/\text{kg})(4)(0.8)(0.6)(0.9)(1.0) \approx 0.74 \text{W}^{-1}$$

That is, the diffraction efficiency is 74% per watt of injected RF power at the input port to the Bragg cell. This is a typical quoted manufacturer value. To obtain the diffraction efficiency at another value of wavelength—that is, 830 nm (GaAlAs laser diode)—we simply calculate the ratio

$$\frac{\eta_{830}}{\eta_{633}} \approx 0.8 \left(\frac{\lambda_{633}}{\lambda_{830}}\right)^2 = 0.34$$

where the extra factor of 0.8 is an empirical conversion factor.

M_2 is the most often used figure-of-merit in the literature. It relates diffraction efficiency *directly* to acoustic power for a given device aspect ratio (L/H) in the small-signal limit, namely,

$$\eta = \frac{\pi^2}{2\lambda^2} M_2 \frac{L}{H} P_s \qquad (8.24)$$

M_2 is given by

$$M_2 = \frac{n^6 p^2}{\rho v_s^3} \qquad (8.25)$$

where n is the index of refraction, ρ is the density of material, P is the strain-optic (or elasto-optic) coefficient, and v_s is the acoustic velocity. P is actually a tensor because crystal orientation plays a role in determining M_2. Typical values assume a crystal orientation has been chosen to maximize M_2, in which case for simplicity in calculating approximate performance the elasto-optic tensor **P** can be approximated by the scalar value P_{max}.

Having introduced diffraction efficiency, we need to reconsider bandwidth because it is intimately related to achieving the Bragg condition and must be traded-off with diffraction efficiency. To get exact Bragg matching, expressed by Equation (8.6), we must change the illumination angle. To achieve this, while holding the incident optical beam fixed, requires that the acoustic wave vectors have a wide angular spectrum. Then

$$\Delta k = 2|\mathbf{k}|\Delta\theta_s \sin\theta_B \qquad (8.26)$$

where $\Delta\theta_s$ is the acoustic beamwidth. $\Delta\theta_s$ obeys diffraction (based on the acoustic aperture length) and thus is given by $\Delta\theta_s = \lambda_s/L$.

Diffraction efficiency, however, is proportional to transducer length [see Equation (8.24)]. Therefore by gaining bandwidth, diffraction efficiency is lost. This is reflected in the family of curves of Fig. 8.9 for various normalized bandwidths. Bandwidths of ~500 MHz and diffraction efficiencies of ~10% can be achieved without special measures. One measure to improve bandwidth and efficiency involves segmenting the acoustic transducers and steering the acoustic beam via relative phase shifting between segments. This technique allows the Bragg condition to be obeyed over a wide bandwidth with a narrow $\Delta\theta_s$ and large interaction length (L) for a high diffraction efficiency (η). Another technique involves using anisotropic media where the idea of "tangential phase matching" can be employed to allow a wide range of k_s values to still satisfy the Bragg condition.

The last performance parameter that we will consider is the dynamic range of Bragg cells. Dynamic range is generally limited by intermodulation products appearing in the diffracted light coming from the Bragg cell. Intermodulation products can be caused by RF amplifier electronics, multiple diffraction within the AO phase grating, and nonlinear piezoelectric, photoelastic, and acoustic effects. The dominant effect is via the two-tone third-order intermodulation products ("intermods") produced by the nonlinear acoustic effect. Acoustic attenuation also reduces dynamic range across the Bragg cell aperture. Under these circumstances the dynamic range is given by

$$\text{DR(dB)} = 10 \log_{10}\left(\frac{\eta_{\text{Fnd}}}{\eta_{\text{IMP}}}\right) - 10 \log_{10}(\alpha T) \tag{8.27}$$

where η_{IMP} is the diffraction efficiency of intermods, η_{Fnd} is the diffraction efficiency of fundamental, α is the acoustic attenuation constant/unit time, and T is the aperture time.

When two frequencies (f_1 and f_2) are restricted to an octave bandwidth, the magnitude of the two-tone third-order product (at $2f_1 - f_2$) is

$$\frac{I}{I_1} = \frac{I_1 I_2}{36} \tag{8.28}$$

where I_1 is the fraction of light diffracted by the acoustic wave at frequency f_1, and I_2 is the fraction of light diffracted at f_2.

EXAMPLE 8.5 Dynamic Range of Bragg Cell Assuming two-tone third-order intermodulation distortion as the primary factor in limiting dynamic range, we can easily calculate a typical baseline dynamic range for a Bragg cell. Assuming that the percentage of light diffracted into frequencies f_1 and f_2 is $I_1 = I_2 = 2\%$, we obtain

$$\frac{I}{I_1} = \frac{(2 \times 10^{-2})^2}{36} = 1.11 \times 10^{-5}$$

Therefore the dynamic range is given by:

$$\text{DR} = \left|10 \log_{10}\left(\frac{I}{I_1}\right)\right| = 49.5 \text{ dB}$$

As it turns out, there is a tradeoff between DR and diffraction efficiency (η). For instance; $\eta = 1\%$ for DR $= 52$–53 dB and $\eta = 5\%$ for DR $= 40$ dB, assuming slow-shear TeO$_2$ as the material and that the two-tone third-order intermod is dominant (as derived from typical manufacturer specifications).

The attenuation constant is also important in determining the aperture time; that is, the maximum aperture time is generally limited by the allowable acoustic attenuation. The attenuation constant is given by

$$\alpha \left[\text{Nepers/unit time}\right] = \frac{\gamma^2 \omega_s^2 K T}{\rho V^4} \tag{8.29}$$

where γ^2 is a proportionality constant, T is the temperature, K is the thermal conductivity, and ω_s is the angular frequency of an acoustic wave.

A representative set of material parameters for the more common AO materials is given in Table 8.2.

EXAMPLE 8.6 Acoustic Attenuation for Typical Bragg Cell For slow-shear TeO$_2$ the attenuation factor is $\alpha' = 17.6$ dB/μsec · GHz^2, and the actual attenuation constant is $\alpha = \alpha' T (f_1^2 - f_2^2)$. For $T = 70$ μsec, $f_1 = 100$ MHz, and $f_2 = 50$ MHz, α is

$$\alpha = \left(17.6 \frac{dB}{\mu sec \cdot GHz^2}\right) 70 \, \mu sec \left[(100 \times 10^{-3} \, GHz)^2 - (50 \times 10^{-3} \, GHz)^2\right]$$

$$\alpha = 9.24 \, dB$$

Thus for propagation across a Bragg cell aperture of 70 μsec the dynamic range (or signal level) is reduced by ~90% due to acoustic attenuation. Thus, if the Bragg cell were imaged using an incident plane wave, the far-end of the image (with respect to the transducer end) would be dimmer by 90%.

Table 8.2 summarizes some of the parameters of AO materials.

Some data on commercially available AO Bragg cells are summarized in Table 8.3. Among these a slow-shear wave TeO$_2$ Bragg cell is shown as an example in Fig. 8.10. It has an aperture of 42 mm, an aperture time of 70 μsec, a center frequency of 75 MHz, a 3-dB bandwidth of 50 MHz and a dynamic range in excess of 50 dB at 5% diffraction efficiency.

8.3 LIQUID CRYSTAL SPATIAL LIGHT MODULATORS

Electro-optical devices can control light by applying a voltage across a crystal that subsequently rotates linearly polarized light that passes through it. The degree of rotation varies with applied voltage. Such a change in polarization can be converted to a change in light amplitude output by using a polarizer on the input side and an analyzer on the output side. Liquid crystal devices

Table 8.2 Acousto-optic Materials

Material	Acoustic Mode and Axis[a]	Figure of Merit (M_2) (× 10^{-18} s^3/g)	Acoustic Velocity (V_S) (× 10^3 m/s)	Acoustic Attenuation (α') (dB/μsec·GHz[b])	Index of Refraction	Optical Band (μm)
GaP	L(110)	44.6	6.32	3.80	3.31	0.6–10
	S(110)	24.1	4.13	1.96	3.31	
LiNbO$_3$	L(100)	7.00	6.57	1.0	2.2	0.5–4.5
	S(100)35°	14.0	3.6	1.0	2.2	
TeO$_2$	L(001)	34.5	4.2	6.3	2.26	0.35–5.0
	S(110)[b]	793.0	0.62	17.6	2.26	0.35–5.0

[a]L, longitudinal; S, shear.
[b]Birefringent.

Figure 8.10 Commercial slow-shear TeO$_2$ Bragg cell.

Table 8.3
Summary of Typical Commercial Grade Bragg Cells

Type	Center Frequency (GHz)	Bandwidth (MHz)	Time Aperture (μsec)	Diffraction Efficiency (%/W)[a]
GaP (L)	1.0–2.0	500–1000	0.15–0.5	18[b]–50
LiNbO$_3$ (L)	1.0	500	2.0	2.5
LiNbO$_3$ (S-35°)	2.3–2.5	1000	0.3–0.4	16[b]–32
TeO$_2$ (SS)	0.045–0.075	24–50	40–80	1–250
TeO$_2$ (L)	0.4	200	1.0–5.0	60–100

[a]Measured at 633 nm (except as noted).
[b]Measured at 815 nm.

operate in just such a fashion. A liquid crystal material is modulated optically or electrically with a signal that creates an image or pattern across its surface. The light to be modulated is passed through the liquid crystal so that the polarization at each point is rotated by an angle that depends on the light intensity or electrical amplitude applied to the liquid crystal at that point. Again a polarizer, used in the exit path of the light, creates an amplitude modulation.

There are two key effects used in certain liquid crystals. These are the twisted nematic effect and optical birefringence, which are important in understanding the principles of operation of the liquid crystal SLM. The liquid crystal layer is fabricated in a twisted alignment configuration, which causes the polarization direction of linearly polarized incident light to rotate exactly through the twist angle. The liquid crystal molecules at the electrodes are aligned with their long axes parallel to the electrode surfaces. The twisted alignment configuration is

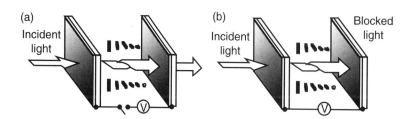

Figure 8.11 Nematic liquid crystal device showing the liquid crystal orientation (a) with and (b) without an applied electric field.

obtained by orienting the two electrodes so that the liquid crystal molecules rotate through a 45° angle as the space is traversed between the two electrodes In the nematic phase there are weak interactive forces between the ends of the elongated molecules that make up the crystal as illustrated in Figs. 8.11(a) and 8.11(b). As a result, applied electric and magnetic fields can modify the orientation of these molecules and rotate the polarization of light passing through according to the angle the twist is "pinned" by brushed-on alignment layers, which induce correct relative alignments at the cell boundaries. Linearly polarized light is rotated by 45° for molecular orientation at 45°, yielding a total rotation of 90°. A cross-polarizer is used to block this light. The application of an electric field also creates birefringence in the liquid crystal. Thus light becomes elliptically polarized before passing through the cross-polarizer, so that a portion passes through as a function of the applied voltage, as illustrated in Fig. 8.12 for a reflective liquid crystal SLM.

The twisted nematic effect is enabled in the off-state (no voltage on the liquid crystal). For the on-state the pure optical birefringence effect of the liquid crystal is manifested. This dual-state operation is referred to as the *hybrid field effect mode*. To further clarify this effect, first consider the off-state. As shown in Fig. 8.12(a) a crossed polarizer/analyzer pair is placed between the light valve and the readout light source. The polarizer is placed in the incident beam parallel to the first electrode twist prealignment layer, and the analyzer is placed in the reflected beam. After its first pass through the liquid crystal layer the direction of polarization of the linearily polarized incident light is rotated through 45°. The light passes a second time through the liquid crystal, after reflection from the dielectric mirror, and its polarization is thus rotated back to the direction of the incident light, where it is blocked by the crossed analyzer. Therefore the off-state is a dark state and is determined entirely by the twisted nematic effect. For the on-state, voltage is applied and the positive dielectric molecules rotate. However, if full voltage were applied, then the molecules would rotate to a perpendicular alignment leaving the polarization of the light unaffected by the liquid crystal and a dark on-state would occur as well. But there exists a voltage regime between full "on" and full "off" where the device

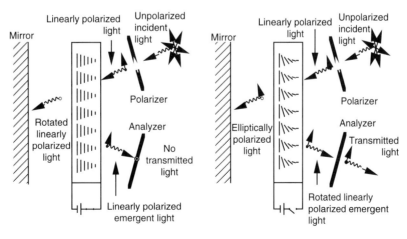

Figure 8.12 Operation of liquid crystal pixels with polarizers used to convert polarization modulation into amplitude modulation. "Off" state is (a) and "on" state is (b).

will transmit light through the crossed analyzer. In this orientation of the molecules, between parallel and perpendicular, the optical birefringence of the molecules can affect the polarization of the light. As a result, at these intermediate voltages the light that emerges from the device after reflection from the mirror is no longer linearly polarized, so that transmission can occur [see Fig. 8.12(b)].

Actual construction of a liquid crystal light valve (LCLV) is shown in Fig. 8.13. It consists of a number of thin film layers sandwiched usually between two high-quality glass substrates. In this case the SLM oprerates in a reflection mode. An electric field is applied across the transparent electrodes using a low-voltage (5- to 10-V rms) audio-frequency power supply. The transparent electrodes are made from indium tin oxide. The write light entering from the left (which can be either coherent or incoherent) is applied to the cadmium sulfide (CdS) photoconductive layer. A dielectric mirror and a cadmium telluride (CdTe) light-blocking layer optically isolate the photoconductor from the readout light. The CdTe layer is needed because the dielectric layer that follows the liquid crystal layer cannot completely block the input light. It also enables simultaneous reading and writing of the SLM without regard to the spectral content of the input and readout light beams. It has more than four orders of magnitude isolation between the input light beam and the readout light. The dielectric mirror consists of alternate quarterwave films of high and low refractive index materials. The mirror can be "tuned" to reflect any portion of the visible spectrum. The dielectric mirror also prevents the flow of DC current through the liquid crystal, thus requiring AC operation. The thickness of the liquid crystal is typically 2 μm.

Typical performance parameters and other features are summarized in Table 8.4.

A systems-level description of the characteristics of the LCLV for optical processing and imaging are provided by the MTF and the input/output (I/O) transfer characteristic of the

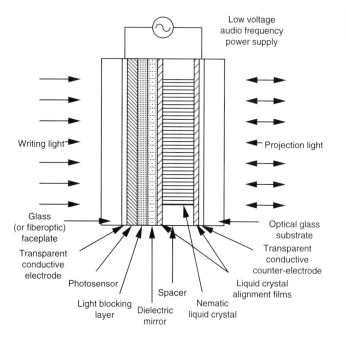

Figure 8.13 Construction of liquid crystal light valve showing the various layers.

Table 8.4
Liquid Crystal Light Valve Parameters

Resolution:	>40 line pairs/mm
Contrast:	>100:1
Speed:	<50 msec (cycle time)
Sensitivity:	~1 erg/cm^2 (at threshold)
Aperture size:	≤50 mm (diameter)
Response time:	40 msec (rise), 30 msec (fall)

device. A simple Gaussian model MTF is plotted in Fig. 8.14 (and a circular aperture MTF is shown for comparison), and the I/O characteristic modeled as a sigmoidal function is plotted in Fig. 8.15.

For all SLMs, contrast decreases as the spatial frequency increases, although when peak contrast values and resolution at the visibility limit are quoted, they may not be obtained simultaneously. The thickness of the SLM material affects the device spatial resolution, with thin materials generally exhibiting higher resolution. Sensitivity is measured in units of microjoules/cm^2 (as the reciprocal of the exposure needed to change the intensity by $1/e$). The required cycle time (write, read, and erase) is generally determined by the source of the input data and the required processing (for example, TV raster). Minimum input signal duration is usually fixed by the sensitivity of the SLM, whereas the response time of the SLM determines when the read cycle can start.

A popular variation on the LCLV is the electrically addressable LCTV, whose pixel structure is shown in Fig. 8.16. An actual commercial device is shown in Fig. 8.17. This device is useful for modest performance optical bench testing of optical processing concepts.

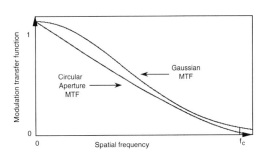

Figure 8.14 Modulation transfer function of a typical LCLV. A typical value for f_c is given in Table 8.4.

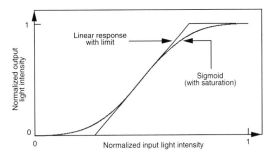

Figure 8.15 Input/output transfer characteristic of a typical LCLV. A linear transfer characteristic is shown for comparison.

216 SPATIAL LIGHT MODULATORS

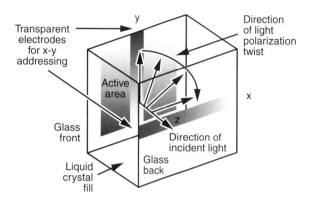

Figure 8.16 Electrically addressed liquid crystal SLM (LCTV) showing details of pixel construction and electrical addressing.

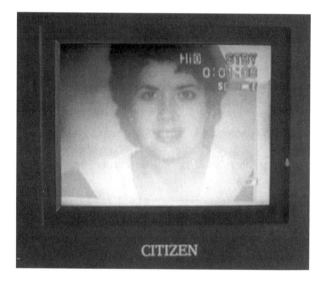

Figure 8.17 Photograph of commercial LCTV device.

It employs an optically transmissive matrix array of x–y electrically addressable pixels to impart modulation to the liquid crystal pixels and hence to a beam of light passing through it. By removing the display unit from a commercially available LCTV module, the device can be used in a transmissive mode to spatially modulate light using standard (television) video input.

8.4 MAGNETO-OPTIC SPATIAL LIGHT MODULATOR

The magneto-optic spatial light modulator (MOSLM) is functionally similar to the LCTV in that it consists of a two-dimensional array of electrically addressable pixels that affect the polarization state of light transmitted through each pixel. It differs from the LCLV in its operating principle in that it employs the magneto-optic effect in single-crystal thin films of bismuth-doped iron garnet, which have a high degree of Faraday rotation under applied

8.4 MAGNETO-OPTIC SPATIAL LIGHT MODULATOR

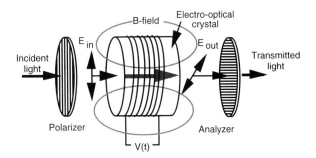

Figure 8.18 Faraday rotation effect whereby the plane of polarization of polarized light is modified by a longitudinal magnetic field.

magnetic fields. The Faraday effect results when an appropriate material (e.g., gadolinium–gallium–garnet or iron–garnet) is subjected to a magnetic field along the direction of light propagation, as depicted in Fig. 8.18. The plane of polarization of incident linearly polarized light is rotated by an angle θ as a linear function of the applied H-field in the crystal, namely, $\theta = \theta_F t = VHt$, where θ_F is the Faraday rotation coefficient, V is the Verdet constant, H is the magnetic field, and t is the thickness of film. Values of V for typical materials such as glass and quartz are 4.64–9.28 and 4.81 radians/Tesla-meter, respectively. The total rotation angle between two states, which can be clockwise ($+\theta_F t$) and counterclockwise ($-\theta_F t$) is $2\theta_F t$ and depends on the magnetization of the film.

By analyzing the polarization state of the transmitted light as shown in Fig. 8.19(a), we can control the transmission according to the relation

$$T(\theta_F) = \sin^2(2\theta_F t) \tag{8.30}$$

Linearly polarized light entering a MOSLM pixel can be rotated in three different ways as shown in Fig. 8.19(b). The material in the appropriate pixel site can be magnetized in two orthogonal directions (saturated state) or left in an unbiased (neutral) state. Orientation of the analyzer polarization relative to the three states is also shown in Fig. 8.19(b). As a result, we can get three intensity levels, which, with the use of polarizers, corresponds to relative electric field values of $+1$, 0, and -1. In essence then the MOSLM is a bipolar three-level modulator. Operation of the MOSLM in a binary (± 1) mode is illustrated in Fig. 8.20. Absorption also occurs in the magneto-optic film. Therefore the total transmission in the on-state is

$$T_{ON} = \exp(-\alpha t) \sin^2(2\theta_F t) \tag{8.31}$$

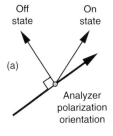
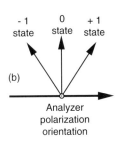

Figure 8.19 (a) Setup to polarize and analyze light passing through a magneto-optic crystal in a binary mode, and (b) bipolar ternary mode.

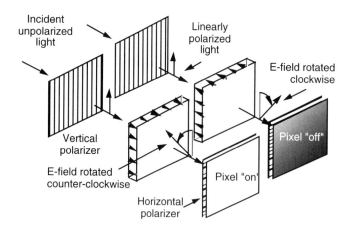

Figure 8.20 Operation of bipolar binary state MOSLM.

To get the overall contrast ratio obtainable with the MOSLM, we have to know the transmission in the off-state, which is

$$T_{\text{OFF}} = R_1 R_2 \exp(-3\alpha t) \sin^2(2\theta_F t) \tag{8.32}$$

where the extra factors of $R_1 R_2$ and $\exp(-3\alpha t)$ are due to multiple reflections and attenuation between the substrate and epitaxial magneto-optic film layers. Then the contrast ratio (CR) is

$$\text{CR} = \exp(2\alpha t)/R_1 R_2 \tag{8.33}$$

showing that the contrast can be strongly influenced by the use of antireflection coatings. Values of CR as high as 50:1 for white light and 1000:1 for coherent light have been obtained with 512×512 arrays switching at ~ 100 times/μsec.

To obtain a 2-D array of pixels the magneto-optic film is etched down to the substrate, and metallic conductors are deposited and patterned to supply drive currents. The spacing between pixels must be two to three times the thickness due to the etching process. Pixels are typically 50×50 μm. The resulting grid structure will decrease optical transmission through the device and causes the output pattern to be replicated due to diffraction. As a result, the fraction of total light incident onto the array that is diffracted into the central diffracted order (which is usually the only one of interest) is given by

$$T_{\text{ON}} = \frac{w^4}{b} \exp(-\alpha t) \sin^2(2\theta_F t) \tag{8.34}$$

where w is the window width and b is the pixel spacing. Typical values of w (56 μm) and b (76 μm) yield a further transmission factor of 29%. To obtain multiple gray levels, multiple cells must be used to form one pixel. Thus, if n cells are used for one pixel, $2n$ gray levels are formed but at the expense of an n-fold reduction in spatial resolution.

8.5 OTHER SPATIAL LIGHT MODULATORS

Several other SLMs will be summarized here. These include the microchannel SLM (MSLM), the deformable mirror device (DMD), the Pockels readout optical modulator (PROM) and a related SLM called PRIZ, and the multiple quantum well (MQW) SLM.

8.5 OTHER SPATIAL LIGHT MODULATORS 219

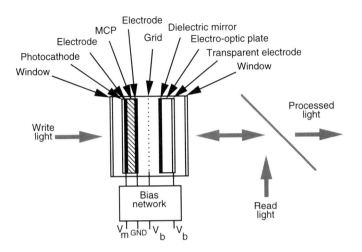

Figure 8.21 Construction of MSLM showing basic components and features.

The MSLM, shown in Fig. 8.21, uses a photocathode to convert incident light into electrons. A microchannel plate (MCP) provides electron gain, and the resulting electron "image" is deposited onto the surface of a crystal ($LiNbO_3$) which modulates the readout light via the longitudinal electro-optic effect. A grid enables the active removal of electrons as a refresh mechanism.

The DMD is an electronically addressed SLM which combines very large scale integration (VLSI) electronics and mechanical motion of pixels to modulate light. Mechanical pixel motion modulates the phase of an incident read beam. VLSI electronics in the form of MOSFET array voltages determines the degree of deflection of mirrors represented as pixel-sized cantilevers, 4 per pixel attached on pedestals as shown in cross section in Fig. 8.22. There is also a variation on this SLM that uses a deformable membrane mirror.

The complete DMD consists of a hybrid integrated circuit containing an array of metalized polymer mirrors bonded to a silicon address circuit. The underlying analog address circuit, which is separated by an air gap from the mirror elements, causes the array of mirrors to be displaced in selected pixels by electrostatic attraction. The resultant two-dimensional displacement pattern yields a corresponding phase modulation pattern for reflected light. This pattern may be converted into analog intensity variations by Schlieren projection techniques and used as the input transducer for an optical information processor.

The organization of the DMD array with electrical addressing is shown in Fig. 8.23. The $x-y$ array of mirror/air-gap capacitors are line-addressed by an underlying array of MOS

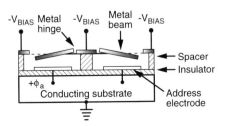

Figure 8.22 Detail of four-petal pixel used in the DMD.

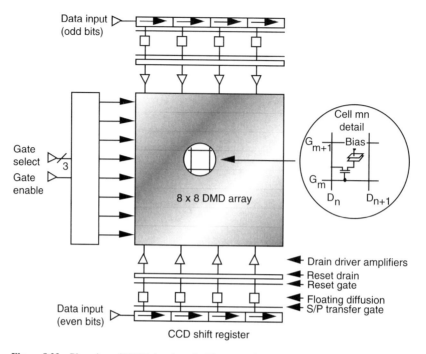

Figure 8.23 Plan view of DMD showing pixel layout and x–y addressing.

transistors. The drains (D_n) of the address transistors are connected to an analog (or digital) serial-to-parallel convertor. The gates (G_m) are connected to a decoder. Although the system described is electrically addressed, optical addressing is also possible.

The PROM is a photorefractive device that uses photoconductive and electro-optic effects and is depicted in Fig. 8.24. It is signal multiplying and nonholographic. It uses a thin (∼500 μm) wafer of photorefractive BSO ($Bi_{12}SiO_{20}$) with a pair of transparent electrodes separated from the crystal by paralene blocking layers (∼5 μm thick). Blue or near-ultraviolet light incident on the device generates a image distribution of electron–hole pairs within the crystal bulk. Mobile electrons drift through the bulk to the electrodes under an applied electric field. The space-charge field setup by the immobile hole pattern and transparent electrodes modulates the refractive index of the crystal and thereby phase-modulates the readout light, which is usually red. This phase modulation is converted to amplitude modulation using polarizers. The relationship between output amplitude and input light intensity is given by the function $\sin[\pi(V(x, y)/2V_\pi]$, where the voltage distribution $V(x, y)$ is induced by an electric field $E(x, y)$ proportional to the charge distribution $\rho(x, y)$ induced by the input exposure $I(x, y)$. (V_π is the voltage required to induce a π phase shift—that is, half-wave delay, hence the term "half-wave" voltage). The PRIZ is very similar to the PROM and originated in the Soviet Union. (Note that PRIZ stands for *Preobrasovatel Izobrazheniy*, which means "image transducer" in Russian.) It employs either BSO ($Bi_{12}SiO_{20}$) or BGO ($Bi_{12}GeO_{20}$) crystals, and is also nonholographic and photorefractive.

Multiple quantum-well (MQW) devices are double heterojunction structures fabricated into ultrathin (∼10 nm) layers of semiconductor material (e.g., GaAs and $Al_xGa_{1-x}As$). These layered structures have unique electrical and optical properties that enable them to be used as lasers, photodetectors, and spatial light modulators, depending on their configuration. They offer high speed, high sensitivity, and large time–bandwidth products.

MQW SLMs are the most advanced SLMs and promise to provide very high-speed capability and excellent sensitivity. As a spatial light modulator, such a device is constructed as shown in Fig. 8.25. When an electric field is applied perpendicular to the well, the electrons and holes are pulled apart to some degree, making it energetically more favorable to form an exciton, which is a favored bound state in quantum wells. Since the exciton manifests itself through optical absorption, the absorption peak (the characteristic signature of incident photons interacting with the exciton bound state) moves down in energy with increasing electric field strength as shown in Fig. 8.26. Since the absorption is changing at a specific wavelength (λ_0), the transmission of light passing through the MQW structure should be modulated accordingly.

In early MQW devices, light propagated perpendicular to these layers. This results in a limited interaction length (between light and the MQWs) equal to the epitaxial thickness, which is typically less than 1 μm. In a waveguide configuration, however, light travels parallel to the layers for much longer interaction distances, giving rise to much higher modulations. Another attractive feature of the waveguide configuration is the integration of the modulator and laser diode on the same structure. MQW materials compromise spectral bandwidth in

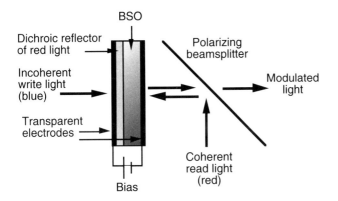

Figure 8.24 Basic structure of the PROM SLM.

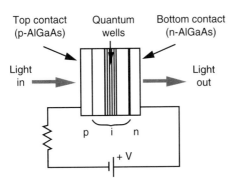

Figure 8.25 Basic diagram of multiple quantum well device used for spatial light modulation.

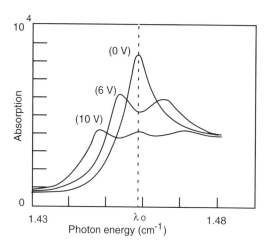

Figure 8.26 Absorption spectra of quantum well diode at different bias voltages. Note the change in absorption at a particular photon energy.

order to achieve a high electro-optic (electroabsorptive) effect at low bias voltages and short (submicrosecond) response times. Charge-coupled devices (CCDs) and QWs can be integrated on the same substrate in a single monolithic structure in which the CCD signal charge controls the optical transmission through a MQW by alternating the electric field across the MQW. This structure shows promise as an electrically addressed high-speed SLM.

Optically controlled MQWs are termed *self-electro-optic effect devices* (SEEDs). They can be used for optical logic operations in optical computing applications. SEED devices incorporate an optical detector and a quantum well modulator. They operate using the Stark effect, and combine optical detection with optoelectronic feedback to achieve bistable operation. As a bistable device the SEED can act as a switch with times ranging from microseconds to seconds or as an oscillator operating up to high frequencies (> 1 MHz). Absorbed power is proportional to the drive current giving self-linearized modulation.

Two-dimensional arrays have been fabricated with a limited number of pixels—for example, 6 × 6 arrays with 30 μm × 30 μm active areas spaced on 90-μm centers. In operation, two light beams are incident upon each device being tested: a visible control beam and a transmitted infrared (IR) beam. A binary thresholded version of an image can be impressed on the transmitted IR beam or an image can be superimposed upon the IR beam, yielding an AND function between the visible and the IR images. Arrays can also be operated linearly such that power subtracted from the IR beam is proportional to the incident visible light, although maximum contrast obtainable is less than 2:1. Thus, one drawback to QW technology is the limited contrast ratio ($\sim$2:1 to 3.5:1). Contrast ratios as high as 8:1 have been achieved with internal reflectors. Further contrast ratio improvements can be obtained by increasing the width and thickness of the QW region as well as fabricating arrays into a waveguide-type configuration.

In summary, it is clear that a number of alternative technologies have been employed to create 2-D SLMs. Only a few of these technologies have matured to practical levels of use in optical processing. Even those SLMs used in practice do not achieve the desired level of throughput, as measured by number of pixels times speed. This short-fall can be related to the fundamental tradeoff between speed of response and electro-optical efficiency. Solid-state modulators may be fast but they lack in diffraction efficiency, whereas liquid crystals are more efficient but slower. Of course this difference is due to the different physical mechanisms employed. As it turns out in liquid crystals, the large electro-optic coefficient results from the fact that an entire molecule is rotated by the applied electric field in a liquid with small elastic constants but slow response. In solids, optical modulation is achieved through relatively small displacements of atoms or ions by the electric field. Because of the much larger elastic constants in solids the optical response is smaller but faster. Thus it can be argued that the product of electro-optic efficiency (i.e., polarizability) and speed of response (i.e., bandwidth) is basically constant for a given *spectral* bandwidth. This tradeoff is directly analogous to the gain–bandwidth product tradeoff in electronic amplifiers. When looking for the best spatial light modulator we should select one that balances the two abovementioned parameters in a way suitable for our needs in optical signal processing. MQW SLMs offer a compromise in this tradeoff potentially optimal for optical signal processing, in which electro-optic efficiency and speed are both maximized but at the expense of narrowed spectral bandwidth.

PROBLEM EXERCISES

1. Normally, laser light incident on a Bragg cell has an amplitude profile which is Gaussian (in the x axis) and has been compressed by a cylindrical lens in the y axis as suggested in the figure below. In addition, the effect of the finite aperture of the Bragg cell is to truncate the laser beam. Neglecting the required Bragg angle incidence and assuming a laser beam centered on the Bragg cell aperture, write an expression for the beam intensity profile $|U_{out}|^2$ after the Bragg cell along the x axis, where x is measured from the input piezoelectric transducer, D is the aperture length, and r_0 is the radius of the laser beam at the $1/e$ points. Sketch the amplitude profile U_{out} and label axes and key values. Show that the Fourier transform of the weighting function is

$$|W(f)|^2 = |W(0)|^2 \frac{A_1}{T^2 \pi^3} |1 + \exp(-\alpha\tau)|^2 \exp[(-2T^2 + \alpha\tau - \alpha^2\tau^2/4T^2)f^2]$$

for large f, where

$$A_1 = 4\left\{\text{erf}\left[1 + \frac{\alpha\tau}{4T^2}\right] + \text{erf}\left[1 - \frac{\alpha\tau}{4T^2}\right]\right\}^2$$

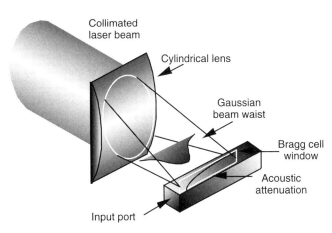

2. In describing the output beam profile U_{out} from a Bragg cell, it is necessary to include the effects of acoustic attenuation, in which the acoustically attenuated light amplitude profile is described

by $U_{in} \exp(-\rho x)$. Combine this effect with the result of Problem 1 to yield the total output beam amplitude. Include the appropriate factor to account for diffraction efficiency. By completing the square on all x-dependent terms, isolate all the x-dependence into the Gaussian and show that the center of the beam is $r_0^2 \rho/2$ closer to the input port and that the beam is attenuated by an overall factor that is independent of x.

3. The diffraction efficiency, accounting for acoustic attenuation is defined by

$$\eta = \frac{\int_0^\infty |U_{out}|^2 \, dx}{\int_0^\infty |U_{in}|^2 \, dx}$$

Show that η is given by

$$\eta = \exp\left[-\rho D \left(1 - \frac{\rho D}{8R^2}\right)\right] \left\{ \frac{\text{erf}[\sqrt{2}(R - \rho D/4R)] + \text{erf}[\sqrt{2}(R + \rho D/4R)]}{2\,\text{erf}[\sqrt{2}R]} \right\}$$

where

$$\text{erf}(x) = \frac{2}{\sqrt{\pi}} \int_0^x \exp(-x'^2) \, dx'$$

and $R = D/2r_0$ is the truncation ratio.

4. In the result of Problem 3, show that for $\rho \ll 1$ and $R > 1$ the diffraction efficiency reduces to $\eta = \exp(-\rho D)$. Under these conditions using a Fourier transform architecture consider the focused intensity profile of the 1st-order diffracted beam, given by

$$I(f_x) = \left| \int_{-\infty}^\infty U_{out} \exp(-j2\pi f_x x) \, dx \right|^2$$

where f_x is $x'/\lambda F$, x' is the coordinate in the focal plane, F is the focal length of the Fourier transform lens, and λ is the wavelength of the (laser) illumination. Evaluate this integral and plot the normalized intensity $I(f_x)/I(0)$ for $D = 25$ mm, $R = 0, 1,$ and 2 and ρ corresponding to 0, 10, and 20 dB attenuation across the aperture.

5. With reference to Problem 4, what effect does the acoustic attenuation have on beamwidth and sidelobe level in the focal plane? How are resolution and dynamic range in the focal plane affected? Specifically, at $R = 10$- and 20-dB attenuation, how much is the resolution reduced (in percent) from its value for 0-dB attenuation, assuming that the resolution element at 0-dB attenuation is defined to be $f_x = 1/D$ (consistent with a 4-dB width of the focused beam profile)? What is the increase in decibels for $R = 2$ at 10- and 20-dB attenuation? Below what level of acoustic attenuation across the aperture is the major effect primarily just a reduction in diffraction efficiency?

6. For wideband operation of a Bragg cell as a spatial light modulator, several (N) resolvable frequencies coexist in the Bragg cell and as a result of acoustic nonlinearities the overall diffraction efficiency η is reduced over that for a single frequency η_1 according to the approximate expression (developed by Hecht)

$$\eta = \eta_1 \left(1 - \frac{1}{3}\eta_1 - \frac{2}{3}\sum_{k=2}^N \eta_k \right)$$

where η_k is the diffraction for the signal at frequency f_k. Calculate the loss in overall diffraction efficiency for the case of two frequencies separated by 50 MHz for the case of slow-shear TeO_2 Bragg cell as described in Example 8.4. Expressed as an equivalent dynamic range, how does this result compare with the result of Example 8.4 with and without acoustic attenuation?

7. For the case of a liquid crystal SLM operated in a controlled birefringence mode with incident light polarized at 45° with respect to the liquid crystal orientation, as illustrated below, the output light intensity is given by

$$I_{out} = I_{in} \sin^2(\delta/2)$$

The phase retardation δ experienced by the light is given by

$$\delta = \frac{2\pi d \Delta n}{\lambda} f(V)$$

where

$$f(V) = \begin{cases} \alpha\left(\frac{V}{V_{th}} - 1\right) & \text{(low fields)} \\ \left(1 - \frac{\beta}{V}\right) & \text{(high fields)} \end{cases}$$

where V is the applied RMS AC voltage. For $d = 10$ μm, $\lambda = 633$ nm, $\Delta n = 0.001$, and $\alpha = 1$, plot I_{out}/I_{in} versus normalized voltage (V/V_{th} or V/β).

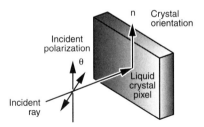

8. In the twisted nematic configuration a liquid crystal SLM is operated as suggested in the figure. The fraction of light intensity exiting the liquid crystal cell with polarization parallel to the liquid crystal alignment is given by

$$\frac{I_{out}}{I_{in}} = 1 - \frac{\sin^2\left[\theta\left(1 - U^2\right)^{1/2}\right]}{1 + U^2}$$

where U is the normalized retardation defined by

$$U = \frac{\delta}{\theta_T} = \frac{2\pi d \Delta n}{\theta_T \lambda}$$

where θ_T is the twist angle (small compared to the classical rotatory power δ), and θ is the angle between the applied field and the liquid crystal orientation as in the previous picture. Plot this versus U for the same parameters as in Problem 7, for $\theta = 0.1$ and 1 radians, and for $\theta_T = 0.001$ radians.

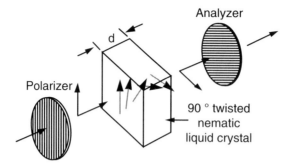

9. In a liquid crystal SLM, dynamic range is set by the maximum $\Delta\phi$ achievable, which is given by

$$\Delta\phi = \frac{2\pi d \Delta n}{\lambda}$$

which is thus proportional to thickness d. Response time is given by

$$\tau(V) = \frac{\tau_0 d^2}{\left(\frac{V}{V_{\text{th}}}\right)^2 - 1}$$

To maximize dynamic range while minimizing response time we must maximize the quantity $\Delta\phi/\Delta t$. First show that

$$\frac{\delta\phi}{\Delta t} \approx \frac{\partial\delta}{\partial V}\frac{\Delta V}{\tau}$$

Using this result and values of δ for low- and high-field cases cited in Problem 7, determine corresponding expressions for $\Delta\phi/\Delta t$. Thus for either case show that $\Delta\phi/\Delta t \propto \Delta n/\tau d$, where $\Delta n/d$ is analogous to gain and $1/\tau$ is equivalent to bandwidth, thus making $\Delta\phi/\Delta t$ equivalent to gain–bandwidth product.

10. The on-state transmission T_{on} of a unipolar operated magneto-optic SLM (MOSLM) is determined by absorption in the material as well as the polarizer-analyzer transmission function. Write an expression for T_{on}. The off-state transmission (T_{off}) includes a second pass through the MOSLM as shown in the figure. Write an expression for T_{off} including effects due to reflection at the surfaces of the MOSLM active medium. From this determine the contrast ratio for the case where the absorption coefficient $\alpha = 0.108$ μm^{-1}, the index of refraction $n = 2.35$ and the thickness of the MOSLM active material $d = 10$ μm. If the reflectivity is improved to 1% by using AR coatings, what is the improved contrast ratio?

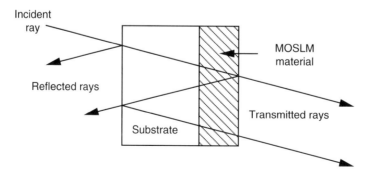

11. A useful model for describing the transfer characteristic of a generic spatial light modulator is the sigmoid function defined by

$$g(f) = \frac{\alpha - \beta}{1 + \exp\left[-\gamma(f - \langle f \rangle)\right]} + \beta$$

where g is the output signal, f is the input (modulating) signal, $\langle f \rangle$ is the mean value of input signal, α is the saturation level, β is the noise floor, and γ is the slope of the linear region. Plot this function and identify the asymptotic values of $g(f)$ for $\exp[-\gamma(f - \langle f \rangle)] \ll 1$ and $\gg 1$. Determine the value of γ in terms of α, β and $df/dx|_{x=\langle x \rangle}$. Using a sinusoidal modulation as the input signal

$$f(x) = f_o + f_1 \cos kx$$

and assuming there are an integer number of cycles spanning the SLM array, determine the output g in terms of x. Determine the Fourier transform of the input modulation. Then determine the Fourier transform of the SLM output signal for small γ to second-order in γ. What is the effect of the nonlinearity on the output spectrum?

12. A high gamma (γ) transfer characteristic is shown where

$$\gamma = \frac{\log(T_1/T_2)}{\log(E_1/E_2)}$$

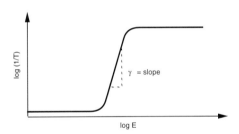

can be achieved with a MSLM by using it as an active birefingent crystal inside a Fabry–Perot cavity. The ratio of the reflected light intensity to incident light intensity is therefore given by

$$T = \frac{4R \sin^2(\phi/2)}{(1 - R)^2 + 4R \sin^2(\phi/2)}$$

where R is the reflectivity and ϕ is the phase shift or retardation Γ, which is linearly proportional to applied voltage; that is, $\Gamma = \pi V/V_\pi$, where V_π is the half-wave voltage (i.e., the voltage required to shift the phase by π). Plot T versus normalized voltage for $R = 0.15$ and $R = 0.80$. In the limit of small R, what is the limiting functional form of T? For large R, how does the transfer characteristic appear? Estimate its gamma.

13. In the previous problem the transfer characteristic can be approximated as a binary characteristic in which below a certain threshold the output is zero and above it the output is unity. Calculate the spectrum of a modulating signal before and after modulating the MSLM, assuming that the input is $f(x) = \cos(kx)$. What has happened to the spectrum? (*Hint*: Use the results of Chapter 2, Section 2.4.)

14. A DMD SLM can be modeled simplistically as a membrane with sinusoidal membrane deflection given by

$$d(x) = d_o + d_1 \sin(2\pi f_x x)$$

in one dimension. The effect of this deformation on a uniform laser beam normally incident on the DMD is described by

$$A(x, y) = T_o A_o(x, y) \exp[j\phi(x, y)]$$

where $A_o(x, y)$ describes the DMD aperture function, T_o is the transmit coefficient, and $\phi(x, y)$ is the phase variation. In one dimension, $\phi(x, y)$ is

$$\phi(x) = \frac{2\pi}{\lambda}(n - 1)\,dx$$

where n is index of refraction of the DMD material. Show that the output wave is given by

$$A(x) = T_o A_o(x) \sum_{m=-\infty}^{\infty} j^m J_m[k(n-1)d_1] \exp[jm 2\pi f_x x]$$

BIBLIOGRAPHY

R. Adler, "Interaction Between Light and Sound," *IEEE Spectrum*, pp. 42–54 (May 1967).

G. Baym, *Lectures on Quantum Mechanics*, W.A. Benjamin, Reading, MA (1974).

N. J. Berg and J. N. Lee, eds. *Acousto-Optic Signal Processing: Theory and Implementation*, Marcel Dekker, New York (1983).

W. P. Bleha et al., "Application of the Liquid Crystal Light Valve to Real-Time Optical Data Processing," *Opt. Eng.* **17**, 371 (1978).

W. P. Bleha et al., "Optical Data Processing Liquid Crystal Light Valve," *Proc. SPIE* **201**, 122 (1979).

G. D. Boyd, D. A. B. Miller, D. S. Chemla, S. L. McCall, A. C. Gossard, and J. H. English, "Multiple Quantum Well Reflection Modulator," *Appl. Phys. Lett.* **50**, 1119 (1987).

D. Casasent, "Spatial Light Modulators," *Proc. IEEE* **65**, 157 (1977).

J. A. Davis and J. M. Waas, "Current Status of the Magneto-Optic Spatial Light Modulator," in *Spatial Light Modulators and Applications III, Critical Reviews of Optical Science and Technology*, U. Efron, ed., **Proc.** SPIE, Vol. 1150, pp. 27–43 (1989).

U. Efron et al., "Silicon Liquid Crystal Light Valves: Status and Issues," *Opt. Eng.* **22**, 682 (1983).

A. D. Fisher and J. N. Lee, "The Current Status of Two-Dimensional Spatial Light Modulator Technology," in *Optical and Hybrid Computing*, SPIE, Vol. 634, pp. 352–372 (1986).

W. D. Goodhue, B. E. Burke, K. B. Nichols, G. M. Metze, and G. D. Johnson, "Quantum-Well Charge-Coupled Device-Addressed Multiple-Quantum-Well Spatial Light Modulators," *J. Vac. Sci. Tech.* **B4**, 76 (1986).

D. L. Hecht, "Spectrum Analysis Using Acousto-optic Devices," *Opt. Eng.* **16**, 461 (1977).

L. J. Hornbeck, "128 × 128 Deformable Mirror Device," *IEEE Trans. Electron Devices* **ED-30**, 539 (1983).

L. J. Hornbeck, "Deformable-Mirror Spatial Light Modulators," in *Spatial Light Modulators and Applications III, Critical Reviews of Optical Science and Technology*, U. Efron, ed., SPIE, Vol. 1150, pp. 86–102 (1989).

T. Y. Hsu and U. Efron, "Multiple Quantum Well Spatial Light Modulators," in *Spatial Light Modulators and Applications III, Critical Reviews of Optical Science and Technology*, U. Efron, ed., SPIE, Vol. 1150, pp. 61–85 (1989).

T. Y. Hsu, U. Efron, W. Y. Wu, J. N. Schulman, I. J. D'Haenens, and Y. C. Chang, "Multiple Quantum Well Spatial Light Modulators for Optical Processing Applications," *Opt. Eng.* **27**, 372 (1988).

T. D. Hudson and D. A. Gregory, "Optically Addressed Spatial Light Modulators," *Optics and Laser Tech.*, Vol. 23, pp. 297–302 (1991).

D. A. Jared and K. M. Johnson, "Ferroelectric Liquid Crystal Spatial Light Modulators," SPIE Critical Reviews Series, *Spatial Light Modulators and Applications III*, Vol. 1150, pp. 46–60.

K. M. Johnson and G. Moddel, "Motivations for Using Ferroelectric Liquid Crystal Spatial Light Modulators in Neurocomputing," *Appl. Opt.* **28**, 4888 (1989).

G. Livescu, D. A. B. Miller, J. E. Henry, A. C. Gossard, and J. H. English, "Spatial Light Modulator and Optical Dynamic Memory Using a 6 × 6 Array of Self-Electro-optic-Effect Devices," *Opt. Lett.* **13**, 297 (1989).

D. A. B. Miller, "Quantum Wells for Optical Information Processing," *Opt. Eng.* **26**, 368 (1987).

D. A. B. Miller, "Optoelectronic Applications of Quantum Wells," *Optics and Photonics News*, pp. 7–15 (February 1990).

D. A. B. Miller, D. S. Chemla, T. C. Damen, T. H. Wood, C. A. Burrus, A. C. Gossard, and W. Wiegmann, "The Quantum Well Self Electro-optic Effect Device: Optoelectronic Bistability and Oscillation, and Self-Linearized Modulation," *IEEE J. Quant. Electr.*, **QE-21**, 1462 (1985).

D. A. B. Miller, J. E. Henry, A. C. Gossard, and J. H. English, "Integrated Quantum Well Self Electro-optic Effect Device: 2 × 2 Array of Optically Bistable Switches," *Appl. Phys. Lett.* **49**, 821 (1986).

J. A. Neff, "Future Directions of Optical Processing," in *Proceedings of NASA Conference on Optical Information Processing for Aerospace Applications II*, August 30–31, 1983, Hampton, VA, pp. 1–18.

J. A. Neff, R. A. Athale, and S. H. Lee, "Two-Dimensional Spatial Light Modulators: A Tutorial," *Proc. IEEE* **78**, 826 (1990).

D. R. Pape, "An Optically Addressed Membrane Spatial Light Modulator," *Proc. SPIE* **465**, 17 (1984).

D. R. Pape et al., "Characteristics of the Deformable Mirror Device for Optical Information Processing," *Opt. Eng.* **22**, 675 (1983).

J. M. Waas and M. Waring, "Spatial Light Modulators: A User's Guide," in *The Photonics Design and Applications Handbook 1989*, pp. 362–365 (1989).

T. H. Wood, C. A. Burrus, R. S. Tucker, J. S. Weiner, D. A. B. Miller, D. S. Chemla, T. C. Damen, A. C. Gossard, and W. Wiegmann, "100 ps Waveguide Multiple Quantum Well (MQW) Optical Modulator with 10:1 On/Off Ratio," *Elec. Lett.* **21**, 693 (1985).

Chapter 9
OPTICAL SPECTRUM ANALYSIS AND CORRELATION

9.1 OVERVIEW

Two of the most basic concepts of signal processing are spectrum analysis and correlation. In this chapter we will describe in detail how these two concepts are implemented in optical signal processing architectures. We will introduce these mathematical concepts in the context of a more general framework appropriate for categorizing different types of optical signal processing architectures. We will distinguish spectrum analyzers and correlators as time- and space-integrating, and we will distinguish correlators as incoherent (e.g., shadow casting) and coherent (e.g., interferometric). Specific examples of each type of processor will be analyzed in detail.

If we confine our attention to two-dimensional functions (or images), which can vary in time, then the input image to an optical system is defined by $f(x, y, t)$ and the output image is defined by $g(x, y, t)$. For linear systems the relationship between input and output is given by the superposition integral

$$g(x', y', t') = \int_{-\infty}^{\infty} \int_{-\infty}^{\infty} \int_{-\infty}^{\infty} f(x, y, t) h(x, y, t; x', y', t') \, dx \, dy \, dt \tag{9.1}$$

where $h(0, 0, 0, x', y', t')$ is the system impulse response (or kernel). This is a quite general result, which is not as interesting as the more specific cases that we will consider.

9.2 TIME- AND SPACE-INTEGRATING ARCHITECTURES

Many optical processors are implemented with one-dimensional acousto-optic Bragg cells, in which case we usually limit the optical processor to two-dimensional (2-D) linear operations that can be performed with one-dimensional (1-D) processors. Two types of kernels can then be considered: shift-invariant and separable. Shift-invariant kernels are of the form $h(x-x', y-y', t-t')$, and separable kernels are of the form $h(x, y, t) = h_x(x) \cdot h_y(y) \cdot h_t(t)$. In many cases, for complex processing tasks, time- and space-integrating processors are used together. Each line of the input image is processed by space integration, and the output image is produced by time integration. Space integrations may introduce approximately 10^{10} samples per second. Time integrations subsequently reduce the throughput by approximately three orders of magnitude.

Alternatively, several useful optical processing architectures are solely space- or time-integrating. We will deal with these first. What do we mean by space- or time-integrating? Obviously, space-integrating means that integration is performed with respect to the spatial coordinate (x or y). This is accomplished for spectrum analysis by allowing the dispersive spatial frequency characteristics of free space propagation to be combined with lenses, which act like space-quadratic phase shifters. Essentially space integration is the diffraction integral. Correlation using a space-integrating architecture yields an output correlation that is a function of time. Detector integration times are short compared to the time it takes to fill the spatial light modulator with the modulation function—for example, for acousto-optic (AO) cell transit times on the order of microseconds. For instance, for a correlator the cross-correlation function, $R_{xy}(t)$, is given by

$$R_{xy}(t) = \int_{\tau} x(\tau) y(\tau - t) \, d\tau \qquad (9.2)$$

where the integration variable (τ) is time delay, which is proportional to the space variable (z) for an AO cell; that is, $\tau = z/v_s$. Time integration obviously means integration with respect to the time coordinate. Here the integration times can be long, on the order of tens of milliseconds. For a correlator the output is now

$$R_{xy}(\tau) = \int_{\tau} x(t) y(t - \tau) \, dt \qquad (9.3)$$

where the integration variable is explicitly time. In practice, this processor uses a light detector to charge a capacitor to do the time integration. Long-duration signals can be processed, and large time–bandwidth products can be achieved. For spectrum analysis the Fourier transform is achieved by integration in time of a signal located at a particular point in space relayed to the output plane by optical imaging.

9.3 COHERENT AND INCOHERENT ARCHITECTURES

Another important distinction needs to be made with regard to the types of optical architectures. Coherent or incoherent light can be used. Although coherent laser light can be obtained from some lasers (e.g., He–Ne lasers) with modest output power, it is more problematic to get both sufficient coherency and power from single laser diodes (as discussed in Chapter 7). Laser diodes are preferred, however, for their high-modulation bandwidths and compact size, so this is an important consideration. If reasonably long coherence lengths can be obtained (e.g., 0.5–5 cm), then interferometric processing can be effectively implemented. In interferometric processing phase modulation of an optical beam is converted to an amplitude modulation. This is accomplished by square-law detection of the linear superposition of two interfering waves: one plane ($A \exp(-jkx)$) and one modulated ($f(x, y)$), namely,

$$g(x, y) = |f(x, y) + A \exp(-jkx)|^2 \qquad (9.4)$$

so that

$$g(x, y) = |f(x, y)|^2 + A^2 + 2A|f(x, y)|\cos[kx - \phi(x, y)]$$

where $f(x, y) = |f(x, y)| \exp[j\phi(x, y)]$.

This is the basis of the VanderLugt matched filter, which we will discuss later. If the temporal or spatial phase stability were not ensured, then interferometric detection would not be guaranteed because the interference (or cross) term shown in Equation (9.4) would be smeared-out by phase fluctuations. When we deliberately make the source illumination incoherent by using diffuse light sources, then we can only implement incoherent or "shadow-casting" correlators. Ambient natural or man-made illumination is often of this type and is used in practice with most electro-optical sensors or cameras. So consideration of incoherent optical processors is important because of the ease of interfacing to these sensors. In addition, the phase is lost from the input light when square-law detection is used, because normally an interference pattern is not established. In many cases, phase is not important for spectrum analysis because the power spectrum is all that is desired. Coherent optical spectrum analyzers, however, achieve greater dynamic range, as will be discussed later.

9.4 SPECTRUM ANALYSIS

The simplest optical architecture for processing signals is the spectrum analyzer. In the setup where the input plane, pupil plane, and output plane are separated by the focal length of the lens (located at the pupil plane), we get the exact Fourier transform, as discussed previously. Let us consider a 1-D Fourier transform architecture implemented with a laser diode, an AO Bragg cell, and a linear self-scanned detector array as shown in simplified schematic form in Fig. 9.1. The laser diode is DC-biased and pulse-modulated, and its beam is expanded, spatially filtered, collimated, and corrected for anamorphic distortion (by a prism pair). Subsequently a cylindrical lens is used to focus the light through the Bragg cell window, whereupon it is recollimated by a second cylindrical lens. The intensity of this beam is given by $I_0 w(x)$. The time-varying signal applied to the Bragg cell is given by $v(t)$. When the signal $v(t)$ is transduced into the Bragg cell, it forms a traveling wave $v(x - v_s t)$, which fills the aperture. At the moment that $v(x - v_s t)$ is centered, it is illuminated by strobing the laser diode for an appropriate dwell time τ. The appropriate (first-order) output of the Bragg cell is now given by $i(x, t) = I_1 w(x) v(x - v_s t)$, where I_1 denotes the intensity of the first-order beam. The applied voltage $v(t)$ induces a modulation on the first-order diffracted beam given by $I_1 = \eta I_0(t)$, where $I_0(t) = I_0 \text{rect}[(t - \tau)/T]$ is the incident laser beam intensity, where T is the Bragg cell aperture time, and the diffraction efficiency η is proportional to the square of the applied voltage, namely $\eta \propto |v(t)|^2$. A second cylindrical lens is used to recollimate the beam. Then a spherical lens can be placed after the second cylindrical lens to Fourier transform $i(x)$. At the output plane we then have

$$I(u) = \int_0^{x_{\max}} I_1 w(x) v(x - vt) \exp(-j2\pi ux)\, dx$$

or

$$I(u) = I_1 W(u) * V(u) \exp(j2\pi v_s t) \tag{9.5}$$

9.4 SPECTRUM ANALYSIS

The space coordinates of this signal are linearly proportional to spatial frequency for small deflection angles.

Recall that when a Bragg cell is modulated with an RF source with several frequencies $\{f_k\}$ present, the Bragg condition must be satisfied for all of them. The corresponding deflection angles for all these frequencies are given by

$$\theta_k = 2 \sin^{-1}(\lambda f_k / 2nv_s) \quad (9.6)$$

which for small angles (≤ 0.1 radians) is $\theta_k \approx \lambda f_k / nv_s$, such that in the transform plane the frequencies are located at positions $x_k \approx \lambda F f_k / v_s$, where F is the Fourier transform lens focal length. Equation (9.6) gives the angle locations of all the frequencies present in $i(x)$.

Normally the Bragg cell is tilted with respect to the optical axis (defined by the input laser beam) so that the $+$ first-order diffracted beam is preferentially enhanced over the $-$ first-order beam. The zero-order beam is blocked by a stop. Then the output beam is detected by the linear detector array, to yield

$$E(u) = I_0 |W(u) * V(u)|^2 \quad (9.7)$$

As can be seen in Equation (9.7) the detected signal is not purely the Fourier transform of the time signal $f(t)$, but is corrupted by the presence of $w(x)$. The resulting $W(u)$ term in Equation (9.7) is caused by several factors: Bragg cell aperture, acoustic attenuation, laser beam profile, and acoustic bandshape. Other considerations may also enter such as multisignal interaction, detection, and post-detection filtering, but we will not deal with these here. Our main interest shall be with the classic resolution versus sidelobe tradeoff of spectral analysis. We can, in fact, model some of the basic effects to this end. The aperture function corresponding to the Bragg cell aperture is defined as

$$w_1(x) = \text{rect}\left(\frac{x - D/2}{D}\right) \quad (9.8)$$

where D is the Bragg cell window length ($= v_s T$). The acoustic attenuation is defined by $w_2(x) = \exp(-\alpha x)$, where α is the frequency dependent loss factor, defined as

$$\alpha = \frac{\alpha' f^2}{20 \log_{10} e} \quad (9.9)$$

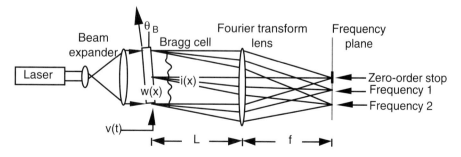

Figure 9.1. Simple one-dimensional acousto-optic spectrum analyzer architecture.

and α' is a material acoustic loss constant with units of $(dB \cdot cm^{-1} \cdot GHz^{-2})$ discussed previously in Chapter 8. The optical beam profile is typically Gaussian (for at least He–Ne lasers), so that it is defined by

$$w_3(x) = I_0 \frac{1}{\sigma_x \sqrt{2\pi}} \exp\left[\frac{-(x - D/2)^2}{2\sigma_x^2}\right] \quad (9.10)$$

where σ_x is the standard deviation of the Gaussian beam, centered at a mean value of $D/2$. These functions are illustrated in Figs. 9.2 (a)–(c). The total convolution window in the spatial frequency domain is now given by

$$W(u) = \int_{-\infty}^{\infty} w_1(x) w_2(x) w_3(x) \exp(-j2\pi u x)\, dx \quad (9.11)$$

where u is the normalized spatial frequency given by $f_x D$. The total apodization function is shown in Fig. 9.2(d). For the ideal case of no attenuation ($\alpha = 0$) and uniform laser beam (σ_x approaching ∞), $W(u)$ reduces to the classic sinc function

$$W(u) = I_i \frac{\sin(\pi u)}{(\pi u)} \quad (9.12)$$

in which case the sidelobes are 13 dB down from the mainlobe.

In addition to the direct windowing effects just described, there is also an acousto-optic bandshaping. Since the Bragg condition cannot be matched (i.e., momentum conserved as described in Chapter 8) over a wide band, the diffraction efficiency is determined for each frequency according to the acoustic transducer radiation pattern in the direction required for the Bragg condition to be met. This effect limits the interaction length and thus the corresponding efficiency for a given bandwidth. For a single uniform transducer the angular spectrum is of

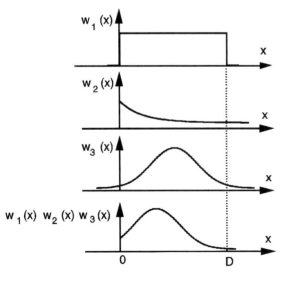

Figure 9.2 Individual apodization curves (a–c), and total apodization function (d).

the sinc² form such that for normal diffraction we have

$$W(f, f_m) = \text{sinc}^2\left[\frac{L}{2L_0}(ff_m - f^2)\right] \qquad (9.13)$$

where f is the frequency normalized to midband, f_m is the frequency at which Bragg angle is matched, and L_0 is the AO characteristic length at midband, given by $\lambda_s^2 n/\lambda \cos\theta_i$. Let us now consider a specific implementation where the modulation bandwidth is narrow.

EXAMPLE 9.1 Narrowband AO Spectrum Analyzer Experiment Figure 9.3 is a more detailed diagram of an experimental laboratory architecture for performing one-dimensional spectrum analysis. This architecture could also be used as the first (range compression) stage of an optical SAR processor, as will be discussed in Chapter 11.

Tests of the architecture were conducted using a laser diode pulsed at a PRF of 26 kHz with pulse widths of 14 nsec. The CID array used for detection integrated 433 pulses per output video frame. The RF signal level applied to the Bragg cell was 16 mW for each of the input waveforms shown in Figs. 9.4 (a)–(c). The detected double-sided Fourier transform is depicted in Figs. 9.4 (d)–(f). The CID was used in lieu of the linear self-scanned array for these figures because it is easier to display the spectra. Note that the mainlobes of the sinc functions (at the harmonic and its image frequency) shown in Figs. 9.4(e) and 9.4(f) are near the saturation level of the CID, while the furthest visible sidelobe is near the noise level. Resolving even higher frequency sidelobes would require an increase in system dynamic range. This might be realized by reducing stray background light and CID noise (through cooling).

An important consideration in this architecture is its overall size and the proper matching of the number of detector pixels to the number of resolvable frequencies in the spectrum. Refer to Fig. 9.5 for this discussion.

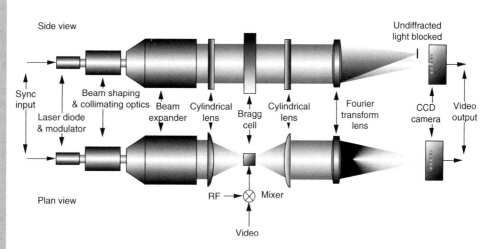

Figure 9.3 Narrow-band acousto-optic spectrum analyzer test architecture.

236 OPTICAL SPECTRUM ANALYSIS AND CORRELATION

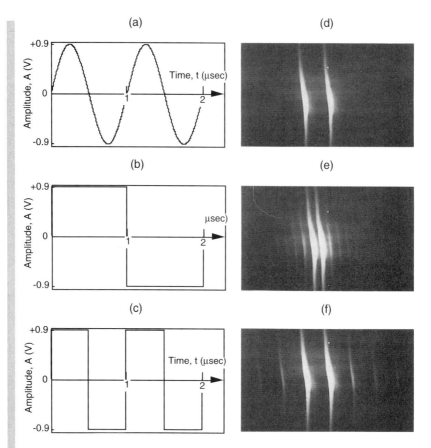

Figure 9.4 Diagram of input test signals and corresponding spectrum displayed on CCD focal plane array.

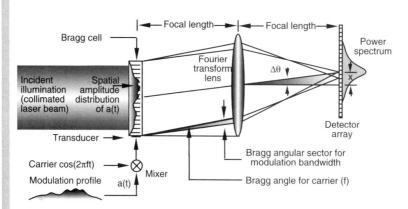

Figure 9.5 Signal input, acousto-optic diffraction from a typical Bragg cell, and Fourier transform architectural geometry for spectrum analysis.

Given that the video bandwidth is 4 MHz and that the laser wavelength λ is 800 nm, the corresponding deflection angle θ is 0.31°. Then using the equation $F = x/\Delta\theta$ [where x is linear displacement in the detector (or Fourier) plane] a focal length (F) of 1.36 m is consistent with a typical detector array size x_{max} of 0.74 cm. The distance in the Fourier plane corresponding to the spatial frequency of the entire Bragg cell aperture (which represents the *lowest* frequency that can be supported by the Bragg cell and the lowest frequency that can therefore be resolved in the spectrum according to Rayleigh's criterion for optical resolving power) is given by $\Delta x = \lambda F/D = 26$ μm in the small-angle approximation, where D is the Bragg cell aperture (42 mm). The number of resolvable frequency components is then given by $x_{max}/\Delta x = 285$. The minimum resolvable frequency is $\sim$14 kHz, a little lower than the standard TV horizontal line rate ($\sim$16 kHz). Another one of the results of this design exercise is the simple fact that the overall length of the spectrum analyzer is at least 2.72 m, unless some optical path folding is done. Of course, the equation $F = x/\Delta\theta$ says that either decreasing the size of the detector array or increasing the deflection angle (or correspondingly the modulation bandwidth) will reduce overall processor size.

EXAMPLE 9.2 Simple LCTV Spectrum Analyzer Architecture As a second example of spectrum analysis, consider the simple Fourier transform architecture shown in Fig. 9.6 in which the input plane is a two-dimensional spatial light modulator (SLM). In this architecture the SLM is served by a liquid crystal TV (LCTV), which is electrically addressed using standard video signals. The input laser is a continuous He–Ne laser, which is beam-expanded and spatially filtered, as shown. The light incident on the spatial light modulator is a plane wave by virtue of the collimating lens (L_1); but by virtue of being a laser beam, it has a Gaussian amplitude profile.

In addition to the rectangular aperture of the LCTV, there is the dead space between the LCTV pixels, which is associated with the control lines to each of the pixels. Hence the transmission of the LCTV is made up of the spatial light modulation $f(x, y)$ and the pixel pattern given by the function

$$s(x, y) = \left[\text{rect}\left(\frac{x}{x_0}\right) * \text{comb}\left(\frac{x}{x_1}\right)\right]\text{rect}\left(\frac{x}{X}\right)\left[\text{rect}\left(\frac{y}{y_0}\right) * \text{comb}\left(\frac{y}{y_1}\right)\right]\text{rect}\left(\frac{y}{Y}\right)$$

where x_0 and y_0 are LCTV pixel dimensions, x_1 and y_1 are pixel spacings, and X and Y are LCTV display dimensions. Hence the total amplitude distribution exiting from the LCTV is $g(x, y) = f(x, y)s(x, y)$.

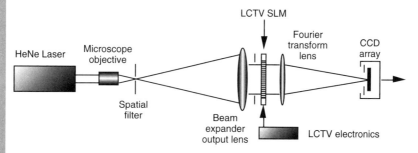

Figure 9.6 Two-dimensional spectrum analyzer employing a liquid crystal TV as the spatial light modulator.

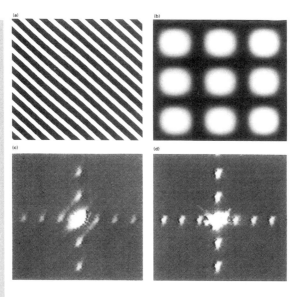

Figure 9.7 Input signals (a, b) electrically addressed to the LCTV, and corresponding spectra (c, d) seen at the output focal plane.

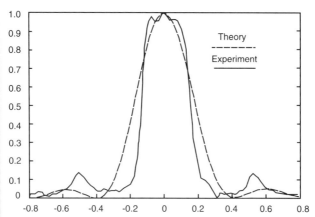

Figure 9.8 Comparison of spectra from LCTV architecture of Fig. 9.6 and theory (1-D).

This amplitude distribution is subsequently Fourier-transformed by the second lens, which yields $G(u, v) = F(u, v) * S(u, v)$, where

$$S(u, v) = C\text{sinc}(x_0 u)\text{comb}(x_1 u) * \text{sinc}(Xx)\text{sinc}(y_0 v)\text{comb}(y_1 v) * \text{sinc}(Yv)$$

where $C = x_0 x_1 y_0 y_1$. This Fourier domain amplitude distribution is then detected by a CCD focal plane array, which square-law detects the complex amplitude to yield the power spectrum $|G(u, v)|^2 = |F(u, v) * S(u, v)|^2$.

This architecture is simple to implement, but it results in a Fourier transform which is corrupted by the pixel deadspace in the LCTV. For the input spatial patterns shown in Figs. 9.7(a) and 9.7(b) the corresponding spectra are shown in Figs. 9.7(c) and 9.7(d). The lower contrast ratio found in commercial grade LCTVs, which is nominally 16:1, is sufficient to allow a reasonably accurate Fourier transform to be derived at the output focal plane (except for pixel artifacts in the spectrum), as shown in Fig. 9.8.

The notion that a low-to-moderate contrast ratio LCTV can support an accurate Fourier transform can be justified on the basis of a simple digital signal analysis described below. A comparison of the Fourier transform of an image, which is digitized to various gray levels (1 through 8 bits), is made using a mean-square error (MSE) measure of likeness between the various Fourier transform images created. As can be seen in Fig. 9.9, when the number of bits reaches 4 the MSE approaches the 8-bit level, which is asymptotically close to the maximum dynamic range of any component in the system (e.g., CCD focal plane array).

An example of a reconstructed image is shown in Figs. 9.10(a)–(d). The premise here is that a given bit-level (or dynamic range) signal can be accurately Fourier-transformed for reconstruction if, after reconstruction, it is reasonably close to the original image as measured by mean-square error (as a global measure of error). Note the aliasing for lower bit-level reconstructions. In some instances, depending on the particular image or class of images, 1-bit of dynamic range is sufficient to yield an accurate Fourier transform. In fact, only the phase (or only the real or imaginary components) are required to get an accurate Fourier transform in some cases, as pointed out by various investigators. The significance of this is that the edges or spatial distribution of object boundaries is more important than the detailed amplitude distribution of each object in the image. This notion can also be exploited in developing binary optical correlators, as will be discussed later.

Now that we have examined some basic experimental 1-D and 2-D spectrum analyzer architectures, let us consider two examples of other designs, one that is space integrating and one that is time integrating. However, we will restrict our attention to 1-D AO spectrum analyzers.

9.5 SPACE-INTEGRATING SPECTRUM ANALYZER

Consider first the space-integrating AO spectrum analyzer (developed by Hecht et al.) which is illustrated in Fig. 9.11. This is an extremely simple and compact design. Its principle of operation is based on the following analysis. Consider the Fourier transform of the function $g(t)$ modified by the truncation effect of the time window (representing the AO cell), $\text{rect}(t/T)$:

$$S_T(f) = \int_{-\infty}^{\infty} s(t) \text{rect}\left(\frac{t}{T}\right) \exp(-j2\pi ft)\,dt \qquad (9.14)$$

Figure 9.9 Mean-square error between ideal images and their reconstruction via digital Fourier transformation for various bit levels.

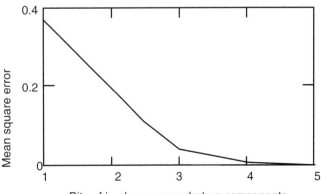

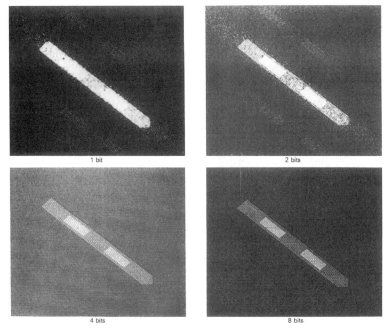

Figure 9.10 Examples of the reconstruction of a ship-like object at various bit levels. Note the aliasing at lower bit-levels.

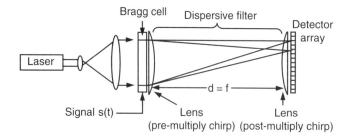

Figure 9.11 Space integrating AO spectrum analyzer.

Now we can simply expand the argument of the Fourier kernal:

$$\exp(-j2\pi ft) = \exp(-j\pi f^2)\exp(-j\pi t^2)\exp[j\pi(f-t)^2] \quad (9.15)$$

Then the Fourier transform can be written

$$S_T(f) = \exp(-j\pi f^2)\int_{-\infty}^{\infty} s(t)\,\text{rect}\left(\frac{t}{T}\right)\exp(-j2\pi t^2)\exp[j\pi(f-t)^2]\,dt \quad (9.16)$$

↑	↑	↑	↑
Postmultiply chirp	Input Window signal	Premultiply chirp	Dispersive delay line (impulse response)

where each term has been identified as to its function. This is the chirp transform, which is equivalent to the Fourier transform. It is more flexible than the (discrete) Fourier transform

9.5 SPACE-INTEGRATING SPECTRUM ANALYZER

and preferred in analog signal processing because it is implemented in the time domain as a convolution with variable input function duration. It rests on the linear relationship between time and frequency; that is, $f = \alpha t$ [where in Equation (9.15), α has been set to 1]. For real symmetric signals the convolution is equivalent to correlation. Without the gating effect of the window function we would get $S(f)$ defined by

$$S(f) = \exp(-j\pi f^2)[s(t)\exp(-j\pi t^2) * \exp(-j\pi t^2)] \tag{9.17}$$

where we have used the convolution theorem in obtaining Equation (9.17). Thus the Fourier transform of the windowed time function is

$$S_T(f) = S(f) * T\operatorname{sinc}(fT) \tag{9.18}$$

We can illustrate the equivalent electrical model for the optical architecture shown in Fig. 9.11 by the block diagram shown in Fig. 9.12.

If we analyze the architecture of Fig. 9.11 in light of the above discussion and the diagram of Fig. 9.12, we can define the amplitude of light in the output plane (before detection) as

$$\begin{aligned}A(x_0) = A_0 &\overset{\text{Second lens}}{\underset{\downarrow}{\frac{\exp(j2\pi d/\lambda)}{j\lambda d}}} \exp\left(-j\frac{\pi x_0^2}{\lambda f_2}\right) \int_{-\infty}^{\infty} [1 + j\alpha s(t - x_i/v_s)]\operatorname{rect}\left(\frac{x_i}{D}\right) \exp\left(-j\frac{\pi x_i^2}{\lambda f_1}\right) \\ &\times \underset{\underset{\text{Optical propagator}}{\uparrow}}{\exp\left[j\frac{\pi(x_0-x_i)^2}{\lambda d}\right]} dx_i\end{aligned} \tag{9.19}$$

where it is clear by analogy with Equation (9.16) that the first lens is equivalent to the premultiply chirp, the optical propagation is equivalent to the dispersive delay line filter, and the second lens is equivalent to the postmultiply chirp. Clearly D is the Bragg cell aperture length, and we have assumed that the signal $g(t)$ converts to the downshifted diffracted beam component at $\nu - \nu_0$, consistent with the analysis of Chapter 8. If we assume $f_1 = f_2 = d = F$, define spatial frequency as $f_x = x_0/\lambda F$, and define the constant B as

$$B = \frac{A_0}{j\lambda F} \exp\left(j2\pi \frac{F}{\lambda}\right) \tag{9.20}$$

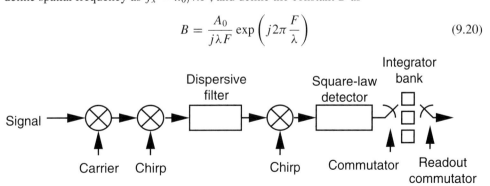

Figure 9.12 Equivalent electrical model of space integrating AO spectrum analyzer.

then Equation (9.19) reduces to

$$A(f_x) = B \int_{-\infty}^{\infty} \left[1 + j\alpha s\left(t - \frac{x_i}{v_s}\right)\right] \text{rect}\left(\frac{x_i}{D}\right) \exp(-j\pi f_x^2)$$
$$\times \exp[j\pi(f_x - x_i)^2] \exp(-j\pi x_i^2) \, dx_i \quad (9.21)$$

which is very close in form to Equation (9.16). Using the definition of the Fourier transform, the similarity, shift, and convolution theorems then yield

$$A(f_x) = BD \, \text{sinc}(f_x D) + j\alpha B \exp(j2\pi f_x v_s t)|v_s|S(v_s f_x) * D \, \text{sinc}(f_x D) \quad (9.22)$$

where $S(v_s f_x)$ is the Fourier transform of $s(t)$. The variable of integration in this result is f_x, but by changing variables to temporal signal frequency $f = f_x v_s$ and converting D to Bragg cell transit time (τ) via $\tau = D/v_s$ we obtain

$$A(f) = BD \, \text{sinc}(f\tau) + j\alpha B \exp(j2\pi f t) S(f) * D \, \text{sinc}(f\tau) \quad (9.23)$$

The first term is the result of the zero diffraction order, which can be blocked by a stop. Thus the desired spectrum of $s(t)$, given by $S(f)$, is smeared by convolution with the Fourier transform of the rectangular window function representing the Bragg cell aperture. Note also that the output field distribution represents the complex linear temporal frequency distribution because of the presence of the term $\exp(j2\pi f t)$.

Finally, square-law detection is usually done in the output plane. The temporal frequency is heterodyned to zero frequency except for those terms resulting from the heterodyning (beating) of adjacent frequencies within the spectrum $S(f)$, which are seen at adjacent detectors in the output plane. If the individual photodetector elements have integration times greater than 1/BW, resulting from their finite spatial resolution, then the finite duration power spectrum of $s(t)$ results (over Bragg cell transmit time τ). Then the voltage at the detector output is

$$V(f) = A_0^2 \Delta x T \left(\frac{D}{\lambda F}\right)^2 \text{sinc}^2(f\tau) * \text{rect}\left(\frac{f}{\Delta f}\right) + A_0^2 T \alpha^2 \left(\frac{D}{\lambda F}\right)^2$$
$$\times \left\langle [S(f) * \text{sinc}(f\tau)]^2 * \text{rect}\left(\frac{f}{\Delta f}\right)\right\rangle \quad (9.24)$$

where Δf is the frequency width corresponding to the detector width in the output plane, such that $\Delta f = v_s \Delta x / \lambda F$. The desired spectrum is buried in the second term of Equation (9.24). It can be isolated from the first term by appropriate filtering.

9.6 TIME-INTEGRATING SPECTRUM ANALYZER

Consider next the time-integrating AO spectrum analyzer shown in Fig. 9.13. This architecture differs from the previous one in that two Bragg cells are used. The optics actually implement a correlation operation in conjunction with the Bragg cells by virtue of carrying-out imaging using two telecentric lens pairs, as shown in the figure. The spectrum of the signal loaded into the first Bragg cell is created by virtue of the fact that time-domain chirp signals are also mixed with the local oscillator signals supplied to both Bragg cells.

9.6 TIME-INTEGRATING SPECTRUM ANALYZER

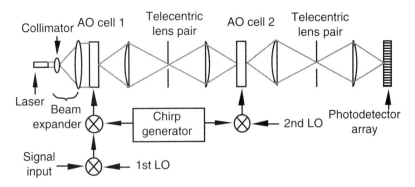

Figure 9.13 Time-integrating AO spectrum analyzer using two Bragg cells.

The first Bragg cell has the following spatial light modulation:

$$A_1(x_1) = A_o \exp[j\alpha s_1(t - x_1/v_s)] \tag{9.25}$$

The first lens pair images $A_1(x_1)$ onto the second cell such that the spatial modulation in front of the second cell is $A_2(x_2) = A_1(-x_1)$ for a magnification of $M_1 = -1$. [This means that $s_1(t - x_1/v_s)$ becomes $s_2(t + x_2/v_s)$]. Note that at the intermediate (Fourier) plane of the first lens pair a spatial filter is used to pass only the $+1$ or -1 diffraction order. The inverted image is thus the spatial analog of two Fourier transforms.

The second Bragg cell provides a spatial modulation to multiply with $A_2(x_2)$ given by $\exp[j\beta s_2(t - x_2/v_s)]$ such that the output spatial modulation function after the second cell is

$$A_2'(x_2, t) = A_o \exp[j\alpha s_1(t + x_2/v_s)] A_o \exp[j\beta s_2(t - x_2/v_s)] \tag{9.26}$$

where we have assumed that both Bragg cells are identical.

If we expand Equation (9.26) in a power series to obtain the spatial modulation in the small signal limit, we obtain

$$\begin{aligned} A_2'(x_2, t) \approx & A_o^2 + A_o^2 j\alpha s_1(t + x_2/v_s) + A_o^2 j\beta s_2(t - x_2/v_s) \\ & - A_o^2 \alpha\beta s_1(t + x_2/v_s) s_2(t - x_2/v_s) \end{aligned} \tag{9.27}$$

The first term occurs at zero spatial frequency. The fourth term is the correlation term and the one of interest in calculating the spectrum. The second lens pair now reimages this result to the output (focal) plane; that is, $A_3'(x_3) = A_2'(-x_2)$.

To actually implement the spectrum analyzer, we must obtain the Fourier transform via correlation. Hence we select the following chirped signals for modulating the Bragg cells:

$$s_1(t) = s_o(t) \exp[-j(2\pi f_o t + \pi a t^2)] \tag{9.28}$$

and

$$s_2(t) = \exp[-j(2\pi f_o t + \pi a t^2)] \tag{9.29}$$

where f_o is the center (chirp) frequency (at $t = 0$), and a is chirp acceleration.

Each photodetector integrates the output for a period of time equal to the dwell time (T). Subsequently a low-pass filter is used to smooth the output. This is expressed by the following equation for the detector/filter output voltage versus x position $[v(x_n)]$ on the detector array:

$$v(x_n) = \left\{ \int_{-T/2}^{T/2} [A(x, t)]^2 \, dt \right\} * |h_D(x)|^2 \tag{9.30}$$

where $A(x, t)$ is $A_3(x_3)$ with an explicit time dependence indicated, and $h_D(x)$ is the postdetection filter impulse response, expressed in spatial coordinates.

If we make a few changes of variables, by defining $\tau_1 = \tau_2 = x_3/v_s$, $\tau = \tau_1 + \tau_2 = 2x_3/v_s$ (assuming Bragg cell velocities are equal) and $t' = t - \tau_1$, then the output voltage becomes

$$v(\tau_n) = \int_{-\infty}^{\infty} \text{rect}\left(\frac{t' - \tau_1}{T}\right) (|s_2|^2 + |s_1|^2 + s_1 s_2^* + g_1^* s_2) \, dt' * |h_D'(\tau)|^2 \tag{9.31}$$

where the discrete time delay variable is $\tau_n = v_s/x_{3n}$. Substituting for Equations (9.28) and (9.29) yields

$$v(\tau_n) = A_o^2 [\alpha^2 \langle s^2(t') \rangle + \beta^2] * |h_D'(\tau)|^2 + A_o^2 \alpha \beta \exp(j2\pi f_o \tau)$$
$$\times \left\{ \int_{-\infty}^{\infty} s_o(t') \text{rect}\left(\frac{t' - \tau}{T}\right) \exp(-j\pi a t'^2) \exp[j\pi a(t' - \tau)^2 \, dt' \right\} * |h_D'(\tau)|^2$$
$$+ \text{Complex Conjugate} \tag{9.32}$$

From inspection it is clear that the first term is a constant that we will denote as V_o. The integral (or cross-term) is the desired form of the spectrum, except for the absence of a postmultiply chirp. Hence we have a legitimate power spectrum of $s_o(t)$. From these examples it is clear that the explicit form of the power spectrum is not obvious. Rather, it is often buried in the result with a host of other terms. In addition, it is clear that the chirp transform and chirp modulation play a key role in obtaining the desired result.

9.7 CORRELATION

The distinction between time and space integrating architectures will be revisited in the case of optical correlators using acousto-optical Bragg cells and coherent laser light sources. Before we do that we can look at the various incoherent optical correlators, which are very simple and highlight the various types of correlator approaches to handling two-dimensional input images, which do not require Fourier transforming the input signal, but keep all operations entirely in the space domain. These variations on the theme of correlation give a more comprehensive view. In this discussion we follow the work of Sprague.

9.8 INCOHERENT OPTICAL CORRELATOR ARCHITECTURES

The most general linear transformation in two-dimensions is the starting point of the discussion, namely,

$$g(x', y') = \int_{-\infty}^{\infty} \int_{-\infty}^{\infty} f(x, y) h(x, y; x', y') \, dx \, dy \tag{9.33}$$

9.8 INCOHERENT OPTICAL CORRELATOR ARCHITECTURES

The basic architecture for incoherent optical correlation is illustrated in Fig. 9.14. This architecture enables us to calculate the zero "lag" of the cross-correlation of two images $f(x, y)$ and $h(x, y)$, denoted as $g(0, 0)$ and given by

$$g(0, 0) = \int_{-\infty}^{\infty} \int_{-\infty}^{\infty} f(x, y) h(x, y) \, dx \, dy \tag{9.34}$$

This is the peak correlation value.

To obtain a full one-dimensional correlation, we must either shift the input signal $f(x, y)$ (e.g., along x) or translate the mask $h(x, y)$, in which case we have a scanning system. The result of correlation is

$$g(x_o, 0) = \int_{-\infty}^{\infty} f(x - x_o, 0) h(x, 0) \, dx \tag{9.35}$$

where x_o is the lag variable. This system is not very useful or practical, but it can be extended to simultaneously correlate the input $f(x, 0)$ with a library of reference functions $h_n(x, 0)$. To achieve this we must use a multichannel mask and astigmatic optics as shown in Fig. 9.15. This architecture then calculates

$$g_n = \int_{-\infty}^{\infty} f(x) h_n(x) \, dx \qquad (n = 1, 2, \ldots, N) \tag{9.36}$$

If we allow the input to translate, we obtain

$$g_n(x_o) = \int_{-\infty}^{\infty} f(x - x_o) h_n(x) \, dx \qquad (n = 1, 2, \ldots, N) \tag{9.37}$$

where x_o is the shift variable. This is the discrete version (in y) of the full 2-D integral

$$g(x_o) = \int_{-\infty}^{\infty} \int_{-\infty}^{\infty} f(x - x_o) h(x, y) \, dx \, dy \tag{9.38}$$

This integral can also represent the correlation function when the input is continuously translated in x, such that $x_o = v_s t$, where the scanning velocity (v_s) can signify a mechanical, electrical or acousto-optic implementation. The correlation output of the detector is then an explicit function of time.

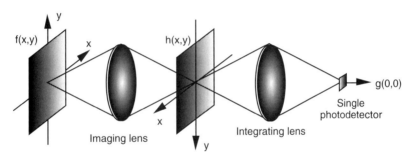

Figure 9.14 Simple incoherent optical correlator architecture corresponding to Equation (9.34).

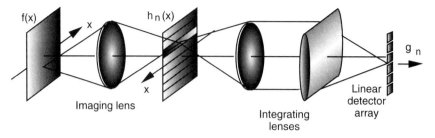

Figure 9.15 Multichannel optical correlator architecture corresponding to Equation (9.36). The input can also be allowed to translate, which is described by Equation (9.37).

Now consider a shift in two dimensions, given by

$$g(x_o, y_o) = \int_{-\infty}^{\infty} \int_{-\infty}^{\infty} f(x - x_o, y - y_o) h(x, y) \, dx \, dy \quad (9.39)$$

Instead of a full 2-D shift as indicated in Equation (9.39), we can implement a 2-D multichannel correlation by using replicating optics. A parallel 2-D array of lenslets for image casting is illustrated in Fig. 9.16. Each individual spherical lens (or "fly's eye") images the entire input image $f(x, y)$ onto the mask array denoted by $h_{mn}(x, y)$, where each (m, n)th submask thereby multiplies a miniature replica of $f(x, y)$. The resulting output at the (m, n)th detector is

$$g_{mn} = \int_{-\infty}^{\infty} \int_{-\infty}^{\infty} f(x, y) h_{mn}(x, y) \, dx \, dy \quad (m, n = 1, 2, \ldots, N) \quad (9.40)$$

Each $h_{mn}(x, y)$ can represent the (n, m)th subimage of a larger discretely sampled image or separate reference functions entirely.

Therefore, within one time interval all N^2 detectors can be read-out in parallel. Alternatively the read-out may be mixed (parallel in m, serial in n) or fully serial, but the resulting detector throughput and matching mask refresh rate will be lower.

The above-described correlators achieve a correlation by spatial shift. Another class of correlators achieves correlation by temporal shift (or scanning) as already intimated in the discussion following Equation (9.36). A typical system employs a modulated light source such as a temporally modulated laser diode as indicated in Fig. 9.17. An optical mask (transparency or SLM), represented by $h_n(x)$, is employed right after a collimating (projection) lens which

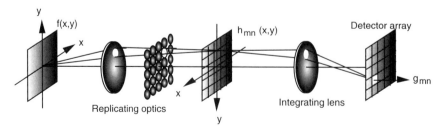

Figure 9.16 Multichannel correlator implemented using a parallel 2-D lenslet array.

illuminates the impulse response $h_n(x)$. The impulse response $h_n(x)$ consists of a library of 1-D reference signals. These are cross-correlated with the temporal laser modulation function $f(t)$. The light transmitted through $h_n(x)$ is the product of $f(t)$ and a time-shifted version of $h_n(x)$. During the time interval over which $f(t)$ is processed (integration period), either the optical mask or detector is translated in the x direction with a relative velocity v such that the output of the nth channel of the multichannel detector array is

$$g_n(x) = \int_{-\infty}^{\infty} f(t) h_n(x - vt) \, dt \qquad (n = 1, 2, \ldots, N) \tag{9.41}$$

Various implementations of scanning mask systems have been tried, including continuous strips of film, rotating disks, cylindrical drums, and scanning mirrors. Alternatively scanning detectors can be used such as translating photographic film or operating CCDs in a time-delay and integrate (or shift and add) mode.

Finally, let us consider nonscanning correlators. These use diffuse light sources and employ shadow casting in which no scanning or shifting is required. The diffusely scattered light source yields the irradiance $f(x, y)$, which is multiplied by $h(x, y)$ and integrated over x and y as depicted in Fig. 9.18. This is expressed by

$$g(x_o, y_o) = \int_{-\infty}^{\infty} \int_{-\infty}^{\infty} f\left(x - \frac{d}{f} x_o, y - \frac{d}{f} y_o\right) h(x, y) \, dx \, dy \tag{9.42}$$

Shadow-casting systems, however, are ultimately limited by diffraction effects, which limit the achievable space–bandwidth product to approximately 10^4.

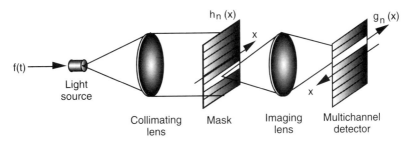

Figure 9.17 Temporal scanning optical correlator architecture employing scanning in one axis.

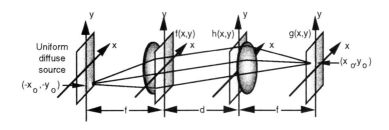

Figure 9.18 Shadow-casting correlator employing a telecentric lens pair.

9.9 COHERENT OPTICAL CORRELATOR ARCHITECTURES

In addition to optical correlators using incoherent light, we will consider time- and space-integrating correlators that use coherent light in conjunction with AO (Bragg) cells for spatial modulation. We will also consider (combined) time–space-integrating correlator architectures. In this discussion we will follow closely the work of Psaltis.

Consider first the space-integrating correlator shown in Fig. 9.19. In this architecture a point source (e.g., laser diode) is used as a source of illumination. A Bragg cell (at P_2) is used to spatially modulate the incident plane wave beam collimated by the lens L_1. The transmitted light distribution $t(x, t)$ is derived from the traveling acoustic wave $s_1(x - v_s t)$ created by the applied voltage $v(t)$. The Bragg cell window acts to truncate this spatial modulation by rect (x/w). The lens L_2 focuses the transmitted light to the Fourier plane P_3 where a spatial filter blocks the 0 and -1 diffracted orders. The $+1$ order beam continues on to lens L_3, which collimates the beam in order to illuminate a mask (or second Bragg cell) at P_4. The light amplitude $u(x, t)$ incident on the mask at P_4 is therefore given by

$$u(x, t) = \frac{\eta}{2} \tilde{s}_1^*(x - v_s t) \text{rect}(x/w) \qquad (9.43)$$

where η is the diffraction efficiency of the first Bragg cell. If we assume that the mask transmittance is given by $t_m(x) = 1 + s_2(x)$, then the light distribution exiting the mask plane (P_4) would be $u'(x, t) = u(x, t) t_m(x)$. (The mask can be replaced with another AO cell if the input signal is time-reversed.)

By employing another Fourier transform lens (L_4), we actually produce the Fourier transform with respect to the spatial coordinate (x) at the focal plane (P_5), which is given by

$$\text{FT}\{u'(x, t)\} = \int_{-\infty}^{\infty} u(x, t) t_m(x) \exp(-j2\pi u x) \, dx$$

$$= \frac{\eta}{2} \int_{-\infty}^{\infty} \tilde{s}_1^*(x - v_s t) \text{rect}\left(\frac{x}{w}\right) [1 + \tilde{s}_2(x)] \exp(-j2\pi u x) \, dx \qquad (9.45)$$

By virtue of the fact that a pinhole spatial filter is employed in the focal plane (P_5), the constant

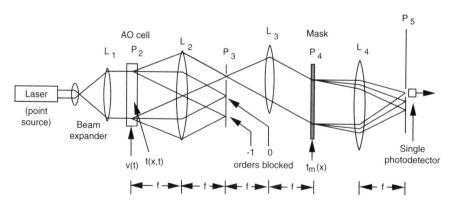

Figure 9.19 Space-integrating AO correlator. This correlator employs coherent light.

term in $t_m(x)$ is blocked and the spatial frequency u can be set to zero. Then the light distribution becomes only a function of time, namely,

$$u(t) = \frac{\eta}{2} \int_{-\infty}^{\infty} \tilde{s}_1^*(x - v_s t) \text{rect}\left(\frac{x}{w}\right) \tilde{s}_2(x) \, dx \tag{9.46}$$

and is equivalent to a correlation of $s_1(x)$ and $s_2(x)$ with lag variable $v_s t$ over a finite region defined by $\text{rect}(x/w)$. Denoting the correlation function as $R_{12}(t)$, the actual detected signal at the detector is a voltage $v_D(t)$ given by $V_D(t) = \mathcal{R} R_{12}^2(t)$, where $\mathcal{R}$ is the responsivity of the detector.

Obviously the correlation function explicitly depends on time, where the integration time is effectively set by either photodetector dwell time (or RC filter time) or by spatial resolution permitted by the Bragg cell to achieve a specified number of resolution cells in the truncation interval w, whichever is smaller.

Next consider the time-integrating correlator shown in Fig. 9.20. In this architecture a point source is again beam expanded and collimated to fully illuminate the Bragg cell located at P_1. In this case, however, the laser diode is not only DC-biased but modulated in time. Thus the signal applied to it is $v_o + v_1(t)$. The other signal $v_2(t)$, to correlate with $v_1(t)$, is applied to the Bragg cell. Thus the laser illumination intensity is $I_L(t) = K[v_o + v_1(t)] = s_o + s_1(t)$, and the diffracted beam after the Bragg cell is

$$u(t) = \eta[s_o + s_1(t)][1 + \tilde{s}_2^*(t - x/v_s)]\text{rect}\left(\frac{x}{w}\right) \tag{9.47}$$

Using two successive Fourier transform lenses (L_1 and L_2), separated from the intermediate Fourier plane P_2 by their common focal length (f), and using a spatial filter at P_2 that passes the first order diffracted beam and shifts the phase of the zero order (undiffracted) beam by $90°$, we get intensity at the detector plane given by

$$I_D(x, t) = \eta I_L(t) \left| 1 + \frac{1}{2}\tilde{s}_2^*(t - x/v_s) \right|^2 \tag{9.48}$$

After postdetection filtering (or integration for a duration ΔT) the detected output voltage versus spatial location on the linear detector array at P_3 is

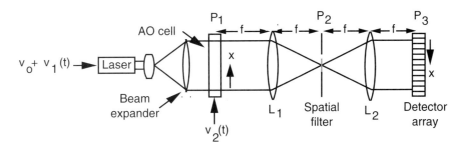

Figure 9.20 Time-integrating AO correlator. The spatial filter at P_2 shifts the zero order by $\pi/2$.

$$V_D(x) = \frac{1}{\Delta T} \int_0^{\Delta T} \eta I_L(t)\, dt + \frac{1}{4\Delta T} \int_0^{\Delta T} \eta I_L(t)|\tilde{s}_2(t - x/v_s)|^2\, dt$$
$$+ \frac{1}{4\Delta T} \int_0^{-\Delta T} \eta I_L(t)\tilde{s}_2^*(t - x/v_s)\, dt \tag{9.49}$$

The first term in Equation (9.49) is simply a bias, and the second term can be neglected in the small signal limit. The cross-term contributes another bias term and the cross-correlation term, namely,

$$V_D(x) = \text{Bias} + \text{Negligible Term} + \frac{\eta}{\Delta T}\int_0^{\Delta T} s_1(t)\tilde{s}_2^*(t - x/v_s)\, dt \tag{9.50}$$

The advantage of this architecture is that the integrating time ΔT can be quite large, leading to a large time–bandwidth product and large signal-to-noise ratio limited only by the sensor or the desired data acquisition time.

Finally, to tie everything together that has been introduced so far, we will combine time- and space-integrating architectures into one system. In order to consider optical processors that are both time- and space-integrating (TSI), we will restrict our attention to 2-D linear operations which can be performed by 1-D processors. This will be possible for two types of kernels: shift-invariant or those separable in two coordinates. Each line of the input signal is processed by space integration, and the output is produced by time integration. The space integration typically yields a very high effective throughput (e.g., $\sim 10^{10}$ samples/sec), and the time integration compresses the result typically by three orders of magnitude.

Time- and space-integrating processors can implement the general linear superposition integral:

$$g(x', y') = \int_{-\infty}^{\infty}\int_{-\infty}^{\infty} f(x, y)h(x, x'; y, y')\, dx\, dy \tag{9.51}$$

where $f(x, y)$ is an image that will generally be sampled in y and scanned in x, and $h(x, y)$ is the system impulse response. Under these circumstances a temporal signal can be processed as sequential raster-scanned video, where the raster line is usually made equal to the AO cell aperture, and the input (laser) light source illuminates the entire aperture. If the laser illuminator is strobed synchronously with the filling of the AO cell aperture (with the acoustic modulation signal), then each raster-line pattern spatially modulates the laser light and sequential lines temporally modulate it. Thus $f(x, y)$ is converted to a function spatially modulated in x and temporally modulated in y, denoted by $f(x, n\Delta y)$. Thus the first stage in the processor is a 1-D multichannel space integration such that

$$g(x', y') \to g_n(x', y') = \int_{-\infty}^{\infty} f(x, n\Delta y)h(x, x'; n\Delta y, y')\, dx \tag{9.52}$$

The output is detected by a 2-D integrating CCD imager where consecutive lines are accumulated. After N lines we have

$$g(x', y') = \sum_{n=0}^{N-1} \int_{-\infty}^{\infty} f(x, n\Delta y)h(x, x'; n\Delta y, y')\, dx \tag{9.53}$$

9.9 COHERENT OPTICAL CORRELATOR ARCHITECTURES

Thus, if the kernel is separable, then $h(x, x'; y, y') = h_1(x, x')h_2(y, y')$ and

$$g(x', y') = \sum_{n=0}^{N-1} \left[\int_{-\infty}^{\infty} f(x, n\Delta y) h_1(x, x') \, dx \right] h_2(n\Delta y, y') \quad (9.54)$$

Separable kernels can therefore be implemented with a cascade of 1-D space and time integrators.

Shift invariant kernels can also be processed. They are of the form

$$g(x', y') = \sum_{n=0}^{N-1} \int_{-\infty}^{\infty} f(x, n\Delta y) h(x + x', n\Delta y + y') \, dx \quad (9.55)$$

which is equivalent to

$$g(x', y') = \sum_{n=0}^{N-1} \left[\int_{-\infty}^{\infty} f(x, n\Delta y) h(x + x', y) \, dx \right] \otimes \delta(y + n\Delta y) \quad (9.56)$$

where the integral in brackets is a 1-D correlation of the nth line of $f(x, y)$ with all lines of $h(x, y)$, and the second operation denoted by $\otimes$ represents a 1-D correlation with lag variable $n\Delta y$ perpendicular (orthogonal) to that of the first correlation. Ultimately the 2-D correlation is obtained by a sum over all partial correlations (or lagged products) and is performed by the time integrating detector array. A shift in y must be accomplished, however, before the sum over n.

Consider the architecture shown in Fig. 9.21 as first described by Psaltis. We will consider this architecture for calculating:

$$g(x', y') = \int_{-\infty}^{\infty} \int_{-\infty}^{\infty} f(x, y) h_1(x; y') h_2(x + x') \, dx \, dy \quad (9.57)$$

where (x, y) are the input variables and (x', y') are the output variables. The input $f(x, y)$ can be expressed as a temporal raster modulation $s(t)$, given by

$$s(t) = \sum_{n=0}^{N-1} \text{rect} \left[\frac{t - (n + 1/2)\tau}{\tau} \right] f\left(t - n\tau, \frac{n}{N}\tau\right) \quad (9.58)$$

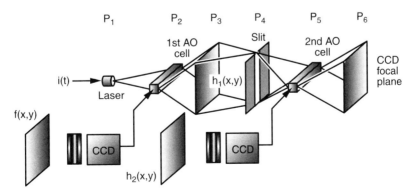

Figure 9.21 Time- and space-integrating AO correlator, in which the input signal is separable in x and y coordinates.

or

$$s(t) = \text{rect}\left(\frac{t - \tau/2}{\tau}\right) f(t, 0) + \text{rect}\left(\frac{t - 3\tau/2}{\tau}\right) f(t - \tau, \tau/N)$$
$$+ \text{rect}\left(\frac{t - 5\tau/2}{\tau}\right) f(t - 2\tau, 2\tau/N) + \ldots \quad (9.59)$$

where τ is the raster line duration, N is the number of raster lines, and $0 < (x, y) < v_s\tau$ and $0 < t < N\tau$. $s(t)$ is illustrated in Fig. 9.22. In the AO cell the temporally modulated signal $s(t)$ becomes a traveling wave $s(t - x_1/v_s)$ given by

$$s(t - x_1/v_s) = \sum_{n=0}^{N-1} \text{rect}\left[\frac{t - x_1/v_s - (n + 1/2)\tau}{\tau}\right] f\left(t - \frac{x_1}{v_s} - n\tau, \frac{n}{N}\tau\right) \quad (9.60)$$

where it is assumed that the raster line completely fills the AO aperture.

The first cell is modulated by a coherent light source with PRF $= 1/\tau$ and ideal (instantaneous) pulses, which are functionally described as

$$i(t) = \sum_{n=0}^{N-1} \delta[t - (n + 1)\tau] \quad (9.61)$$
$$= \delta(t - \tau) + \delta(t - 2\tau) + \delta(t - 3\tau) + \ldots \quad (9.62)$$

and are illustrated in Fig. 9.23.

Right after the AO cell at P_2 the modulated light intensity is therefore $i(t) \cdot s(t - x_1/v_s)$, neglecting the zero and -1 diffraction orders. The output (after the first cell) is imaged in x_1 and uniformly expanded in the y direction. At P_3 a 2-D mask (or SLM) is placed with an intensity transmittance $h_1(-x_1, y)$, which is recorded on a spatial carrier in y to synthesize an effective complex transmittance. The light intensity at P_3 is therefore $i(t) \cdot s(t - x_1/v_s) \cdot h_1(-x_1, y)$.

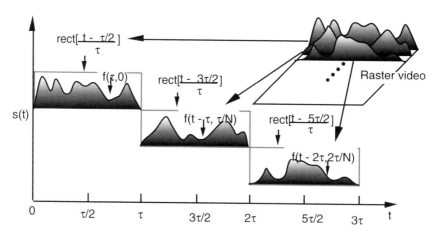

Figure 9.22 Temporal raster modulation signal which represents the 2-D input signal to the TSI architecture shown in Fig. 9.21.

Figure 9.23 Ideal laser diode temporal modulation impulse train.

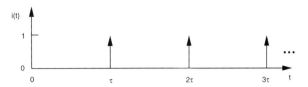

This result is imaged onto plane P_5 where the second AO is located. At the intermediate plane (P_4) a slit is placed which acts as a low-pass filter in the x (image) direction but does not affect the y direction. If the slit is narrow enough, the result at plane P_5 will have no variations in amplitude versus x, which can be expressed as a spatial integration. Then

$$s_1(t, y) = \int_0^{v_s \tau} o(t) s(t - x_1/v_s) h_1'(-x_1, y) \, dx_1 \quad (9.63)$$

Thus, the first part of the architecture is a 1-D space integrating processor, where the original image $f(x, y)$ is filtered in the x direction.

To realize filtering in the y direction, it is necessary to compress the scale of $h_2(y)$ by a factor of two (as will be seen below) and apply it periodically to the second AO cell. Thus the intensity modulation at the second cell is

$$h_2(t - x_2/v_s) = \sum_{n=0}^{N-1} h_2 \left\{ 2[t - x_2/v_s - (n + 1/2)\tau] + \frac{n}{N}\tau \right\} \quad (9.64)$$

where $0 < y < \tau/v_s$, and $0 < x_2 < \tau/v_s$, where x_2 is the acoustic wave direction. Writing Equation (9.64) out for $x_2 = 0$ yields the time dependence $h_2(t)$, namely,

$$h_2(t) = h_2(2t - \tau) + h_2(2t - 3\tau + \tau/N) + h_2(2t - 5\tau + 2\tau/N) + \ldots$$
$$= h_2[2(t - \tau/2)] + h_2[2(t - 3\tau/2 + \tau/2N)] + h_2[2(t - 5\tau/2 + \tau/N)] + \ldots \quad (9.65)$$

which is illustrated in Fig. 9.24. As can be seen, $h_2(t)$ is centered at every $(\tau - \tau/N)$ interval, which is a period slightly smaller than the period of the pulsed source. When the pulsed source is on, at times $t = (n + 1)\tau$, the modulation in the second AO cell is

$$h_2[(n + 1)\tau - x_2/v_s] = h_2[\tau - 2x_2/v_s + n\tau/N] \quad (9.66)$$

Therefore, at successive pulse times the modulation in the second AO cell is shifted in x_2 by $\tau/2N$.

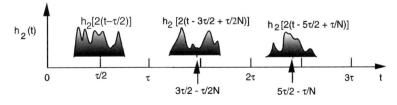

Figure 9.24 Intensity modulation at the second Bragg cell of the TSI architecture described in Fig. 9.21.

The spatial modulation at plane P_5 is then imaged onto a 2-D time-integrating CCD at plane P_6 with integration time $N\tau$. When imaging from P_5 to P_6 a half-plane stop is used to block the 0 and -1 diffracted orders, a technique known as Schlieren imaging, which will be described later in Chapter 10. The total light intensity integrated at each point in space on the CCD is then

$$g'(x_2, y) = \int_0^{N\tau} s_1(t, y) h_2(t - x_2) \, dt \tag{9.67}$$

Substituting Equations (9.60)–(9.65) into Equation (9.67) therefore yields

$$g'(x_2, y) = \sum_{n=0}^{N-1} \int_0^{N\tau} \delta[t - (n+1)\tau] h_2 \left\{ 2[t - x_2/v_s - (n + \frac{1}{2})\tau] + \frac{n}{N}\tau \right\} \tag{9.68}$$

$$\times \left\{ \int_0^\tau \text{rect} \left[\frac{t - x_2/v_s - (n+1/2)\tau}{\tau} \right] f\left(t - x_2/v_s - n\tau, \frac{n}{N}\tau\right) h'_1(-x_1, y) \, dx_1 \right\} dt$$

By virtue of the delta function the integral over t is eliminated and Equation (9.68) reduces to

$$g'(x_2, y) = \sum_{n=0}^{N-1} \int_0^\tau f\left(\tau - x_2/v_s, \frac{n}{N}\tau\right) h'_1(-x_1, y) h_2\left(\tau - x_2/v_s + \frac{n}{N}\tau\right) dx_1 \tag{9.69}$$

Letting $\tau - x_2/v_s = \tau'$, $n\tau/N = t'$, $\tau' = 2x_2/v_s = x/v_s$ and $h'_1(\tau - \tau') = h_1[v_s(\tau - \tau'), y]$ yields

$$g'(x, y) = \sum_{t'=0}^{\tau} \int_\tau^0 f(\tau', t') h_1(\tau', y) h_2(t' + x/v_s) \, d\tau' \tag{9.70}$$

where t' takes on discrete values corresponding to the integer values of n. Thus the input image $f(x, y)$ is processed by continuous spatial integration along the x direction (related to τ') and discrete temporal integration along the y direction (related to t' indexed by n). The utility of this result is that it can be applied to SAR image formation and the calculation of moments as will be described in the Chapters 11 and 12.

In most correlators the critical performance factor is the time–bandwidth product (τB), since it determines overall (output) signal-to-noise ratio (SNR_o) according to $SNR_o = \tau B \cdot SNR$, where SNR is the signal-to-noise ratio of the received signal. In radar applications, such as pulse-compression radar, bandwidth should also be large to ensure a narrow correlation peak since the received signal mainlobe width (which is a measure of range resolution) is inversely proportional to bandwidth.

For spatially integrating correlators the maximum achievable time–bandwidth product is indeed τB, but for time-integrating correlators it is the square of dynamic range (since it is not fundamentally limited by integration time). In fact, the minimum detectable signal level is determined by the reciprocal of τB, except for the time-integrating correlator, where it is limited by RMS detector noise, or equivalently the reciprocal of dynamic range squared.

Incoherent optical correlators employ light intensity and therefore must incorporate a bias level to represent bipolar numbers. The subsequent bias at the detector means that dynamic range is reduced according to

$$DR_o = DR \left(\frac{SBR}{1 + SBR} \right) \tag{9.71}$$

where DR is the dynamic range of the CCD, DR_o is the dynamic range of the overall system, and SBR is the ratio of the maximum output signal-to-bias level. Typical values of DR for CCDs are limited to approximately 10^3, and SBR is defined to be between 0 and 1. Overall dynamic range can also be reduced by additive interference. (Incoherent systems, however, are insensitive to coherent noise, such as speckle and interference.) Values of 100:1 are sufficient for reliable peak detection, but values below 10:1 are not.

Incoherent systems are most suitable when the signal is much greater than the interference, which is often true in pattern recognition applications. In general, however, coherent architectures are preferred not only because they eliminate the need for bias (or defer it until after detection) but because they increase output SNR with processing gain (G) according to $SNR_o = SNR \cdot G^{3/2}$.

PROBLEM EXERCISES

1. Consider the space-integrating correlator shown below. The signal to be correlated, $s_1(t)$, produces a phase modulation such that the amplitude of the transmitted beam is

$$a(x, t) = a_o \exp\{j[\omega t + \alpha s_1(x + v_s t)]\}$$

(a) Write down the form of $a(x, t)$ in the small signal limit to first order in α.

(b) Derive an expression for the light distribution at the first transform plane (P_2).

(c) Assuming the zero-order light is blocked, what is the form of the light distribution function at the second transform plane (P_3)?

(d) Assuming that a mask of the form $t(x) = 1 + s_2(x)$ is placed at P_3, what is the overall transmitted amplitude?

(e) The final transform lens produces a light distribution in plane P_4 that is detected by a single photodiode. What is this light distribution?

(f) If the photodetector is a square-law device, what is the final expression for the photodiode current $i(t)$ and how is it related to the cross-correlation function between s_1 and s_2?

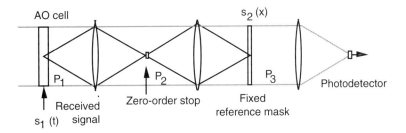

2. Consider the simple correlator architecture shown below. A noisy signal of time duration T is injected into the AO cell of length D as a modulation on a carrier at a center frequency ω_c. Assume that the mask is close enough to the Bragg cell to neglect diffraction effects and that it has the functional form given by

$$\{b_o = b_1 a(x/v_s) \cos[(\omega_o + \omega_B) x/v_s + \phi(x/v_s)]\} \text{rect}(x/v_s)$$

Determine the light amplitude at the focal plane of the Fourier transform lens.

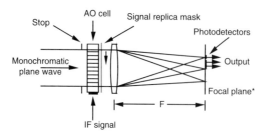

3. For the space-integrating AO spectrum analyzer described in Fig. 9.11, assume that the input is a two-tone signal given by

$$s(t) = a_1 \exp(j\omega_1 t) + a_2 \exp(j\omega_2 t)$$

Calculate the resultant amplitude distribution in the output plane denoted by $A(f_x)$, where $f_x = x_o/\lambda F$. Then calculate the photodetector response $V(x_n)$.

4. For the time-integrating AO spectrum analyzer described in Fig. 9.13, assume that the input signal $s(t)$ is the same as that in Problem 3. Determine the output light distribution after the second AO cell to first order in each of the exponential terms $\exp(j\alpha s_1)$ and $\exp(j\beta s_2)$. Then calculate the photodetector response given by

$$V_n(\tau_n) = \int_{-\infty}^{\infty} \text{rect}\left(\frac{t' - \tau_1}{T}\right) [|s_1|^2 + |s_2|^2 + s_1 s_2^* + s_1^* s_2] \, dt' * |h'_D(\tau)|^2$$

where

$$|h'_D(\tau)|^2 = \frac{2}{v_s} \text{rect}\left(\frac{\tau}{\Delta x/2\pi}\right)$$

5. Compare the results of Problems 3 and 4 by plotting the resulting normalized spectra on a common set of axes. How do sidelobe levels and frequency resolution compare?

6. In the nonheterodyne correlator of Fig. 9.19 the incident field amplitude at plane P_4 is given by

$$E_{\text{inc}}(x, t) = \tfrac{1}{2} \tilde{s}_1^*(x - v_s t) \text{rect}(x/D)$$

and the mask transmission is given by

$$t_m(x) = 1 + s_2(x) = 1 + \tfrac{1}{2}[\tilde{s}_2(x) + \tilde{s}_2^*(x)]$$

(a) Write down the form of the transmitted field amplitude $E_{\text{trans}}(x, t)$.

(b) Neglecting a quadratic phase factor and looking at the signal at zero spatial frequency, derive the amplitude at the output plane (P_5) and show that it is given by

$$E(t) = \tfrac{1}{4} \int \tilde{s}_1^*(x - v_s t) \tilde{s}_2(x) \text{rect}(x/D) \, dx$$

(c) Derive the photodetector output, assuming that it is a square-law detector. Note that the detected output is the envelope of the cross-correlation of s_1 and s_2.

7. In the architecture of Problem 6 (Fig. 9.19), instead of placing the first pinhole at plane P_3 at the $+1$ diffracted order and the photodiode at the zeroth order, place a half-plane stop so that both the $+1$ and 0 orders pass at plane P_3 and place the photodiode at an off-axis position. Also allow a $\pi/2$ phase shift to occur for the $+1$ component at plane P_3. Derive an expression for the transmitted field $E_{\text{trans}}(x, t)$. Assuming that only singly diffracted components survive, that is,

$$\tfrac{1}{2}\tilde{s}_1^*(x - v_s t) \quad \text{and} \quad \tfrac{1}{2}\tilde{s}_2^*(x)$$

   show that the photodetector current

$$i_D(t) = \int |E_D(x, t)|^2 \, dx$$

   has a cross-term given by

$$E(t) = \tfrac{1}{2} \int \tilde{s}_1^*(x - v_s t)\tilde{s}_2(x)\text{rect}(x/D) \, dx$$

   which is the desired cross-correlation term. What is the temporal frequency at which this term is located? How can it be isolated from the other squared terms in the output photodetector current?

8. Calculate the spectrum of a two-tone signal $v(t) = a_1 \cos \omega_1 t + a_2 \cos \omega_2 t$ input to the AO cell of Fig. 9.1. Assume that $W(u)$ is derived from Equation (9.11) using the models for $w_1(x)$, $w_2(x)$, and $w_3(x)$ given by Equations (9.8), (9.10), and $w_2(x) = \exp(-\alpha x)$.

9. Using Fig. 9.21 and following the discussion of the time- and space-integrating correlator operation in the text, complete the design by defining the lenses required to enable this architecture to perform the indicated processing steps. Be sure to specify the focal lengths, size, position, and orientation of each element.

BIBLIOGRAPHY

B. G. Boone, S. A. Gearhart, O. B. Shukla, and D. H. Terry, "Optical Processing for Radar Signal Analysis," *John Hopkins APL Tech. Digest* **10** (1), 14–28 (January–March 1989).

J. R. Fienup, "Phase Retrieval Algorithms: A Comparison," *Appl. Opt.* **21**, 2758 (1982).

I. Glaser, "Information Processing with Spatially Incoherent Light," in *Progress in Optics*, Vol 24, E. Wolf, ed. Elsevier Science Publishers, New York, pp. 389–509 (1987).

P. S. Guilfoyle, D. L. Hecht, and D. L. Steinmetz, "Joint transform time-integrating acousto-optic correlator for chirp spectrum analysis," *Optical Engineering* **20**, 556 (1981).

M. Hayes, "The Reconstruction of a Multidimensional Sequence from the Phase or Magnitude of Its Fourier Transform," *IEEE Trans.* **ASSP-30**, 140 (1982).

D. L. Hecht, "Spectrum Analysis Using Acousto-optic Devices," *Opt. Eng.* **16**, 461 (1977).

D. L. Hecht and P. S. Guilfoyle, "Acousto-optic Spectrum and Fourier Analysis Techniques,"

M. A. Monahan, K. Bromley, and R. P. Bocker, "Incoherent Optical Correlators," *Proc. IEEE* **65**, 121 (1977).

D. Psaltis, "Optical Image Correlation Using Acoustooptic and Charge-Coupled Devices," *Appl. Opt.* **21**, 491 (1982).

D. Psaltis, E. G. Paek, and S. Venkatesh, "Acousto-optic/CCD Image Processor," pp. 54–58, Appendix Vb in "Acousto-Optic Processing of 2-D Signals Using Temporal and Spatial Integration," Report for AFOSR Contract 82-0128, May 31, 1983.

W. T. Rhodes, "Acousto-optic Signal Processing: Convolution and Correlation," *Proc. IEEE* **69**, 65 (1981).

Chapter 10
IMAGE AND MATCHED SPATIAL FILTERING

10.1 OVERVIEW

In this chapter we will address the concept of correlation from a filtering perspective. In the previous chapter we considered it in the context of its equivalencey to spectrum analysis, a connection established by the Wiener–Khintchine theorem. As we will see, viewing correlation from a filtering perspective is also a productive approach since it leads eventually to other types of filtering and to feature extraction for pattern recognition. We will set the stage by reviewing the classic VanderLugt filter and then go on to consider improvements on the basic matched filter that improve its performance and make it less susceptible to distortions of the input image. Finally, we will consider a novel alternative to standard (Cartesian) correlation, known as *angular correlation*.

10.2 VANDERLUGT FILTER

A historically important optical processing architecture involved with correlation is the VanderLugt filter. This is a matched filter created by a holographic method in which a reference image, consisting of a uniform plane wave of amplitude A and oblique incidence θ, is added to the Fourier transform of an input function $h(x, y)$ at the output plane of a simple Fourier transform architecture as shown in Fig. 10.1. The result is

$$H'(u, v) = H(u, v) + A_o \exp(j\alpha x_2) \qquad (10.1)$$

where $(u, v) = (x_2/\lambda F, y_2/\lambda F)$, $\alpha = (2\pi \sin\theta)/\lambda$, and $H(u, v)$ is the Fourier transform of $h(x, y)$. The recording of the linear superposition [Equation (10.1)] by a square-law detector (photographic or CCD) yields

$$|H_M(u, v)|^2 = |H(u, v)|^2 + A_o^2 + HA_o \exp(-j\alpha x_2) + H^* A_o \exp(j\alpha x_2) \qquad (10.2)$$

This is the VanderLugt (or matched) filter for $h(x, y)$ and can be placed into the Fourier plane of the so-called coherent optical (matched filter) correlator architecture shown in Fig. 10.2.

10.2 VANDERLUGT FILTER

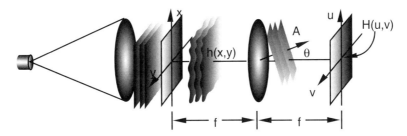

Figure 10.1 Optical setup for preparing a VanderLugt matched filter.

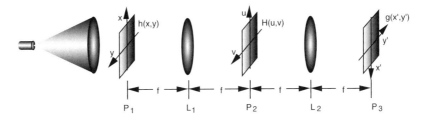

Figure 10.2 Coherent optical (matched filter) correlator architecture.

The first lens of this architecture creates the Fourier transform of the input (object) distribution denoted by $f(x, y)$. Thus the distribution right after the Fourier plane (P_2) is

$$F(u, v)|H_M(u, v)|^2 = F(u, v)[A_o^2 + |H(u, v)|^2 + A_o H \exp(-j\alpha\lambda F u) \\ + A_o H^* \exp(j\alpha\lambda F u)] \quad (10.3)$$

The second lens also performs a Fourier transform to yield the output distribution $g(x, y)$ given by

$$g(x, y) = A_o^2 f(x, y) + f(x, y) * [h(x, y) \otimes h^*(x, y)] \\ + A_o f(x, y) * h(x + F \sin\theta, y) + A_o f(x, y) \otimes h(x - F \sin\theta, y) \quad (10.4)$$

where $*$ denotes convolution and $\otimes$ denotes correlation (but superscript $*$ denotes complex conjugate). A general result for $g(x, y)$ is illustrated in Fig. 10.3 showing the presence of four terms. The first term gives an image of $f(x, y)$ centered on-axis. The second term is also

Figure 10.3 Plan view of a general output plane response for the coherent optical (matched filter) correlator.

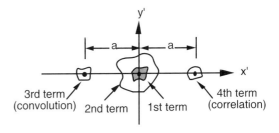

centered on-axis and is the convolution of the input function $f(x, y)$ with the autocorrelation of the matched filter. The third term is located at $x = -F \sin \theta$ and corresponds to the convolution of $f(x, y)$ with $h(x, y)$, whereas the fourth term is located at $x = +F \sin \theta$ and represents the correlation of $f(x, y)$ with $h(x, y)$. The last term is of interest to us because it represents correlation or matched filtering. The matched filter $h(x, y)$ for $f(x, y)$ is indeed $f(x, y)$ and as such will have the maximum possible correlation value at $x = F \sin \theta$. This result can be readily extended to two dimensions by using a reference beam given by $A_1 \exp(j\alpha x_2) + B_1 \exp(j\beta y_2)$.

10.3 IMAGE SPATIAL FILTERING

The notion of a Fourier transform correlator can be applied to many other signal processing functions, including spatial filtering, such as high-pass and low-pass filtering. In fact, we have already discussed low-pass filtering of laser beam noise (see Chapter 7). It can also be used to perform contrast reversal, Schlieren imaging, and, with a slight modification of the classical matched filter, binary phase-only correlation. We will consider these as examples of the utility of the Fourier-transform-based (or matched-filter) correlator architecture.

EXAMPLE 10.1 Low-Pass and High-Pass Filtering In the matched-filter correlator architecture assume that the input plane (P_1) consists of an SLM with the following amplitude transmittance:

$$t(x, y) = 1 + \frac{1}{2}\cos(2\pi x/a) + \frac{1}{2}\cos(2\pi x/2a)$$

which is illustrated in Fig. 10.4(a). (We also require the input light source to be a monochromatic coherent plane wave.) The corresponding Fourier transform in plane P_2 is then

$$T(u, v) = \left[\delta(u) + \frac{1}{4}\delta\left(u - \frac{1}{a}\right) + \frac{1}{4}\delta\left(u + \frac{1}{a}\right) + \frac{1}{4}\delta\left(u - \frac{1}{2a}\right) + \frac{1}{4}\delta\left(u + \frac{1}{2a}\right)\right]\delta(v)$$

which is illustrated in Fig. 10.4(b).

Ignoring the effects of lens apertures, we would expect to exactly reconstruct $t(x, y)$ in the output plane P_3. If, however, we place a spatial filter at plane P_2 with a circular aperture of radius between $1/2a$ and $1/a$, as illustrated in Fig. 10.5(a), then we could modify the output plane distribution by eliminating the higher frequency components $\frac{1}{4}\delta(u \pm 1/a)$. Then the resulting output distribution would be $1 + \frac{1}{2}\cos(2\pi x/2a)$ as shown in Fig. 10.5(b), which is low-pass filtering. If, however, we place a circular stop in plane P_2 as shown in Fig. 10.5(c),

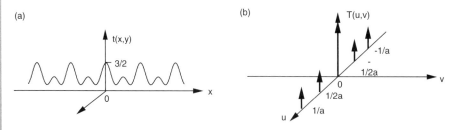

Figure 10.4 (a) Amplitude transmittance function $t(x, y)$ for the SLM, and (b) corresponding Fourier transform $T(u, v)$.

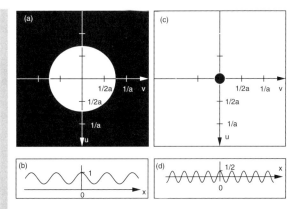

Figure 10.5 (a) Functional form of ideal low-pass filter in plane P$_2$, (b) corresponding output plane distribution, (c) functional form of circular stop (high-pass filter), and (d) corresponding output plane distribution.

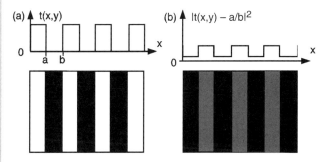

Figure 10.6 (a) Input square-wave intensity pattern input signal to matched filter correlator, and (b) corresponding output intensity pattern showing contrast reversal and attenuation.

then we pass the two high-frequency components yielding $\frac{1}{2}\cos(2\pi x/a)$, which is a bipolar signal, the result of high-pass filtering. It is shown in Fig. 10.5(d).

EXAMPLE 10.2 Contrast Reversal Consider the input signal shown in Fig. 10.6(a) as the input to the matched-filter correlator architecture shown in Fig. 10.2, where the intermediate (P$_3$) plane has a stop located at the center: $(x_3, y_3) = (0, 0)$, as shown in Fig. 10.5(c). This stop blocks out any DC component in the Fourier transform of the input function as was discussed in the previous example. Assume that the input light distribution (or object transmission) for an incident plane wave is $t(x, y)$ given by $t(x, y) = t(x) \cdot 1$, where $t(x)$ is the bar pattern shown in Fig. 10.6(a) in plan view (located at P$_1$). We can describe this pattern as

$$t(x) = \text{rect}\left(\frac{x}{a}\right) * \text{comb}\left(\frac{x}{b}\right)$$

where $b > a$. The Fourier transform of this pattern, which is accomplished by the lens (L$_1$) at plane P$_2$, is

$$T(f_x, f_y) = \int_{-\infty}^{\infty}\int_{-\infty}^{\infty} \text{circ}\left(\frac{\sqrt{x^2+y^2}}{R}\right) t(x, y) \exp\left[-j2\pi(xf_x + yf_y)\right] dxdy$$

The effect of the pupil function of the lens—that is, $\text{circ}[(x^2 + y^2)^{1/2}/R]$—will be neglected. Then the above equation yields $T(f_x, f_y) = T(f_x)T(f_y)$, which follows from the fact that $t(x, y)$ is separable. The preceding equation is quite general, but in this case $T(f_x) = b\,\text{sinc}(bf_x) \bullet a\,\text{comb}(af_x)$ using the convolution theorem, the definition of $t(x)$, and the fact that $T(f_y) = \delta(f_y)$.

In any case, $T(f_x, f_y)$ is multiplied by the Fourier plane stop, which is given by

$$S(f_x, f_y) = 1 - \text{circ}\left(\frac{\sqrt{f_x^2 + f_y^2}}{f_o}\right)$$

Thus the distribution of light in the output (or image) plane P_5 is

$$i(x, y) = \int_{-\infty}^{\infty}\int_{-\infty}^{\infty} T(f_x)T(f_y)\left[1 - \text{circ}\left(\frac{\sqrt{f_x^2 + f_y^2}}{f_o}\right)\right]$$
$$\exp\left[-j2\pi(f_x x + f_y y)\right] df_x df_y$$

since lens L_2 at plane P_4 acts to Fourier transform the distribution of light in plane P_3 (again neglecting aperture effects). In the ideal case where the DC stop blocks only $f_x = f_y = 0$, then

$$\text{circ}\left(\frac{\sqrt{f_x^2 + f_y^2}}{f_o}\right) = \delta(f_x)\delta(f_y)$$

Therefore the output image becomes

$$i(x, y) = \text{rect}\left(\frac{x}{a}\right) * \text{comb}\left(\frac{x}{b}\right) - \int_{-\infty}^{\infty}\int_{-\infty}^{\infty} T(f_x)T(f_y)\delta(f_x)\delta(f_y)$$
$$\cdot \exp\left[-j2\pi(f_x x + f_y y)\right] df_x df_y$$

which reduces to

$$i(x, y) = t(x, y) - T(f_x)T(f_y)|_{f_y=0}^{f_x=0}$$

Now, since $T(f_y) = \delta(f_y)$, then $T(f_y = 0) = 1$ and

$$T(f_x) = \text{FT}\left\{\text{rect}\left(\frac{x}{a}\right) * \text{comb}\left(\frac{x}{b}\right)\right\}\bigg|_{f_x=0}$$

Since the function $t(x, y)$ is a periodic square wave, when we evaluate the last equation at $f_x = 0$ we are looking for the DC term. The DC term is given by the first term in the Fourier series, namely,

$$a_o = \frac{1}{b}\int_{-b/2}^{b/2} \text{rect}\left(\frac{x}{a}\right) dx = \frac{a}{b}$$

Therefore,

$$i(x, y) = t(x) - \frac{a}{b}$$

The corresponding intensity distribution is

$$|i(x, y)|^2 = \left|t(x) - \frac{a}{b}\right|^2$$

which is shown in Fig. 10.6(b). Note that the output is attenuated and contrast-reversed from the input.

EXAMPLE 10.3 Schlieren Imaging As the next example of the application of the matched-filter correlator architecture, consider the arrangement where a phase object defined by $U_o(x_o, y_o)$ is placed in plane P_1 (the object plane of Fig. 10.2) and illuminated with a uniform plane wave. In the Fourier plane (P_2) a half-plane stop is put, otherwise known as a *knife-edge*. The resulting image at the output plane is $U_i(x_i, y_i)$. The light distribution incident on plane P_3 is defined as $U_o(x', y')$, where $U_o(x', y')$ corresponds to the Fourier transform of $U_i(x_i, y_i)$. The half-plane stop at P_2 is defined as

$$t(x', y') = \frac{1}{2}\left[1 + \text{sgn}(x')\right] \cdot 1$$

Thus the light distribution exiting plane P_3 is $U_o(x', y')[1 + \text{sgn}(x')]/2$, and the resulting light distribution in the image plane (P_3) is given by

$$U_i(x_i, y_i) = \int_{-\infty}^{\infty}\int_{-\infty}^{\infty} U_o(x', y')\frac{1}{2}\left[1 + \text{sgn}(x')\right]\exp\left[-j2\pi(x'x_i + y'y_i)\right] dx'dy'$$

If we use the product theorem and the fact that

$$\text{FT}\left\{\frac{1}{2}\left[1 + \text{sgn}(x')\right]\right\} = \frac{1}{2}\left[\delta(x_i) - \frac{j}{\pi}\frac{1}{x_i}\right]\delta(y_i)$$

and that $\text{FT}\{U_o(x', y')\} = U_o(x_i, y_i)$, then we will obtain

$$U_i(x_i, y_i) = \frac{1}{2}\left[\delta(x_i) - \frac{j}{\pi}\frac{1}{x_i}\right]\delta(y_i) * U_o(x_i, y_i)$$

$$U_i(x_i, y_i) = \frac{1}{2}U_o(x_i, y_i) - \frac{j}{\pi}\int_{-\infty}^{\infty}\frac{U_o(x', y_i)}{x_i - x'} dx'$$

where the second term is the Hilbert transform of $U_o(x', y_i)$.

As we pointed out earlier, $U_o(x_o, y_o)$ can be merely a phase object; so we will define it to be $U_o(x_o, y_o) = \exp[j\phi(x_o, y_o)]$. If we assume that $\phi(x_o, y_o) \ll 1$, then $U_o(x_o, y_o) \approx 1 + j\phi(x_o, y_o)$, and the resulting amplitude in the image plane is

$$U_i(x_i, y_i) = \frac{1}{2}\left[1 + j\phi(x_i, y_i) - \frac{i}{\pi}\int_{-\infty}^{\infty}\frac{1 + j\phi(x', y_i)}{x_i - x'} dx'\right]$$

Thus the intensity in the image plane can be shown to be

$$I_i(x_i, y_i) \approx 1 + \frac{2}{\pi}\int_{-\infty}^{\infty}\frac{\phi(x', y_i)}{x_i - x'} dx'$$

If we input a particular phase modulation $\phi(x_o, y_o)$, for example a finite extent, finite thickness glass plate (with transmission = 1), we would describe it as

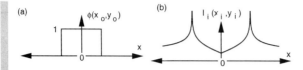

Figure 10.7 (a) Rectangular phase modulation input to Schlieren architecture and (b) output plane intensity distribution, showing edge enhanced image of the phase object.

$$\phi(x_o, y_o) = \phi_o \text{rect}\left(\frac{x_o}{X}\right)$$

in one axis, as shown in Fig. 10.7(a). Then the image plane intensity is

$$I_i(x_i, y_i) \cong \frac{1}{4}\left[1 - \frac{2}{\pi}\phi_o \left|\log_e\left|\frac{x_i - X/2}{X_i + X/2}\right|\right|\right]$$

which is plotted in Fig. 10.7(b). Thus the edges of the phase object are made visible and appear as enhanced edges of a rectangular pillbox. It is clear from the previous examples that the basic matched-filter correlator architecture is useful not only for matched filtering but for various functions implemented as binary spatial filtering. Actually, the binarization of the sensed and matching (reference) images for the classic matched filter (correlator) improves its performance. Furthermore, when the amplitude is binarized, a binary spatial light modulator can be used in the Fourier plane, which is easier to achieve in practice. (In digital correlators the required throughput rate for a given gray-level image is less when using binary signals.) The benefit in either optical or digital correlators is that performance, in terms of output signal-to-noise ratio, may also dramatically improve. In addition, if phase-only or binary phase-only images are used, diffraction efficiency is improved to essentially 100%. Hence the output signal-to-noise ratio (peak-to-sidelobe ratio) at the output detector plane is improved.

10.4 MATCHED SPATIAL FILTER AND BINARY PHASE-ONLY CORRELATORS

Much attention has been devoted to phase-only and binary phase-only correlators implemented with the matched-filter correlator architecture shown in Fig. 10.2. In this architecture two SLMs are used in a transmissive mode, so that they spatially modulate a linearly polarized coherent light beam as in planes P_1 and P_2 in Fig. 10.2. One SLM represents the sensed image, and the other SLM represents the reference image. After the input (sensed) image is Fourier transformed by lens L_1, it multiplies the reference Fourier transform image at the internediate plane P_2. The product is then Fourier-transformed by lens L_2 and square-law-detected by the CCD at plane P_3.

If the reference image is binarized, then

$$H_B(u, v) = \begin{cases} 1, & \text{if } H(u, v) \geq 0 \\ 0, & \text{if } H(u, v) < 0 \end{cases} \quad (10.5)$$

where $H_B(u, v)$ is the binarized matched spatial filter in the Fourier plane, corresponding to $H(u, v)$. The threshold for binarization can be set by first high-pass filtering the reference image and then using a zero threshold nonlinear filter on the high-passed filtered image. Any filtered image value above zero is set to one, and any value below is set to zero. (Alternatively, the histogram of the interference image is calculated, and the median is made the threshold level.)

EXAMPLE 10.4 Full-Dynamic-Range Matched-Filter Correlator As an example consider the correlation of the sensed and reference images shown in Figs. 10.8(a) and 10.8(b). Using a 128 × 128 2-D FFT algorithm the resulting full dynamic range correlation is shown in Fig. 10.9, and the same correlation with additive noise is shown in Fig. 10.10 (with some smoothing of the correlation to make the peaks more visible). The correlation surface shows the location of two objects that match the reference object. The peak of the correlation is not very confined spatially, and the peak-to-sidelobe ratio (PSR) or zero-lag (variance) to RMS value (elsewhere), which is a measure of the matched filter's output signal-to-noise ratio, is not very large. More closely spaced correlation peaks would be less distinguishable.

In practice, an SLM used for introducing the reference function has a grid pattern resulting from the addressing structure. This pattern multiplies the Fourier plane distribution before the last Fourier transform is taken, to yield the correlator output. As a result, the desired correlation peak may be confused with replicas produced by the grid structure. In addition, the effects of thresholding (and binarization) in the Fourier plane can produce harmonics resulting from aliasing, which can also be confused with the desired correlation peak under some circumstances.

It is not necessary to fully binarize the intermediate (Fourier) plane image in the matched-filter correlator. A kth-law nonlinearity can be employed as described by Javidi. However, by virtue of incorporating a nonlinearity in the Fourier plane, whether kth-order or binary, a much higher and narrower autocorrelation peak, smaller sidelobes, and better discrimination performance can be obtained.

The notion of binarizing the matched filter for improving correlation efficiency and PSR can be extended by looking only at the phase and binarizing it. This editing of the matched filter to yield a phase-only matched filter $H^{*\prime}(u, v)$ is expressed as

Figure 10.8 (a) Sensed "image" and (b) reference "image" used in joint transform correlator.

Figure 10.9 Full dynamic range correlation and using "sensed" (or input) and reference images in Fig. 10.8.

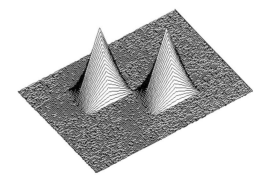

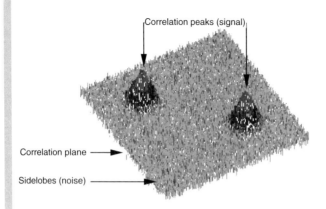

Figure 10.10 Full dynamic range correlation using "sensed" (or input) and reference images in Fig. 10.8 with noise added to original images.

$$H^{*\prime}(u, v) = \frac{H^*(u, v)}{|H^*(u, v)|} = \exp[j\phi(u, v)] \tag{10.6}$$

The potential improvement in matched filtering using this type of filter is well known historically from image processing theory and from work on kinoforms. From Equation (10.6) it is clear that the application of a phase-only restriction is equivalent to applying the inverse filter: $1/|H^*(u, v)|$. The function $|H^*(u, v)|$ usually has strong peak structure in which the amplitude is high for low spatial frequencies but low for high spatial frequencies. Therefore the inverse filter $1/|H^*(u, v)|$ can be interpreted as a high-pass filter. Thus, invoking a phase-only restriction is loosely equivalent to high-pass filtering. High-pass filtering can be implemented with spatial differentiation. When an image is spatially differentiated the operation is equivalent to the gradient

$$\vec{\nabla} f(x, y) = \frac{\partial f(x, y)}{\partial x} \hat{i} + \frac{\partial f(x, y)}{\partial y} \hat{j} \tag{10.7}$$

which has a magnitude (e.g., after square-law detection) given by

$$\left|\vec{\nabla} f(x, y)\right| = \sqrt{\left(\frac{\partial f}{\partial x}\right)^2 + \left(\frac{\partial f}{\partial y}\right)^2} \tag{10.8}$$

Thus an intuitive manner in which correlation of binary spatial filters can be understood is by resorting to using spatial differentiation and detection to emulate the high-frequency filtering effects of such image editing. If we apply to the image of a circular pill box, shown in Fig. 10.11(a), the magnitude of the gradient given by Equation (10.8) and then low-pass filter it (to smooth the result), we get the outline of the pill box shown in Fig. 10.11(b). This is equivalent to placing a stop in the center of a circular pupil function and calculating the OTF. If we were to let the "wall" thickness of the annular pupil function vary between the full circular

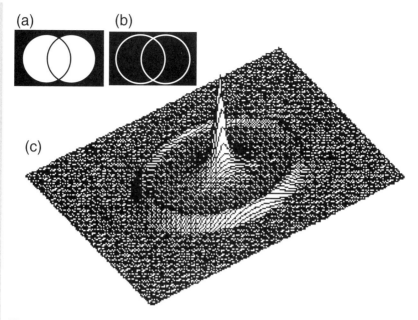

Figure 10.11 (a) Circular pillbox, (b) outline of pillbox resulting from differentiation and low-pass filtering, and (c) corresponding autocorrelation (analogous to the OTF of a circular annulus).

pill box and no "wall" thickness at all, the correlation function would vary so that at some point when the wall thickness gets small, the autocorrelation (or OTF, equivalently) would look like that shown in Fig. 10.11(c). Notice that the PSR varies considerably between the two extremes of this case and is obviously better for the thin-walled annulus because virtually all the image "mass" has been removed so that only the edges are preserved and thus overlap.

EXAMPLE 10.5 Binary Phase-Only Matched Filter As an example in which a matched spatial is restricted to phase-only, consider the binary phase-only correlation of the images in Figs. 10.8(a) and 10.8(b). If we restrict the matched filter to binary phase-only, we set

$$\phi(x, y) = \begin{cases} +1, & \phi \geq 0 \\ -1, & \phi < 0 \end{cases}$$

The correlation result is shown in Fig. 10.12. In effect, binarizing the phase of the images removes the image "mass" or bias, which therefore substantially reduces the cross-correlation contribution to the correlation surface sidelobe levels.

In addition to binarizing the phase of the matched filter, we can also binarize the amplitude of the input (sensed) image and/or the reference image. This will, in some cases, improve the correlator performance as measured by PSR. However, when we binarize an image we actually introduce higher spatial frequencies than were originally in the unbinarized signal. In digital correlators these higher spatial frequencies thus give rise to sampling artifacts because the original sampling may be inadequate. Another, more fundamental drawback of restricting the matched filter to phase-only is a greater sensitivity to offsets or distortions in angle or scale.

268 IMAGE AND MATCHED SPATIAL FILTERING

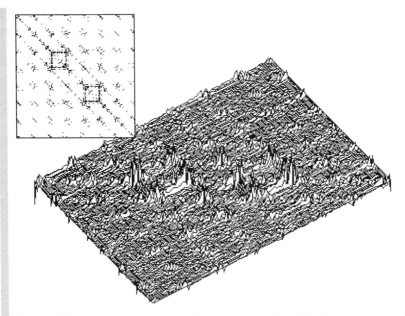

Figure 10.12 Binary phase-only correlation for images in Fig. 10.9. The inset shows a thresholded version of correlation surface showing the outline (magnitude of derivative) of detected objects.

This is intuitively obvious by noting the narrow width of the correlation peak in the binarized phase-only case. Since the important structural information for correlation lies in the edges, which have inherently high frequency content, this is not surprising.

10.5 TECHNIQUES FOR CIRCUMVENTING GEOMETRIC DISTORTIONS

In the application of correlation to pattern recognition or navigation the sensitivity to offsets in position, scale, and rotation are, in fact, crucial. There are generally two methods used to circumvent this:

1. Spatially multiplex or store in parallel all possible matched filters (or reference functions) of the actual desired object.
2. Transform the reference function(s) to a domain where correlation is invariant with respect to position, scale, and rotation offsets.

We will treat this aspect of correlation here because it is germane to practical implementations.

10.6 SPATIAL MULTIPLEXING

Consider first the spatial multiplexing of matched filters corresponding to all possible offset (or distorted) replicas of an object we wish to locate or recognize. The simplest way in which to *spatially* store all possible matched filters representing an object is to place them in parallel in the intermediate plane as suggested by Fig. 10.13(a). The difference in the views presented at each spatial location in the matched filter depend on translation, scale, and rotation and the sensitivity of the matched spatial filter to variations in these parameters. The more sensitivity there is, the more views that must be stored. The input image must be replicated in order to address simultaneously all possible matched filters. A lenslet array (L_1) can do this, as shown. The filter with the best match will yield the maximum correlation in the output detector plane. An imaging lens (L_2) will place the maximum correlation at a location related to the identity of the matching reference (if the matched filter encodes identity according to location). This architecture can use incoherent light, and thus it is an incoherent correlator. If, instead, a coherent correlator is desired, the lenslet array is replaced with a holographic lens matrix generated by a computer so that it replicates the Fourier transform of the input image. Then the second lens (L_2) merely takes the Fourier transform of the product and yields the correlation in the output plane. This version is shown in Fig. 10.13(b), and is called a *spatially multiplexed matched spatial filter*. If, however, individual matched filters are implemented using a reference wave at a different angle of incidence for each filter, then we have a *frequency multiplexed matched spatial filter* (not shown here since it is embodied in the VanderLugt filter of Figs. 10.1 and 10.2).

This latter technique [Fig. 10.3(b)] implements a linear superposition of all possible matched filters. This technique can be analyzed mathematically using synthetic discriminant functions (SDFs) introduced by Casasent et al. SDFs can accommodate distortions (i.e., be made invariant with respect to), translation, scale, in-plane and out-of-plane rotation, and class (as in pattern recognition). If we denote an SDF as $s(x, y)$, it is given by

$$s(x, y) = \sum_{n=0}^{N-1} a_n s_n(x, y) \tag{10.9}$$

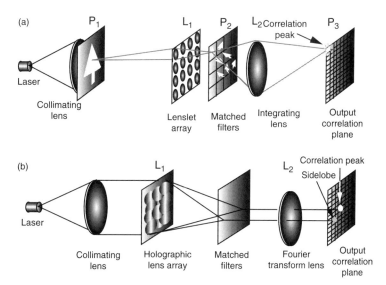

Figure 10.13 (a) Spatially parallel matched filters implemented using incoherent light, and (b) spatially multiplexed matched filter used with coherent light.

where the set of functions $\{s_n(x, y)\}$ represents all possible images corresponding to an object at various offsets or distortions. Thus the SDF is a linear superposition of all possible replicas of $s_n(x, y)$ with weights $\{a_n\}$ essentially representing the holographic lens array (L_1). Possible replicas include images of the object at all translations, scales, and rotations. We only require that the correlation peaks (at zero lag) between $s(x, y)$ and all possible (replicas) images $s_n(x, y)$ be constant, namely, $s_n(x, y) \otimes s = c_n$.

To determine the weight coefficients a_n, we must substitute Equation (10.9) into $s_n(x, y) \otimes s = c_n$. Then

$$s_n(x, y) \otimes s = s_n(x, y) \otimes \sum_{m=0}^{M-1} a_m s_m(x, y) \qquad (10.10)$$

$$s_n(x, y) \otimes s = \sum_{m=0}^{M-1} a_m r_{nm} = c_n \qquad (10.11)$$

where r_{nm} are the elements of a correlation matrix for the data set $s_n(x, y)$. In matrix form, Equation (10.10) is $\mathbf{RA} = \mathbf{C}$, where $\mathbf{R}$ is the correlation matrix and $\mathbf{C}$ is a vector of constants. Thus the coefficients are given by $\mathbf{A} = \mathbf{R}^{-1}\mathbf{C}$. Thus the weight of each basis function in the filter is determined by inverting the correlation matrix $\mathbf{R}$. The vector $\mathbf{C}$ is determined by the particular application of correlation. If every image in the set should yield a correlation peak of equal intensity, then the vector $\mathbf{C}$ should consist of a sequence of 1's. If images occurring in the training set are to be undetectable, then 0's are included in the vector $\mathbf{C}$ where appropriate. To construct a spatial filter that will be used in the matched-filter correlator the coefficients a_n are substituted into Equation (10.9) to generate the SDF $s(x, y)$. Hence

$$s(x, y) = \sum_{n=0}^{N-1} (r_{nm}^{-1} c_n) s_n(x, y) \qquad (10.12)$$

Then a Fourier transform is taken of $s(x, y)$ to actually yield the SDF matched spatial filter. This filter would then be placed in the Fourier plane of the matched-filter transform correlator architecture.

10.7 DISTORTION-INVARIANT TRANSFORMATIONS

In order to make the matched-filter correlator invariant to the usual image distortions of translation, scale, and rotation (in the image plane), many investigators have resorted to appropriate nonlinear and linear transformations, as well as geometrical (or coordinate) transformations. The need for such transformations is supported by the results of sensitivity studies of the correlation peak-to-sidelobe ratio (PSR) (or signal-to-noise ratio) versus percentage scale changes and rotation angle offsets.

Sensitivity to shifts in the input image $f(x, y)$ can be handled by taking the magnitude of its Fourier transform, $|F(u, v)|$. A rotation of $f(x, y)$ rotates $|F(u, v)|$ by the same amount, and a scale change in $f(x, y)$ coordinates (x, y) by an amount a scales the magnitude of the Fourier transform, $|F(u, v)|$, by $1/a$. The effects of scale and rotation can be separated by performing a polar transformation on coordinates (u, v) to yield coordinates (ρ, ϕ). Since

10.7 DISTORTION-INVARIANT TRANSFORMATIONS

$\phi = \tan^{-1}(v/u)$ and $\rho = (u^2 + v^2)^{1/2}$, a scale change by a does not affect ϕ and scales ρ to $a\rho$. Thus a 2-D scaling of the input function is reduced to a scaling in 1-D (ρ coordinate) in the resulting function $F(\rho, \phi)$. If now a 1-D Mellin transform is performed on $F(\rho, \phi)$ in ρ, a completely scale-invariant transform results, a result shown in Chapter 3.

Recall the 1-D Mellin transform is given by

$$M(\omega_\rho, \phi) = \int_{-\infty}^{\infty} F(\rho, \phi) r^{-j\omega_\rho - 1} \, d\rho \tag{10.13}$$

where $\rho' = \ln \rho$. Therefore the Mellin transform of the scaled function $F(a\rho, \phi)$ is

$$M'(\omega_\rho, \phi) = a^{-j\omega_\rho} M(\omega_\rho, \phi) \tag{10.14}$$

where it is clear that the magnitudes of the two transforms are equal. Thus the Mellin transform of the scaled function $F(a\rho, \phi)$ is

$$M(\omega_\rho, \phi) = \int_0^\infty F[\exp(\rho, \phi)] \exp(-j\omega_\rho \rho) \, d\rho \tag{10.15}$$

Thus, Equation (10.15) shows that a logarithmic scaling of the ρ coordinate followed by a 1-D Fourier transform is needed to realize the required optical Mellin transform; that is, $M(\omega_\rho, \phi)$ is the Fourier transform of $F(\exp \rho, \phi)$.

To explicitly delineate the effects of rotation of the input function $f(x, y)$, consider the following discussion. If an angular sector of the function $F(u, v)$ in the (u, v) plane as shown in Fig. 10.14(a) is labeled $F_2(u, v)$ and the remainder is labeled $F_1(u, v)$, then the corresponding polar forms for these are denoted by $F_2(\rho, \phi)$ and $F_1(\rho, \phi)$ as shown in Fig. 10.14(b). If the angular extent of $F_2(u, v)$ is ϕ_o and the entire function is rotated clockwise by ϕ_1, the rotated versions of $F(u, v)$ and $F(\rho, \phi)$ are as shown in Figs. 10.14(c) and 10.14(d), where for simplicity we will assume $\phi_1 = \phi_2 = \phi_0$. It is clear that a rotation shifts the different portions by different amounts; that is, $F_1(\rho, \phi)$ shifts up to $F_1'(\rho, \phi)$ by an amount ϕ_0 and $F_2(\rho, \phi)$ shifts down to $F_2'(\rho, \phi)$ by an amount $2\pi - \phi_0$. The important point here is that the

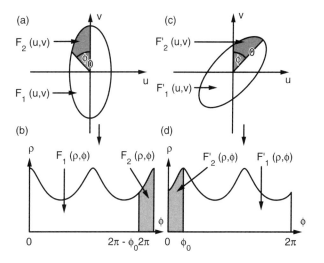

Figure 10.14 (a) Angular sectors of the function $F(u, v)$ in the (u, v) plane, and (b) their corresponding polar maps $F_2(\rho, \phi)$ and $F_1(\rho, \phi)$. Rotated versions of $F(u, v)$ and the two $F(\rho, \phi)$ maps are shown in (c) and (d).

resulting shifts can be expressed as phase factors in the Fourier transform of $F(\rho, \phi)$, which can be explicitly employed in correlation. Denoting the 1-D Fourier transform of $F(\rho, \phi)$ as $G(\rho, \omega_\phi)$, then the 1-D transform of the unrotated input is

$$G(\rho, \omega_\phi) = G_1(\rho, \omega_\phi) + G_2(\rho, \omega_\phi) \tag{10.16}$$

and the transform of the rotated input is

$$G'(\rho, \omega_\phi) = G_1(\rho, \omega_\phi) \exp(-j\omega_\phi \phi_0) + G_2(\rho, \omega_\phi) \exp[-j\omega_\phi(2\pi - \phi_0)] \tag{10.17}$$

Combining all the transformations—that is, Fourier transform magnitude, polar transform, log scaling, and Fourier transform—yields the process shown in the block diagram of Fig. 10.15. Essentially what is done is that a 2-D Fourier transform is produced by taking a Fourier transform in ρ (which is equivalent to a scale invariant Mellin transform) and a Fourier transform is taken in ϕ in order to convert shifts by ϕ_0 into phase terms, as just described. In fact, if we write the result explicitly as a Mellin transform, the complete transformation of $f(x, y)$ is given by

$$M(\omega_\rho, \omega_\phi) = M_1(\omega_\rho, \omega_\phi) + M_2(\omega_\rho, \omega_\phi) \tag{10.18}$$

and the corresponding transformation for the scaled and rotated function $f'(x, y)$ is

$$M'(\omega_\rho, \omega_\phi) = M_1(\omega_\rho, \omega_\phi) \exp\left[-j(\omega_\rho \ln a + \omega_\phi \phi_0)\right] \\ + M_2(\omega_\rho, \omega_\phi) \exp\left\{-j\left[\omega_\rho \ln a - \omega_\phi(2\pi - \phi_0)\right]\right\} \tag{10.19}$$

Now that a shift, scale, and rotation-invariant transformation has been established, we can consider the final correlation given by the Fourier transform of.

$$M^*M' = M^*M_1 \exp\left[-j(\omega_\rho \ln a + \omega_\phi \phi_0)\right] \\ + M^*M_2 \exp\left\{-j\left[\omega_\rho \ln a - \omega_\phi(2\pi - \phi_0)\right]\right\} \tag{10.20}$$

If we take the Fourier transform of M^*M', which is the result of utilizing the matched-filter correlator architecture shown in Fig. 10.16, then we obtain two terms in the output plane:

1. The cross-correlation term $F_1(\exp \rho', \phi) \otimes F(\exp \rho', \phi)$ located at $\rho' = \ln a$ and $\phi' = \phi_0$.
2. The cross-correlation term $F_2(\exp \rho, \theta) \otimes F(\exp \rho', \theta)$ located at $\rho' = \ln a$ and $\phi' = -2\pi + \phi_0$, where (ρ', ϕ') are the coordinates in the output plane.

If these two cross-correlation terms are summed, we obtain the autocorrelation $F(\exp \rho', \phi)$, despite the fact that the two terms are scaled and rotated versions of each other. From the locations of the two peaks the scale factor a and rotation angle ϕ between the two functions can be determined. To recover positional information, which was lost when the magnitude

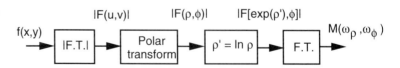

Figure 10.15 Block diagram of processes that yield shift, scale, and rotation invariance.

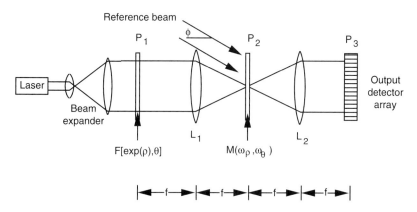

Figure 10.16 Matched-filter correlator used in scale-invariant optical correlation.

of the Fourier transform was taken, it is necessary to either scan the input or use the scale and rotational information from the correlation expressed by Equation (10.20) to perform conventional correlation.

Recalling Fig. 10.16, it should be noted that the function $M^*(\omega_\rho, \omega_\phi)$ appearing in Equation (10.20) is produced by putting $F(\exp \rho, \phi)$ in the input plane (P$_1$) and interfering its transform $M(\omega_\rho, \omega_\phi)$ generated by L$_1$ and a plane wave reference beam (at angle ϕ) in plane P$_2$. Following the conventional holographic spatial filtering synthesis technique (described previously), one of the four terms created at P$_2$ will be proportional to $M^*(\omega_\rho, \omega_\phi)$. Now, if the plane wave is blocked and $F'(\exp \rho, \phi)$ is input at P$_1$, then the light incident on plane P$_2$ will be $M'(\omega_\rho, \omega_\phi)$ so that one term after P$_2$ will be $M'M^*$, which will be subsequently Fourier transformed to yield the two desired cross-correlation peaks at plane P$_3$.

10.8 ANGULAR CORRELATION

In all our discussions up to this point, optical correlation has been discussed in terms of Cartesian correlation—that is, correlation that depends on shifts in x and y coordinates. Correlation can also be performed with respect to polar angle (measured about an optical axis perpendicular to the image plane). This type of correlation, termed *angular correlation*, can also be performed optically and yields useful results for feature extraction in pattern recognition applications. Although we will consider feature extraction and pattern recognition in the next chapter, it is appropriate to include angular correlation here for the sake of completeness.

As with Cartesian correlation, there are two cases of angular correlation: autocorrelation and cross-correlation. In angular autocorrelation we obtain a periodic correlation function which is symmetric over one period (given that the original object is symmetric with respect to azimuth angle ϕ over 2π radians). Optical angular autocorrelation can be used directly to determine object primitives without being sensitive to scale. In general, the angular autocorrelation function $R(\phi)$ is calculated according to the following prescription. In polar coordinates,

$$R_{ff}(\phi) = \int_0^\infty \int_0^{2\pi} f(r, \phi') f(r, \phi + \phi') \, r \, dr \, d\phi' \quad (10.21)$$

where $f(r, \phi)$ is the same as the object (or image) function $f(x, y)$ discussed earlier. For simplicity we are interested in the case where the sensed and reference images are both "binarized" so that the silhouette of the desired object is used. This also enables us to recover ultimately the boundary and descriptors of the overall object size and shape. The advantage for implementation is that binary SLMs can be used to implement the angular correlator in an incoherent (shadow-casting) mode. If the image is properly binarized, the image intensity distribution described by $f(r, \phi)$ reduces to rect$[r/r(\phi)]$, where $r(\phi)$ is the radial extent of the binarized object measured from its centroid. Then the angular autocorrelation is given by

$$R_{rr}(\phi) = \int_0^{2\pi} \int_0^{\infty} \text{rect}\left[\frac{r}{r(\phi')}\right] \text{rect}\left[\frac{r}{r(\phi' + \phi)}\right] r \, dr d\phi' \qquad (10.22)$$

where the integrand is the binarized object with a boundary that circumscribes the area of overlap between $r(\phi')$ and $r(\phi' + \phi)$, as shown in Fig. 10.17(a). This result is consistent with Green's theorem. The resulting angular correlation curve is shown in Fig. 10.17(b). The advantage of angular autocorrelation is, of course, that it is scale-invariant.

The angular cross-correlation function is given by

$$R_{rr}(\phi) = \int_0^{\infty} \int_0^{2\pi} f(r, \phi') f'(r, \phi + \phi') r \, dr d\phi' \qquad (10.23)$$

where the second term, $f'(r, \phi)$, in the integrand is, in general, not equal to the original function $f(r, \phi)$. By appropriate selection of $f'(r, \phi)$ interesting results can be obtained, such as direct recovery of the boundary function of $f(r, \phi)$. This will be discussed in Chapter 12.

The simplest architecture to enable angular correlation is illustrated in plan view Fig. 10.18. Here the concept of video feedback is used. In this processor the basic function of feedback is enabled by viewing a miniature video display through an appropriate transfer lens with a CCD camera. The output of this camera is fed back through video processing electronics to the display unit. Initial input video is interrupted after the first frame using a

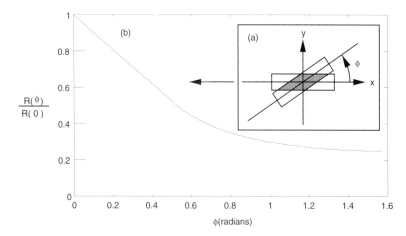

Figure 10.17 (a) Schematic plan view of angular correlation process for a simple rectangular shape, and (b) corresponding angular autocorrelation.

Figure 10.18 Angular correlator including image rotation architecture employing feedback.

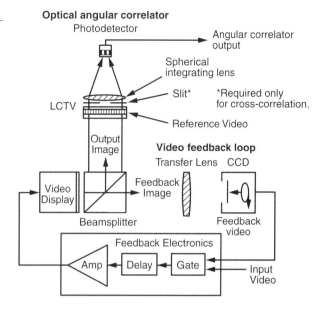

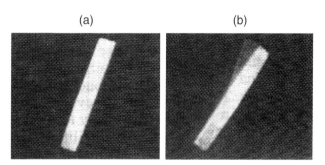

Figure 10.19 Rotated replicas of input image after (a) first and (b) second iterations from video feedback architecture. Note secondary or ghost image after second iteration due residual of first iteration image.

video gate circuit. Subsequent frames of video from the CCD are iteratively passed through the feedback loop, which compensates for losses in the beamsplitter used to output the results to the optical angular correlator. The CCD camera is tilted by a small angle about its optical axis so that each successive image displayed and output through the beamsplitter rotates incrementally at the video frame rate. The feedback process yields a rotated image as indicated in Fig. 10.19. By employing a spherical lens at the output of the feedback loop, light can be focused to a single photodetector. By placing a transmissive SLM between the output of the beamsplitter and the spherical lens the reference function $f'(r, \phi)$ can be displayed. The output image $f(r, \phi + \phi')$ successively multiplies $f'(r, \phi)$ in a shadow-casting mode for each value of ϕ'. The result is a 1-D time-dependent signal output from the photodetector, which can be digitized for subsequent digital processing.

This feedback technique serves to illustrate one possible implementation of angular correlation in which image rotation is enabled without mechanical methods. It is, however, a single-channel temporally multiplexed method. In Chapter 12 we will revisit the concept of angular correlation for feature extraction using an entirely different implementation using lenslet arrays. This approach is a spatially multiplexed method and, as such, is faster than the feedback approach.

PROBLEM EXERCISES

1. Consider the case where the input function to the 2-D Mellin transform system is a rectangular function between between x_1 and x_2 and y_1 and y_2.

   (a) Write an expression for this aperture function.

   (b) Calculate its Mellin power spectrum $|M(u, v)|^2$.

   (c) Show explicitly that the Mellin power spectra for a square aperture of width W is the same as for an aperture of width $2W$.

2. Show that the Mellin transform is not invariant to shift.

3. Consider the matched spatial filter of Fig. 10.2 in which the input is a 2-D mesh grating at plane P_1 illustrated below.

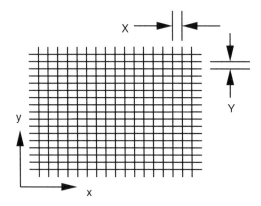

   (a) Write a mathematical expression for this grating.

   (b) Determine the expression for the corresponding Fourier transform at plane P_2.

   (c) What is the spatial filter required at P_2 to permit only the horizontal bars of the mesh to be reconstructed at plane P_3? Write an expression for this filter and for the resulting pattern at plane P_3.

4. A phase distribution $f(x, y)$ illuminated with coherent light is placed in the object plane (P_1) of an imaging system consisting of a spherical lens at plane P_2. It has an amplitude distribution given by

$$t(x, y) = \exp[j\phi(x, y)]$$

   At the focal plane (P_3) of the system an attenuating plate is placed, defined by

$$a(x, y) = a_o(x^4 + 2xy + y^4)$$

   What is the resulting image intensity in plane P_4 and how is it related to object phase?

5. In the matched spatial filter architecture shown in Fig. 10.2 a spatial filter $H(u, v)$ is created as defined by

$$H(u, 0) = \frac{1}{2}[1 + \sin(2\pi a u)]$$

where $2\pi a$ is the separation between two object functions $f_1(x, y)$ and $f_2(x, y)$ placed side-by-side in the input plane.

(a) Calculate the complex light amplitude right behind the filter plane P_2.

(b) Derive the output plane (P_3) light amplitude distribution.

(c) Sketch the result and specify what mathematical operation is performed by this processor.

6. In the coherent matched-filter correlator architecture of Fig. 10.2 an input transparency is placed, given by $f(x, y) + n(x, y)$, where $f(x, y)$ is the signal and $n(x, y)$ is white Gaussian noise. Assume that $f(x, y)$ and $n(x, y)$ are uncorrelated; that is, $n(x, y) \otimes f(x, y) = 0$ and that the matched filter is given by

$$H(u, v) = c_1 + 2c_2|F(u, v)|\cos[x_o u + \phi(u, v)]$$

Calculate the light distribution, $g(x, y)$, in the output plane given that $f(x, y) = \text{rect}(y/y_o)$. Use MATLAB to plot your results and vary the SNR defined as $1/\sigma_n^2$. Does the noise affect the desired (correlation) output?

7. In the "joint" transform correlator shown below, the output of the first stage is often a square-law detector.

(a) Draw a schematic analog processing diagram describing what operations the correlator performs.

(b) Derive an expression for the amplitude at the square-law detector plane.

(c) Derive an expression for the irradiance induced voltage at the detector terminals.

(d) If this is subsequently read-out by a coherent plane wave at P_3, show that the light distribution is given by

$$g(x, y) = f_1(x, y) \otimes f_1^*(x, y) + f_2(x, y) \otimes f_2^*(x, y)$$
$$+ f_1(x, y) \otimes f_2^*(x - 2x_o, y) + f_1^*(x, y) \otimes f_2(x + 2x_o, y)$$

Identify the cross-correlation terms.

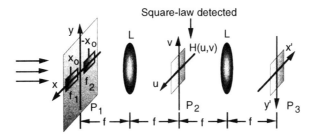

8. (a) If additive white Gaussian noise appears in the $f_1(x - x_o, y)$ channel of Problem 7, what is the complex light amplitude at the square-law detector?

(b) Derive the irradiance.

(c) Derive $g(x, y)$ in this case.

(d) Define each term explicitly.

9. In the Schlieren imaging system described in Example 10.3, derive explicitly the intensity in the image plane, that is,

$$I_i(x_i, y_i) = \frac{1}{4}\left[1 - \frac{2}{p}\left|\log_e\left|\frac{x_i - X/2}{x_i + x/2}\right|\right|\right]$$

for the rectangular phase input object defined by

$$\phi(x_o, y_o) = \phi_o \text{rect}(x_o/X)$$

10. Recall the standard holographic matched-filter correlator architecture as shown in Fig. 10.2 in which the input plane contains a scene that is trans-illuminated by a collimated laser beam and the intermediate plane contains a Fourier-domain wavelet filter. The wavelet-filtered output image of the corresponding input scene is read-out through a CCD in the output plane. Sketch this architecture with appropriate labeling. Using the Mexican-hat form of a wavelet in two dimensions,

$$h(x, y) = \left[1 - \left(\frac{a_x x}{N_x}\right)^2\right]\left[1 - \left(\frac{a_y y}{N_y}\right)^2\right]\exp\left\{-\left(\frac{a_x x}{2N_x}\right)^2 + \left(\frac{a_y y}{2N_y}\right)^2\right\}$$

determine $H(u, v)$. Using the values $N_x = N_y = 128$, $a_x = 0.5$, $a_y = 0$, and $a_x = a_y = 0.25$ and the letter "E" for $s(x, y)$, calculate the 2-D wavelet transform in this architecture. It is recommended that you use MATLAB.

BIBLIOGRAPHY

D. Casasent and D. Psaltis, "Scale Invariant Optical Transform," *Opt. Eng.* **15**, 258 (1976).

D. Casasent and D. Psaltis, "Position, Rotation and Scale Invariant Optical Correlator," *Appl. Opt.* **15**, 1795 (1976).

D. Casasent, W. Rozzi, and D. Fetterly, "Projection Synthetic Discriminant Function Performance," *Opt. Eng.* **32**, 716 (1984).

J. Cederquist and S. H. Lee, "The Use of Feedback in Optical Information Processing," *Appl. Phys.* **18**, 311 (1979).

N. Collings, *Optical Pattern Recognition Using Holographic Techiques*, Addison-Wesley, Reading, MA (1988).

J. P. Crutchfield, "Space-Time Dynamics in Video Feedback," *Physica* **10D**, 229 (1984).

S. R. Curtis, A. V. Oppenheim, and J. S. Lim, "Signal Reconstruction from Fourier Signal Information," *IEEE Trans.* **ASSP-33**, 643 (1985).

J. A. Davis, E. A. Merrill, D. M. Cottrell, and R. M. Borsuk, "Effects of Sampling and Binarization in the Output of the Joint Transform Correlator," *Opt. Eng.* **29**, 1094 (1990).

G. R. Gindi and A. F. Gmitro, "Optical Feature Extraction via the Radon Transform," *Opt. Eng.* **23**, 499–506 (1984).

J. S. Harris, "State of the Art Review: Optical Processing," GACIAC SOAR 87–01 Technical Report (January 1989).

M. H. Hayes, J. S. Lim, and A. V. Oppenheim, "Signal Reconstruction from Phase or Magnitude," *IEEE Trans.* **ASSP-28**, 672 (1980).

J. L. Horner, ed., *Optical Signal Processing*, Academic Press, San Diego, CA (1987).

B. Javidi, "Comparison of Binary Joint Transform Correlators and Phase-Only Matched Filter Correlators," *Opt. Eng.* **28**, 267 (1989).

B. Javidi and C.-J. Kuo, "Joint Transform Correlation Using a Binary Spatial Light Modulator at the Fourier Plane," *Appl. Opt.* **27**, 663 (1988).

A. V. Oppenheim and J. S. Lim, "The Importance of Phase in Signals," *Proc. IEEE* **69**, 529 (1981).

A. V. Oppenheim, J. S. Lim, and S. R. Curtis, "Signal Synthesis and Reconstruction from Partial Fourier-Domain Information," *JOSA* **73**, 1413 (1983).

D. Psaltis, "Incoherent Electro-optical Image Correlator," *Opt. Eng.* **23**, 12 (1984).

Y. Sheng, T. Lu, D. Roberge, and H. J. Caulfield, "Optical N4 Implementation of a Two-Dimensional Wavelet Transform," *Opt. Eng.* **31**, 1859 (1992).

A. VanderLugt, "Signal Detection by Complex Spatial Filtering," *IEEE Trans. Inf. Theory*, **IT-10**, 139 (1964).

F. T. S. Yu and I. C. Khou, *Principles of Optical Engineering*, John Wiley & Sons, New York (1990).

Chapter 11
RADAR SIGNAL PROCESSING APPLICATIONS

11.1 OVERVIEW

In this chapter we will consider a representative set of applications of optical signal processing to radar. Among these radar signal processing applications we will discuss the use of optics in determining the ambiguity function, which is a mathematical transformation that expresses the tradeoff between time and frequency resolution in the measurement of objects using modulated waveforms. A particular class of waveforms of interest in high-range-resolution radar and synthetic aperture radar (SAR) are frequency-coded waveforms, especially chirp waveforms. In fact, one of the earliest and most popular applications of Fourier optics has been to SAR imaging. We will discuss this extensively.

11.2 RADAR SIGNAL PROCESSING

In radar signal processing (and, for that matter, sonar) an active sensor operates by transmitting a particular waveform and reflecting it from an object or distribution of objects. After receiving the reflected signal the waveform is processed in a way to recover range to the object (or a distribution of ranges), as well as range rate (or velocity) in many instances. Simple signal theory would say that the range to a single object (e.g., point scatterer) could be measured to arbitrary accuracy with a sufficiently fast risetime pulse for sufficient signal-to-noise ratio. In practice, however, pulses do not have arbitrarily short risetimes and there may be more than one scatterer. In this case, the ability to resolve two or more scatterers is directly related to pulse width. The narrower the pulse width, the finer the range resolution. Given a transmitted waveform defined by $s_T(t)$, the received signal from two scatterers is

$$s_R(t) = A_1 s_T(t - t_0) + A_2 s_T(t - t_0 - \tau) \quad (11.1)$$

where $s_T(t)$ is the transmitted waveform, t_o is the time delay of first scatterer, τ is the scatterer time separation, and $A_{1,2}$ are factors to account for the relative reflectivity of the two scatterers and for two-way propagation path losses, which we will assume are constant;

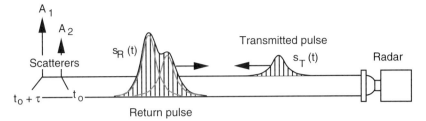

Figure 11.1 Typical pulse radar geometry.

that is, $A_1 = A_2 = A$ over distances corresponding to τ. The more general case where $A_1 \neq A_2$ is diagrammed in Fig. 11.1.

We have assumed no noise in the measurement process for the sake of exposition. Equation (11.1) implies that in the process of reflection the transmitted waveform convolves with the amplitude distribution $\{A_1, A_2\}$ of the object. The trick is to select a transmitted waveform $s_T(t)$ so that the signals received from the two scatterers are as different as possible, which can be measured by the mean-squared error (MSE). MSE is defined as

$$\text{MSE} = A^2 \int_{-\infty}^{\infty} |s_t(t - t_0) - s_T(t - t_0 - \tau)|^2 \, dt \qquad (11.2)$$

which reduces to

$$\text{MSE} = 2A^2 \int_{-\infty}^{\infty} |s_t(t)|^2 \, dt - 2A^2 \, \text{Re} \int_{-\infty}^{\infty} s_T^*(t - \tau) s(t) \, dt \qquad (11.3)$$

where we have shifted the time origin to $t = t_0$ (equivalent to range tracking with respect to t_0). The first term is equivalent to the energy transmitted and can be assumed constant. The second term is the important one for our discussion of the ambiguity function and is equivalent to the autocorrelation function of $s_T(t)$, defined at $R_{ss}(\tau)$. In order to make the MSE as large as possible (for fixed energy) we must make $R_{ss}(\tau)$ a minimum except when $\tau = 0$, where it will, by necessity, be a maximum. This implies that $R_{ss}(\tau)$ should be impulse-like, corresponding to a infinite bandwidth pulse waveform $s_T(t)$—that is, $\delta(t)$. In practice, such a waveform cannot be created, but the effective time (or range) resolution is the inverse of the bandwidth of $s_T(t)$. More importantly, however, we must recognize that the receiver must act like a matched filter in order to best resolve the signal in range from two separate scatterers. This will be considered below.

11.3 AMBIGUITY FUNCTION PROCESSING

In the preceding discussion we overlooked one important characteristic of the received waveform: that the signal will have an extra phase factor given by $\exp(j\omega_D t)$, which accounts for the relative motion between the transmitter and scatterer or between two scatterers, where ω_D is radian Doppler frequency given by $\omega_D = \pm 4\pi v f/c$, where v is the range rate (or relative velocity), f is the radar frequency, and c is the velocity of light. The $\pm$ sign corresponds to an approaching or receding object, respectively.

Thus, for the two-scatterer case just considered we will modify the situation to consider one scatterer as stationary and one moving with respect to the transmitter. The received signal now is

$$s_R(t) = A_1 s_T(t) + A_2 s_T(t) \exp(j\omega_D t) \tag{11.4}$$

Again assuming $A_1 = A_2 = A$ and forming the MSE yields

$$\text{MSE} = 2A^2 \int_{-\infty}^{\infty} |s_T(t)|^2 \, dt - 2A^2 \int_{-\infty}^{\infty} |s_T(t)|^2 \cos \omega_D t \, dt \tag{11.5}$$

The first term is the (constant) energy as before, but the second term is equivalent to the real part of the Fourier transform of $|s_T(t)|^2$. Using the correlation theorem, the second term is also equivalent to the real part of the autocorrelation of the Fourier spectrum of $s_T(t)$. Thus

$$\text{MSE} = 2A^2 \int_{-\infty}^{\infty} |s_T(t)|^2 \, dt - 2A^2 \, \text{Re}\left[S_T(\omega_D) \otimes S_T^*(-\omega_D)\right] \tag{11.6}$$

To maximize this MSE term requires that the second term be as near zero as possible except at $\omega_D = 0$. Thus the *spectral* autocorrelation term should now be an impulse function. A time function that corresponds to an impulsive spectrum is a constant (or DC) signal. This is just the opposite of the requirement for high range resolution. Thus there must be a tradeoff between range and Doppler resolution if their requirements are to be met simultaneously in the same waveform. This means that range and velocity *cannot* be measured simultaneously with arbitrary accuracy using the same waveform. This result leads us ultimately to conclude that an uncertainty principle operates for simultaneous range and velocity resolution. This uncertainty or ambiguity is expressed by the so-called ambiguity function. (It should be noted that uncertainty is the potential range of a variable, whereas ambiguity also means the potential location when one or more locations are possible. They are therefore not strictly the same.)

Thus, in the context of the previous discussion the ambiguity function, in fact, corresponds to the cross-term in each case. It represents the response of a matched filter to a signal for which it is matched as well as one to which it is mismatched (by virtue of a Doppler shift). In matched filtering, of course the received signal is matched with a replica of the transmitted waveform such that

$$R_{RT}(\tau) = \int_{-\infty}^{\infty} s_R(t) s_T^*(t - \tau) \, dt \tag{11.7}$$

where $s_T(t) = m(t) \exp(j 2\pi \omega_0 t)$, $m(t)$ is the complex modulation function, $|m(t)|$ is the (real) envelope of the modulation, and ω_0 is the carrier frequency.

Without loss of generality we can assume that the received (echo) signal and transmitted signal are the same except for the range delay and Doppler shift ω_D. Then

$$s_R(t) = m(t - \tau_0) \exp\left[j 2\pi (\omega_0 + \omega_D)(t - \tau_0)\right] \tag{11.8}$$

Thus Equation (11.7) now becomes a 2-D function given by

$$R_{RT}(\tau, \omega_D) = \int_{-\infty}^{\infty} m(t - \tau_0) m^*(t - \tau) \exp\left\{j \left[(\omega_0 + \omega_D)(t - \tau_0) - \omega_0(t - \tau)\right]\right\} dt \tag{11.9}$$

Again without loss of generality we can simplify the above result by assuming $\tau_o = 0$ and $\omega_o = 0$; that is, range tracking and homodyne detection (with respect to the carrier frequency) occur. Then

$$R_{RT}(\tau, \omega_D) = \chi(\tau, \omega_D) = \int_{-\infty}^{\infty} m(t) m^*(t - \tau) \exp(j\omega_D t) \, dt \qquad (11.10)$$

where the 2-D correlation function is defined as the ambiguity function, $\chi(\tau, \omega_D)$. It is clear that Equation (11.10) expresses the properties discussed earlier leading up to Equations (11.3) and (11.6). That is, the ambiguity function is a correlation along the τ axis (range or delay) where $\omega_D = 0$, and it is a power spectrum along the ω_D axis (range rate or Doppler) where $\tau = 0$. $\tau > 0$ implies that the object measured is beyond the reference delay (τ_o), and $\omega_D > 0$ implies that the object is incoming. Often the ambiguity function is defined as $|\chi(\tau, \omega_D)|^2$.

Another, more general form of the ambiguity function is the cross-ambiguity function given by

$$\chi(\tau, \omega_D) = \int_{-\infty}^{\infty} s_1(t + \tau/2) s_2^*(t - \tau/2) \exp(j\omega_D t) \, dt \qquad (11.11)$$

which can also be implemented optically and is related directly to the Wigner distribution discussed later.

The ambiguity function is used not only to assess the properties of a transmitted waveform regarding resolution as just discussed, but measurement accuracy and response to clutter as well. Although optimum waveforms can in principle be selected using the ambiguity function, often in practice the ambiguity function is used to determine the suitability of a selected waveform after the fact.

Some properties of the ambiguity function are summarized below. The volume enclosed by the surface of the ambiguity function is equal to the square of twice the energy contained in the received signal—that is,

$$\int_{-\infty}^{\infty} \int_{-\infty}^{\infty} |\chi(\tau, \omega_D)|^2 d\omega_D \, dt = (2E_s)^2 \qquad (11.12)$$

This means that energy is conserved, since no matter what shape $\chi(\tau, \omega_D)$ takes Equation (11.12) is constant. If the peak is narrowed in one region to improve accuracy and resolution, then the ambiguity function must be raised elsewhere. This is the so-called radar ambiguity principle. Therefore there is a limit to the combined resolution that can be achieved simultaneously in range and velocity, as measured by the width of $|\chi(\tau, \omega_D)|^2$ centered at $\tau = 0$ and $\omega_D = 0$. This was alluded to earlier in reference to the "uncertainty principle." A single pulse or a continuous waveform has an ambiguity function dominated by a single, central peak, often called a "thumbtack" ambiguity. A repetitive waveform, on the other hand, has periodically repeating peaks at other values of time and frequency, which indeed imply *ambiguity* (not just uncertainty) in measurement of object parameters. If a waveform is expected to have good clutter rejection, then its ambiguity function should be small in regions of time–frequency where clutter power is found. To illustrate the properties of the ambiguity function consider the following examples.

EXAMPLE 11.1 Ambiguity Function of a Gaussian Waveform Consider the Gaussian waveform defined as

$$m(t) = \frac{1}{\left(\sqrt{2\pi}\sigma\right)^{1/2}} \exp(-t^2/2\sigma^2)$$

which has energy equal to unity and is plotted in Fig. 11.2(a). The corresponding Fourier transform is

$$M(\omega) = \sqrt{\frac{\sigma}{\pi}} \exp(-\omega^2\sigma^2/2)$$

and is plotted in Fig. 11.2(b). The value of $\chi(\tau, \omega)$ is then

$$\chi(\tau, \omega) = \exp(-\tau^2/4\sigma^2) \exp(-\omega^2\sigma^2/4) \exp(j\omega\tau/2)$$

such that

$$|\chi(\tau, \omega)|^2 = \exp\left[-\frac{1}{2}\left(\frac{\tau^2}{\sigma^2} + \omega^2\sigma^2\right)\right]$$

which is plotted in Fig. 11.3(a). If this function is sliced with a plane such that

$$-\frac{1}{2}\left(\frac{\tau^2}{\sigma^2} + \omega^2\sigma^2\right) = 1$$

then a plan-view 2-D ambiguity diagram is created as shown in Fig. 11.3(b). Note that the semimajor and semiminor axes correspond to $1/\tau_o$ and $1/B_o$ along the ω and τ axes, respectively, where τ_o is the root mean square (RMS) time duration and B_o is the RMS bandwidth of a Gaussian pulse. For a Gaussian pulse the area bounded by the ellipse in Fig. 11.3(b) is constant independent of τ_o and B_o and equal to unity. The height of the ellipse is a measure of range resolution (or uncertainty) for a given velocity, and the width is a measure of Doppler resolution (or uncertainty) for a given range.

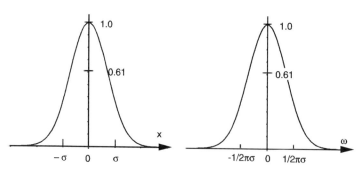

Figure 11.2 (a) Gaussian pulse and (b) corresponding Fourier transform. Plots are normalized to a peak value of one.

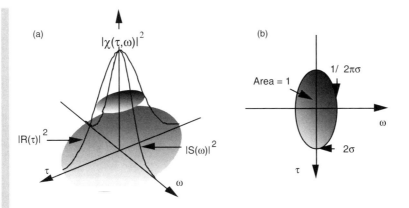

Figure 11.3 (a) Ambiguity function for Gaussian pulse, and (b) plan view cross section (ellipse).

EXAMPLE 11.2 Ambiguity Function of a Rectangular Pulse Train. To develop the ambiguity function for the rectangular pulse train, first consider a general pulse train described by

$$u(t) = \frac{1}{\sqrt{N}} \sum_{n=0}^{N-1} u_n(t - nT)$$

where T is the pulse repetition period and $u_\tau(t)$ is an individual pulse of duration $\tau \ll T$. The total signal $s(t)$ includes the pulse train as an amplitude modulation as well as the carrier such that

$$s(t) = \text{Re}[u_n(t) \exp(j\omega_c t)]$$

The ambiguity function is given in general by

$$\chi(\tau, f) = \frac{1}{N} \sum_{n=0}^{N-1} \sum_{m=0}^{N-1} \int_{-\infty}^{\infty} u_n(t - nT) u_m^*(t - mT - \tau) \exp(j2\pi f t) \, dt$$

By changing variables of integration (i.e., $t_1 = t - nT$), we can express the total ambiguity function as

$$\chi(\tau, f) = \frac{1}{N} \sum_{n=0}^{N-1} \exp(j2\pi f nT) \sum_{m=0}^{N-1} \chi_{nm}[\tau - (n-m)T, f]$$

where χ_{nm} is the cross-ambiguity function between pulses. If all pulses are the same, then the above equation can be written as

$$\chi(\tau, f) = \frac{1}{N} \sum_{n=0}^{N-1} \exp(j2\pi f nT) \sum_{m=0}^{N-1} \chi_c[\tau - (n-m)T, f]$$

where $\chi_c[\tau - (n-m)T]$ is the ambiguity function for a single pulse. As pointed out by Rihaczek, it is expedient to rewrite the double sum in a form that collects all terms with the same delays. By using $p = n - m$ we can exploit the following relationship

$$\sum_{n=0}^{N-1}\sum_{m=0}^{N-1} = \sum_{p=-(N-1)}^{0}\left[\sum_{n=0}^{N-1-|p|}\right]_{m=n-p} + \sum_{p=1}^{N-1}\left[\sum_{m=0}^{N-1-|p|}\right]_{m=n+p}$$

For a specific value of N the above relationship can be analyzed in the form of a matrix whose rows and columns are labeled as m and n, respectively. Using the above relationship in the expression for the total ambiguity function and resolving all expressions such as

$$\sum_{n=0}^{N-1-|p|}\left[\exp(j2\pi fT)\right]^n$$

by noting that they are finite geometric series, which can be resolved into closed form, we arrive at the following

$$|\chi(\tau, f)| = \left|\frac{1}{N}\sum_{p=-(N-1)}^{N-1}|\chi_c(\tau - pT, f)|\left|\frac{\sin[\pi f(N-|p|)T]}{\sin(\pi fT)}\right|\right|$$

provided that the separation between pulses is somewhat greater than the duration of individual pulses.

For monochromatic constant amplitude pulses the resulting ambiguity function versus the lag variable (where $f = 0$) is

$$|\chi(\tau, 0)| = \sum_{p=-(N-1)}^{N-1}\left(1 - \frac{|p|}{N}\right)\left(1 - \frac{|\tau - pT|}{\tau_p}\right), \qquad |\tau - pT| < \tau_p$$

where τ_p is the pulse length, τ is the lag variable, p is the pulse index, and N is the number of pulses for a finite pulse train. In the delay axis the ambiguity corresponds to the correlation function, which is shown in Fig. 11.4(a). The corresponding frequency dependence is given by

$$|\chi(0, f)| = \frac{1}{N}\left|\frac{\sin(\pi f\tau_p)}{\pi f\tau_p}\frac{\sin(\pi fNT)}{\pi fNT}\right|$$

In the frequency axis the ambiguity corresponds to the power spectrum, which is also shown in Fig. 11.4(a). For reference the detailed structure of the ambiguity function for a single pulse, [i.e., $\chi_c(\tau, f)$], is shown in Fig. 11.4(b). Although this pattern is a triangle function along the lag axis and a sinc function along the frequency axis, off both axes it is more complex than the product of these two functions alone. The ambiguity function of the finite pulse train is then a replication (or superposition to be exact) of this basic single-pulse ambiguity function over the entire delay-Doppler plane at locations $\tau \pm pT$, where $p = 0, 1, 2, 3, \ldots$ up to $N - 1$, where each copy is weighted by the discrete sinc functions $\sin(2\pi|p|fT)/\sin(\pi fT)$.

An interesting and useful relative of the ambiguity function that also captures time and frequency information is the Wigner–Ville distribution. It is the 2-D Fourier transform of the

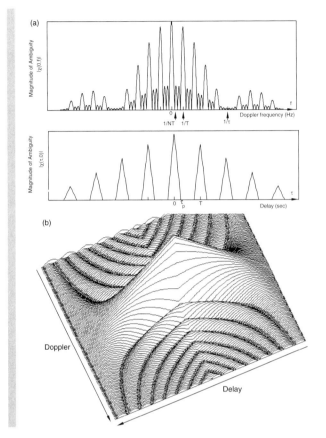

Figure 11.4 (a) Doppler frequency and delay cuts (at $t = 0$ and $f = 0$, respectively) of the ambiguity function of a coherent finite pulse train, where $T = 4\tau_p$ and $N = 5$, and (b) the detailed isometric view of the corresponding single pulse ambiguity function.

ambiguity function. Given the general symmetric form of the cross-ambiguity function defined by

$$A(\tau, \omega_D) = \int_{-\infty}^{\infty} s_1^*(t - \tau/2) s_2(t + \tau/2) \exp(-j\omega_D t) \, dt \tag{11.13}$$

its 2-D Fourier transform is

$$W(t, \omega) = \frac{1}{2\pi} \int_{-\infty}^{\infty} \int_{-\infty}^{\infty} A(\tau, \omega_D) \exp[j(\omega_D t - \omega\tau)] \, d\tau d\omega_D \tag{11.14}$$

which is the Wigner–Ville distribution and is given explicitly as

$$W(t, \omega) = \int_{-\infty}^{\infty} f_1^*(t - \tau'/2) f_2(t + \tau'/2) \exp(-j\omega\tau') \, d\tau' \tag{11.15}$$

This function is very useful because it takes into account for most realistic signals that the frequency content varies with time; that is, most real signals are nonstationary. One might think that appropriately translated short-time power spectra might be just as good, but they suffer from reduced resolution in time and frequency by virtue of windowing. To show how the Wigner distribution carries time-dependent frequency information, consider the following example.

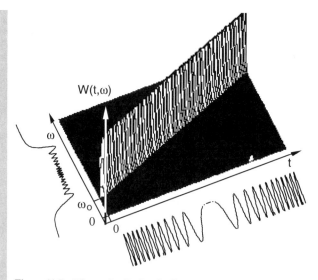

Figure 11.5 Wigner distribution for linear FM chirp-modulated carrier.

EXAMPLE 11.3 Wigner Distribution of a Complex Linear FM Signal The complex linear FM signal of infinite length is given by

$$f(t) = \exp\left[j2\pi\left(f_o t + \frac{1}{2}at^2\right)\right]$$

with center frequency ω_o and chirp rate α. First, determining the required integrand, we get

$$f(t+\tau/2)f^*(t-\tau/2) = \exp[j(\omega_o + \alpha t)\tau]$$

Then, substituting this result into Equation (11.17) yields

$$W(t,\omega) = 2\pi\delta(\omega - \omega_o - \alpha t)$$

as shown in Fig. 11.5. It is immediately obvious that the Wigner distribution tracks instantaneous frequency for a nonstationary signal, an immediate advantage over the Fourier transform. It is also real, which is helpful in optical implementation. Other useful properties are that the Wigner distribution is time- and frequency-invariant; that is, shifts in time and frequency [i.e., $x(t) \to x(t-\tau)$ and $x(t) \to x(t)\exp(j\omega_o t)$], yield Wigner distributions given by $W(f,t) \to W(t-t_o, f)$ and $W(f,t) \to W(t, f-f_o)$, respectively.

In order to calculate the ambiguity function using optics, we will approach the subject from a general viewpoint. We can look at an architecture intended to calculate a triple product. The basic schematic of this architecture is shown in Fig. 11.6. A laser diode is electronically modulated with the function $f_1(t)$. Its optical output illuminates the first AO cell oriented in the vertical (y') direction, which is addressed by the modulation function $f_2(t)$ such that the spatial modulation is $f_2(t - y'/v_s)$. Hence the output at P$_2$ is $f_1(t)f_2(t - y'/v_s)$ so that the light entering the second AO cell at P$_3$ is

$$g(y') = \int f_1(t)f_2(t - y'/v_s)\ dt \qquad (11.16)$$

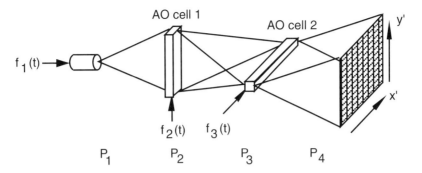

Figure 11.6 Triple product processor architecture employing AO Bragg cells.

The second AO cell is perpendicular to the first AO cell and both are imaged onto plane P_4 where a focal plane a distance of f (focal length) away would be located. The second AO cell operates on the output from the first cell with the kernel function $f_3(t - x'/v_s)$, such that the final output is

$$g(x', y') = \int f_1(t) f_2(t - y'/v_s) f_3(t - x'/v_s) \, dt \qquad (11.17)$$

This is the triple product processor output. Notice that the transfer functions (or kernels) for the horizontal (x') and vertical (y') axes are separately controllable. Although integration times are limited by the dynamic range of the output focal plane detector, the full 2-D space–bandwidth product is available as measured by the number of pixels (e.g., 512^2).

The previous architecture is also interesting because it can be the basis of several types of optical processors, including folded spectrum as well as ambiguity function processors. In fact, there is a large class of linear transformations described by the more general triple product given by

$$g(\xi, \eta) = \int f(\zeta) h_1(\zeta; \xi) h_2(\zeta; \eta) \, d\zeta \qquad (11.18)$$

The most notable example, which we are considering here, is the ambiguity function processor, where $\zeta = t, \xi = x'$, and $\eta = y'$; and $f(\zeta) = f(t), h_1(\zeta; \xi) = h_1(t; x') = f^*(t + x'/v_s)$, and $h_2(\zeta; \eta) = h_2(t; y') = \exp(-jty')$, where x'/v_s corresponds to the time lag variable for correlation and y' corresponds to the frequency variable for the Fourier transform. A large time–bandwidth product results from the long integration time. The limited space–bandwidth product in both output variables results from the limited pupils corresponding to the optical apertures in the architecture and the output detector plane. If $\zeta = x$ were chosen as the input variable (rather than t), then the space–bandwidth product would be reduced, but then t could be used as the output variable. Then the time–bandwidth product could be large and y could be used as the second output variable, in which case arbitrary linear transformations could be realized. For the ambiguity function we then have $\zeta = x, \xi = t, \eta = y'$, or $\eta = x'$. Then the triple product processor output is

$$g(x', t) = \int f(x) h_1(x; t) h_2(x; x') \, dx \qquad (11.19)$$

At fixed t this is a 1-D system with output variable x', and at each x' the system is 1-D with output variable t.

A practical implementation architecture for Equation (11.19) is illustrated in Fig. 11.7. Here the light leaving P$_2$ is $f(x)h_2(x - v_s t')$. Light propagates through the plane at P$_4$ with impulse response, in general, given by $h_1(x; x')$. For a Fourier transform $h_1(x; x') = \exp(jxx')$ and for correlation $h_1(x; x') = h(x - x')$. The output at plane P$_5$ varies in space (x') and time (t). Thus Equation (11.19) becomes

$$g(x', t) = \int f_1(x) f_2^*(x - v_s t') \exp(-jxx') \, dx \tag{11.20}$$

where x' corresponds to Doppler shift and t corresponds to range delay.

A variety of processing architectures can realize the ambiguity function and related processes (correlation and spectrum analysis) as described by Casasent and Psaltis by permuting the input and output variables. Input time–bandwidth (or space–bandwidth) product for these variants is typically 10^3 (for 1-D systems), up to 10^6 (for 2-D systems). The variety of potential processing operations are summarized in Table 11.1.

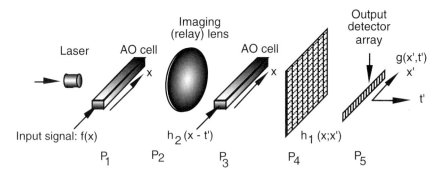

Figure 11.7 Ambiguity function optical processor architecture.

Table 11.1
Signal Processing Architectures

Input Variable(s)	Output Variable(s)	Realizable Kernels	Commonly Used Systems
x	y'	Arbitrary	Vector–matrix multiplier and image plane correlator
x	t'	$h(x - v_s t')$	Space-integrating correlator
x	x'	$h(x - x'), \exp(jxx')$	Frequency plane correlator and Fourier transform lens
t	x'	$h(t - x'/v_s)$, Chirp-z	Time-integrating correlator
x	y', t'	$h_1(x; y')h_2(x - v_s t')$	Ambiguity function
x	x', t'	$h_1(x; x')h_2(x - v_s t')$	Ambiguity function processor
x	x', y'	$h_1(x; x')h_2(x; y')$	Ambiguity function processor
t	x', y'	$h_1(t - x'/v_s)h_2(t - y'/v_s)$	Ambiguity function processor and spectrum analyzer
x, y	x', y'	$h(x - x', y - y')$	Folded spectrum analyzer
		$\exp[j(xx' + yy')]$	Image processor
x, t	x', y'	$\exp[j(xx' + yy')]$	Spectrum analyzer

EXAMPLE 11.4 Design Example of an Optical Ambiguity Function Processor Consider a detailed design example from the technical literature of an ambiguity function processor as shown in Fig. 11.8. This is a space-integrating architecture. Monochromatic plane wave illumination is focused in the vertical dimension by a cylindrical lens (L_1) onto the first AO cell, which is placed in the horizontal (x) direction. The signal $f(t)$ is applied to the first AO cell. Light diffracted from here (plane P_1) is recollimated vertically and focused horizontally by a spherical lens L_2 onto plane P_2 where the second AO cell is oriented in the vertical (y) direction. The second cell is driven by $h(t)$. In the horizontal direction the transmitted light is doubly diffracted and subsequently collimated by a spherical lens L_4. The combination of lenses L_2 and L_4 form an image at plane P_3 of light diffracted from the first AO cell. In the vertical direction a cylindrical lens (L_3) with the spherical lens (L_4) form an image located at P_3 of light diffracted by the second AO cell located at P_2. The amplitude of the 2-D light distribution at P_3 is thus modulated by $f(t - x/v_s)h(t - y/v_s)$. Then the astigmatic pair of lenses L_5 and L_6 image plane P_3 along one dimension and form a Fourier transform along the orthogonal dimension. The direction in which the cylindrical lens L_5 has focusing power is at 45° with respect to the x and y axes, and thus the Fourier transform is performed along a direction that is also at 45° with respect to the (x, y) axes; that is, for $\theta = 45°$ the linear transformation

$$\begin{bmatrix} x' \\ y' \end{bmatrix} = \begin{bmatrix} \cos\theta & -\sin\theta \\ \sin\theta & \cos\theta \end{bmatrix} \begin{bmatrix} x \\ y \end{bmatrix} \tag{11.21}$$

becomes

$$\begin{bmatrix} x' \\ y' \end{bmatrix} = \begin{bmatrix} \frac{1}{\sqrt{2}}(x - y) \\ \frac{1}{\sqrt{2}}(x + y) \end{bmatrix} \tag{11.22}$$

Thus we can define (x', y') as a new set of coordinates with y' parallel to the axis of the cylinder of lens L_5 and x' perpendicular to it. Thus the amplitude of the light distribution at P_4 is a 1-D Fourier transform of the light at plane P_3, namely,

$$g(u, y', t) = \int_{-\infty}^{\infty} f\left[t - (x' + y')/\sqrt{2}v_s\right] h\left[t - (x' - y')/\sqrt{2}v_s\right] \tag{11.23}$$
$$\exp(j2\pi u x') \, dx'$$

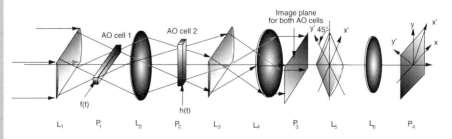

Figure 11.8 Ambiguity function processor with rotated astigmatic optics and orthogonal AO Bragg cells.

where $u = x''/\lambda F_6$, x'' is the spatial coordinate at the output plane, and F_6 is the focal length of lens L_6. If we choose $f(t)$ to be

$$f(t) = h(t - \tau)\exp(j2u_o t) \quad (11.24)$$

then $x'' = u_o \lambda F_6/\sqrt{2} \cdot v_s$ (at time t when both $f(t)$ and $h(t)$ are entirely within the apertures of the AO cells). $g(u, y', t)$ is the autocorrelation with respect to the lag variable y' of h, with a peak that occurs at $y' = v_s \tau/\sqrt{2}$.

Ambiguity function processing is of interest in designing waveforms for high-bandwidth radars, in which a common processing step in the receiver is correlation. The motivation for AO processing of ambiguity functions stems from the high computational burden in typical systems and the desire to do it in real time. Correlation of received signals with reference waveforms in some of these radars requires bandwidths as high as 500 MHz and sample times of 10^3 samples per microsecond. For each lagged product, 1000 multiplications are required. To produce a full correlation therefore requires 10^{12} multiplications per second, far in excess of special-purpose digital architectures. For Doppler radar systems each Doppler bin requires this computation. Only AO approaches can adequately support these requirements. The usual limitation on AO processors, however, is dynamic range, which must be commensurate with the time–bandwidth product to achieve full processing gain. Since time–bandwidth products $\geq 10^6$ are desired, a dynamic range of 60 dB is required but has been achieved only in coherent time-integrating optical architectures. Systems that use large arrays have many signal channels of narrow bandwidth that must be processed repeatedly at high computation rates to achieve such high processing bandwidths.

In ambiguity processing, large computational loads can result even from modest time–bandwidth products and large numbers of Doppler bins. For instance, a signal analyzed over 1000 Doppler channels with a time–bandwidth product of 200 requires an ambiguity surface for each channel. The number of channels therefore requires calculation of 1000 ambiguity surfaces per second. Using a fast Fourier transform to do Doppler analysis requires 10^4 multiplications per second for each range bin. The number of range bins should equal the time–bandwidth product. Therefore, the number of multiplications per second $= 200 \times 1000 \times 10^4 = 2 \times 10^9$, which is a very large number for state-of-the-art digital implementations.

11.4 SYNTHETIC APERTURE RADAR

Synthetic aperture radar (SAR) was among the earliest applications of optical signal processing. In the early days of development (1950s–1960s), film-based systems were developed, but they were large, expensive, and slow. Before explaining the fixed-film optical techniques and describing in detail the latest programmable architectures, it is necessary to give a short tutorial on SAR operation, in particular SAR signal processing and image formation.

For a synthetic aperture radar the transmitted waveform is given by $v(t) = v_o(t)\cos\omega_o t$, where $v_o(t)$ is the pulse envelope amplitude modulation and ω_o is the transmitted radian frequency. Since we know that SAR exploits the Doppler effect, we can explicitly indicate that the range, R, to the target has a velocity term, namely,

$$R = R_o + \frac{dR}{dt}t \quad (11.25)$$

where R_o is the initial range to target, and dR/dt is the relative velocity of the target along the line of sight.

The received signal reflected from a point target is then given by

$$v(t) = v_o(t) \cos\left[\left(\omega_o - \frac{2\omega_o dR/dt}{c}\right)t - \frac{2\omega_o R_o}{c}\right] \quad (11.26)$$

This equation indicates that the received signal has a frequency f_r shifted from the transmitted frequency f_o by the amount

$$f_D = f_r - f_o = \frac{-2f_o dR/dt}{c} = \frac{-2dR/dt}{\lambda} \quad (11.27)$$

where f_D is the Doppler frequency shift and λ is the wavelength. The received phase angle is $\phi_r = -4\pi R_0/\lambda$. Hence, phase is proportional to range and Doppler frequency is proportional to range rate.

From the typical SAR platform trajectory shown in Fig. 11.9, where a SAR is imaging point scatterers along a line on the earth's surface, the resolution of the SAR can be derived. In Fig. 11.9 the resolution in cross range (along the y axis), which is proportional to the Doppler frequency shift, Δf_D, between two scatterers, is given by

$$\Delta R_y = \frac{D}{2} = \frac{\lambda R \Delta f}{2v \sin \theta} \quad (11.28)$$

where

$$\Delta f = \frac{2v}{\lambda} \Delta\theta \sin\theta \quad (11.29)$$

for small $\Delta\theta$, where v is the aircraft velocity and θ is the squint angle to the scatterer 2 ($\theta + \Delta\theta$ is the squint angle to scatterer 1).

For a properly designed SAR the aperture time should be $T = 1/\Delta f_D$, so that the distance the SAR platform travels in that time is

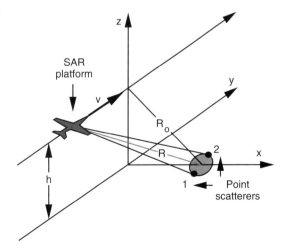

Figure 11.9 Typical SAR imaging geometry showing object illuminated and platform trajectory. The SAR antenna is "squinted" slightly forward from a sidelooking direction, i.e., the so-called "spotlight" mode.

$$L = vT = \frac{\lambda R}{2L \sin \theta} \tag{11.30}$$

Using this result and Equation (11.28), the resolution is

$$\Delta R_y = \frac{D}{2} = \frac{\lambda R}{2L \sin \theta} \tag{11.31}$$

Now consider the case of a focused SAR. As the SAR aircraft flies over a distance L, a point on the ground appears at ranges R_1, R_o, and R_1 (again) as shown in Fig. 11.10. Since the phase is proportional to range, it will change with time. If a phase shift of 90° (or $\pi/4$) is acceptable as the SAR moves through distance L, then

$$R_o^2 + (L/2)^2 = (R_o + \lambda/8)^2 \tag{11.32}$$

where $\lambda/8$ is the acceptable one-way phase change. Solving for L and neglecting terms quadratic in λ yields $L = \sqrt{\lambda R_o}$. Substituting this in Equation (11.31) yields $\Delta R_y = \sqrt{\lambda R_o}/2$. Hence, resolution for a focused SAR depends on range and can be predicted for specific scenarios. In this case the one-way phase shift is quadratic in time for L small compared to R_o; that is,

$$\phi = \frac{\pi y^2 t^2}{\lambda R_o} \tag{11.33}$$

This phase factor is generally corrected by a so-called SAR focusing algorithm.

We can characterize a SAR sensor as a linear system. Hence we can ascribe to it an impulse response. There are several approaches for deriving the impulse response of a SAR. Of course, two dimensions are considered—that is, slant range and cross range. Along the slant-range direction, pulse compression is employed to get better range resolution than simple pulse modulation. This is accomplished using a linear frequency modulation (or chirp) waveform. For the cross-range dimension the Doppler effect produces a natural chirp. Hence, both image dimensions are treated in the same way upon reception by the SAR; that is, they are "de-chirped."

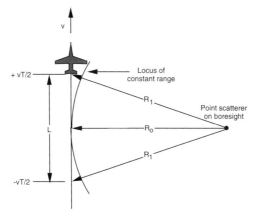

Figure 11.10 Plan view of SAR transit of ground scene, illustrating the "sidelooking" mode.

11.4 SYNTHETIC APERTURE RADAR

Consider a linear frequency-modulated waveform as in pulse compression. The transmitted signal is

$$v(t) = p_\tau(t) \cos\left(\omega_0 t + \frac{1}{2}\mu t^2\right) \quad (11.34)$$

where

$$p_\tau(t) = \begin{cases} 1, & -\tau/2 \leq t \leq \tau/2 \\ 0, & \text{elsewhere} \end{cases} \quad (11.35)$$

The receiver is a matched filter with impulse response $h(t)$ given by

$$h(t) = \sqrt{\frac{2\mu}{\pi}} p_\tau(t) \cos\left(\omega_0 t - \frac{1}{2}\mu t^2\right) \quad (11.36)$$

If there is a Doppler shift, the received signal is given by

$$x(t_o) = \cos\left[(\omega_o + \omega_D)t + \frac{1}{2}\mu t^2\right] \quad (11.37)$$

and the output of the matched filter is given by convolution integral; that is,

$$y(t) = \int_{t_1}^{t_2} x(\tau) h(t - \tau) \, d\tau \quad (11.38)$$

For this situation, Equation (11.38) is

$$y(t_o, \omega_D) = \sqrt{\frac{2\mu}{\pi}} \int_{-\tau/2}^{\tau/2} \cos\left[(\omega_o + \omega_D)t + \frac{1}{2}\mu t^2\right] \cos\left[\omega_o(t_o - t) + \frac{\mu}{2}(t_o - t)^2\right] dt \quad (11.39)$$

The close-form solution to this convolution integral is

$$y(t_o, \omega_D) = p_\tau(t) \sqrt{\frac{\mu}{2\pi}} \cos\left[(\omega_o + \omega_D/2)t\right] \frac{\sin\left[\frac{\omega_D + \mu t_0}{2}(\tau - |\tau_o|)\right]}{\frac{\omega_D + \mu t_0}{2}} \quad (11.40)$$

Note the sinc function response of the matched-filter output.

Figure 11.11 Synthetic array corresponding to effective SAR aperture.

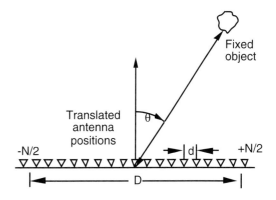

A similar response can be derived for the cross-range axis of the SAR. The cross-range impulse response of synthetic aperture processing can be derived from general considerations of the receiving array. In SAR processing a small sensor is sequentially stepped through small increments in space over a distance called the *synthetic aperture*. A sample of the received wavefront is taken at each time step, and the quadratic phase variation discussed previously (and hence antenna spatial position versus time) is coherently tracked and compensated for using an inertial measurement system. The coherent sum of the stored sample values is the same as the signal that would be received from the corresponding large real aperture. Figure 11.11 depicts such an array of sensor locations that sequentially transmit and receive. The response of the array is then given by

$$y(R_n) = \sum_{n=1}^{N} \exp\left(-j\frac{4\pi R_n}{\lambda}\right) \tag{11.41}$$

If each element of the array is identical and uniformly spaced, then the response of the SAR is

$$y(\theta) = \sum_{n=-N/2}^{N/2} \exp\left(-j\frac{4\pi n d \sin\theta}{\lambda}\right) \tag{11.42}$$

or

$$y(\theta) = \frac{\sin\left[\frac{2\pi(N+1)d\sin\theta}{\lambda}\right]}{\sin\left(\frac{2\pi d \sin\theta}{\lambda}\right)} \tag{11.43}$$

Note that this cross-range (or azimuth angle) response is also a sinc function like the slant range response discussed previously. Remember that the synthetic aperture is formed by moving a small sensor sequentially through an equivalent real aperture of receiving points.

In summary, we can describe the operation of SAR as an imaging process as illustrated in Fig. 11.12. In this illustration a point scatterer at position (x_o, y_o) in the object plane is illuminated by the radar. The received signal resembles a Fresnel zone pattern, which is described analytically as a chirp in x and y coordinates. The final step of imaging involves 2-D correlation of the received signal (Fresnel zone pattern) with a reference function, which is also a chirp function. The final image of this point object is, in fact, the point spread function (or impulse response) as illustrated in Fig. 11.13. Of course, the received Fresnel zone pattern from many point scatterers is a linear superposition of many individual Fresnel zone patterns. This range/azimuth response can be interpreted as two orthogonal time variables: "fast time" corresponding to range and "slow time" corresponding to azimuth (or Doppler). "Fast time" connotes that this axis is generated in the radar transmitter quickly by a deliberate chirping of the waveform, whereas "slow time" connotes that this axis is generated by virtue of the relatively slow radar platform motion with respect to the scatterer, which generates a "natural" azimuth chirp. The resulting amplitude and phase response functions are similar for both axes. The phase "history" of a given point scatterer is quadratic in time, which corresponds to linear FM.

Optical signal processing techniques are very appropriate for SAR image formation because the coherent detection of radar echoes received from the object field in SAR is analogous to the interferometric detection of the Fresnel diffraction pattern of an object that is

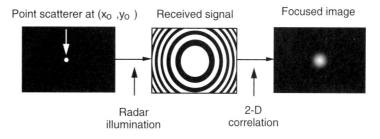

Figure 11.12 SAR imaging process described as a matched filtering (correlation) process.

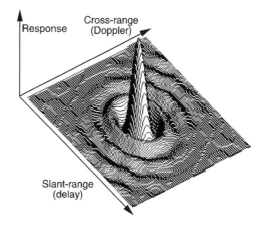

Figure 11.13 SAR imaging point spread function versus "fast" (range) and "slow" (Doppler) axes.

illuminated with coherent light. Thus the synthesis of the image of a single scatterer in SAR is equivalent to the optical reconstruction of a Fresnel hologram of a single point.

The mechanism for optically recording the SAR phase history of an object on the ground in the traditional side-looking SAR geometry is illustrated in Fig. 11.14. A film roll is exposed to light focused from a cathode ray tube (CRT), modulated by the received signal. This is the first step of the SAR processing. As the film is scanned by the CRT, a range swath is scanned in the CRT perpendicular to the film scan direction. The received signal is a bipolar signal that is mixed down to baseband in order to modulate the CRT. The intensity of the light spot on the CRT is therefore modulated as the range is swept. A bias is usually added to the bipolar signal to get the appropriate film exposure.

By virtue of the fact that the phase history is brought to focus on the SAR film roll at different focal lengths in range and azimuth, we must use astigmatic (or cylindrical) optics to subsequently form the image in the second (image forming) step of the process. Separate focal lengths occur because the phase history is different in each axis. The slant range component has a phase given by $\pi a(t - \tau)^2$, where a is the chirp rate; and the azimuth (cross-range) component has a phase given by $-2\pi(vt - y)^2$, where v is the SAR platform velocity. The slant-range term is recorded in the range dimension of Fig. 11.14 at the sweep velocity v_r. The azimuth term is recorded in the azimuth dimension at the film transport velocity v_a.

By virtue of the quadratic distributed phase delay in the film, it acts like a lens when illuminated with coherent collimated light. The corresponding phase function in either axis is given in general by

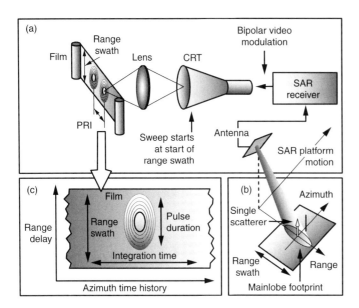

Figure 11.14 Optical technique for recording SAR phase history using photographic film. The Fresnel Zone pattern represents the received signal from a single scatterer and single pulse.

$$\phi(x) = -\frac{2\pi}{\lambda} \frac{x^2}{2F} \tag{11.44}$$

where F is the effective focal length in each axis. The choice of F for each axis depends on the following analysis. First note that the x and y dimensions of the recorded signal (film Fresnel superposition) are $x = v_r(t_1 - \tau)$ in range and $y = v_a(t_2 - y/v)$ in azimuth. Thus the two phase terms equivalent to Equation (11.44) are

$$\phi(t_1) = -\frac{2\pi}{\lambda} \frac{v_r^2(t_1 - \tau)^2}{2F_r} \tag{11.45}$$

and

$$\phi(t_2) = -\frac{2\pi}{\lambda} \frac{v_a^2(t_2 - y/v)^2}{2F_a} \tag{11.46}$$

F_r and F_a can be expressed in terms of radar parameters by equating optical and RF phase components. Thus $F_r = v_r^2/\lambda a$ in range, and $F_a = v_a^2 \lambda R/2\lambda a \, v^2$ in azimuth.

The optical architecture designed to correctly focus the SAR phase history is shown in Fig. 11.15. Cylindrical lenses are used to correct the basic astigmatism as shown in Fig. 11.15. A tilted cylindrical (or an untilted conical lens) is used to correct for the azimuth focal length linear dependence on range R. The first plane (P_2) after the input data plane (P_1) is the location of the focal plane in range (vertical axis). The second (tilted) plane (P_3) is the azimuth focal plane. A tilted cylindrical lens is oriented vertically so that its input focal plane coincides with plane P_2, and another cylindrical lens is oriented orthogonally to focus on plane P_3. As a result, both axes are collimated. Therefore the final spherical lens focuses the collimated light onto the output plane, thus forming the SAR image.

Figure 11.15 Optical processor for focusing a SAR phase history on film.

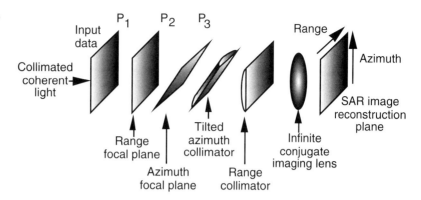

This architecture represents the simplest and one of the earliest methods of optical SAR image formation. The output plane usually employs photographic film. Thus these systems are not real-time, nor are they rapidly programmable. Because not all SAR platform trajectories are so simple as the side-looking case, it is often necessary to refocus the light for different geometries. To do this requires SAR optical architectures that employ programmable SLMs to accommodate changes in geometry and waveforms. We will discuss these next.

Recall the geometry of the SAR platform viewing an object on the ground as was shown in Fig. 11.9. The figure depicts a slightly forward-looking field of view, which in practice could be more pronounced and is termed a spotlight mode rather than a side-looking mode because it is squinted toward the indicated point scatterer and is usually a more confined angular sector. The range to the point scatterer is

$$R(t) = \left(R_o^2 + x^2\right)^{1/2} \tag{11.47}$$

which, for small displacement along y (compared to R_o), is

$$R(t) = R_o\left[1 + v^2(t^2 - t_o^2)^2/2R_o^2\right] \tag{11.48}$$

This expression (minus the leading term on the right-hand side, which is just the range offset to the target) appears in the phase factor attributed to the natural Doppler chirp in the received signal, as discussed below.

In the following discussion we will follow closely recent technical developments, especially those published by Pslatis, Wagner, and Haney. The transmitted signal is a FM or chirped pulse train:

$$s(t) = \sum_n \underbrace{\text{rect}\left(\frac{t-nT}{\tau}\right)}_{\substack{\uparrow \\ \text{Pulse modulation} \\ \text{envelope}}} \cdot \underbrace{\exp\left[ja(t-nT)^2\right]}_{\substack{\uparrow \\ \text{Linear FM} \\ \text{chirp}}} \cdot \underbrace{\exp(j\omega t)}_{\substack{\uparrow \\ \text{Carrier}}} \tag{11.49}$$

where τ is the pulse duration, T is the pulse repetition period, a is the chirp rate, and ω is the transmission frequency. (The "rect" pulse envelope implies infinite bandwidth modulation, which is not met in practice, but is convenient for this explanation.) The patch on the earth's surface is illuminated with this waveform. For each point scatterer in the patch, the received signal is $r(t) = A(t)s(t - 2R/c)$, where R is the instantaneous range to the point scatterer, and $A(t)$ is the far-field pattern of the antenna. Considering each received pulse and assuming that $A(t)$ does not vary enough to be significant, the received pulse train is

$$s\left(t - \frac{2R}{c}\right) = \sum_n \underbrace{\text{rect}\left(\frac{t - (2R/c) - nT}{\tau}\right)}_{\substack{\text{Return pulse}\\ \text{modulation envelope}}} \underbrace{\exp\left[jb\left(t - \frac{2R}{c} - nT\right)^2\right]}_{\text{Linear FM chirp}}$$

$$\times \underbrace{\exp(j\omega t)}_{\text{Carrier}} \underbrace{\exp\left[\frac{j\omega_o v(nT - t_o)^2}{R}\right]}_{\text{Natural azimuth chirp}} \quad (11.50)$$

The SAR received signal can be shown, from Equation (11.50), to be separable in range and Doppler. Hence, it can be processed with two sequential Bragg cells. The 1-D signal can also be interpreted as a 2-D unfocused image by defining $\tau_o = t - nT$. Therefore Equation (11.50) becomes

$$s(t - 2R/c) = \text{rect}\left(\frac{\tau_o - 2R_o/c}{\tau}\right) \exp\left[jb(t - 2R_o/c)^2\right]$$

$$\times \exp\left[j\omega_o v(t - t_o)^2\right] \exp\left[j\omega_o \tau_o\right] \quad (11.51)$$

(for a single member of the pulse train) where ground (delay) coordinates $\tau_o = 2R_o/c$ and $t = t_o$ are the center of the received modulation, which can be interpreted as a nonsymmetric Fresnel zone plate, as discussed earlier.

A schematic diagram of a real-time AO SAR processor is shown in Fig. 11.16. The system is illuminated with a beam from a pulsed laser diode. The light amplitude modulation is

$$P(t) = \sum_n \text{rect}\left(\frac{t - nT}{\tau}\right) \quad (11.52)$$

where τ is the duration of each light pulse. The spherical lens, L_1, collimates the light from the source, and the cylindrical lens, L_2, focuses the light in the vertical (y) direction so that it can

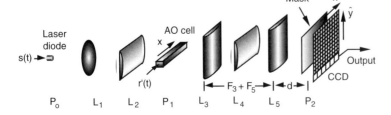

Figure 11.16 Real-time AO SAR processor architecture.

pass through the aperture of the AO device in plane P_1. The received radar signal is heterodyned to the center frequency v_1 of the AO device; and a reference signal, $B \exp(j2\pi v_2 t)$, is added to it. The frequency v_2 is chosen such that the difference $(v_2 - v_1)$ is equal to the bandwidth of each radar pulse; that is, $v_2 - v_1 = b\tau/\pi$. The resulting signal $r'(t)$, is related to $r(t)$ (the signal received by the antenna) by the following equation:

$$r'(t) = r(t) \exp[-j2\pi(v_o - v_1)t] + B \exp(j2\pi v_2 t) \tag{11.53}$$

where $r'(t)$ is the signal applied to the transducer of the AO device and B is the strength of the reference signal. The light modulation of the beam diffracted by the AO device is

$$S_1(t, x) = \text{rect}\left(\frac{x}{X}\right) P(t) r'\left(t + \frac{x}{v_s}\right) \tag{11.54}$$

where x is along the direction of propagation of the acoustic wave, X is the aperture of the AO device, and v_s is the speed of sound in the AO device. The undiffracted light is blocked in the focal plane of the cylindrical lens, L_3. The combination of lenses L_3 and L_5 accomplishes two tasks. First, since the two lenses are separated by the sum of their focal lengths, a single plane wave incident on L_3 will be recollimated when it exits lens L_5. Second, the system impulse response in the x direction from plane P_1 to P_2 is made to be equal to

$$h(x, \hat{x}) = \exp\left[jb_1 (x - \hat{x})^2\right] \exp(-j2\pi v_1 x/v_s) \tag{11.55}$$

where $\hat{x}$ is the horizontal coordinate in plane P_2. The constant b_1 can be set by appropriately choosing the focal lengths of L_3 and L_5 and the distance d. The term $\exp(-j2\pi v_1 x/v_s)$ reflects the fact that the optical system following the AO device is tilted so that its optical axis coincides with the wave diffracted at the Bragg angle. In the vertical direction y, the light is recollimated by lens L_4, and therefore the amplitude of light intering plane P_2 does not vary along y. At plane P_2, a transparency is placed immediately in front of the CCD dectector. The intensity transmittance $T(x, y)$ of this mask is

$$T(x, y) = \frac{1}{2} + \frac{1}{2} \cos\left[2\pi u_o \hat{x} + \frac{b_2 \hat{y}^2}{\hat{x}}\right] \tag{11.56}$$

where

$$u_o = \frac{(v_2 - v_1)}{v_s} \tag{11.57}$$

and

$$b_2 = \frac{4 v_o v^2 T^2 v_s}{c^2 \Delta y^2} \tag{11.58}$$

where v_s = speed of sound in the AO device, and Δy = pixel size in the CCD.
    The instantaneous intensity incident on the CCD is

$$I(\hat{x}, \hat{y}, t) = T(\hat{x}, \hat{y}) \left| \int S_1(t, x) h(x, \hat{x}) \, dx \right|^2 \tag{11.59}$$

The CCD detector is operated in the shift-and-add mode, in which it is exposed to light periodically. The photogenerated charge is accumulated on the CCD for a short time, and after each exposure the entire charge pattern is shifted by one pixel along one of the dimensions of the CCD. The charge resulting from the next exposure is added to the charge that is already stored in each pixel, and the process is repeated. The charge transference is done in synchronism with the PRF of the radar; that is, the integration period of the CCD is set equal to the period T of the radar. The charge generated on the CCD during the nth radar pulse is

$$Q(\hat{x}, \hat{y}, n) = \int_{t-nT/2}^{t+nT/2} I(\hat{x}, \hat{y}, t) \, dt \qquad (11.60)$$

After N exposures, the charge pattern is shifted along the y direction of the CCD by $(N - n)$ pixels. Then the total charge that accumulates at each pixel located at coordinates (x, y) is

$$Q(\hat{x}, \hat{y}) = \sum_{n}^{N} Q\left[\hat{x}, \hat{y} + (N - n)\Delta y, n\right] \qquad (11.61)$$

$$= \frac{By_s \tau \tau_o}{2} \operatorname{sinc}\left[\frac{b\tau}{\pi v_s}\left(\hat{x} - \frac{2R_o v_s}{c}\right)\right]$$

$$\times \operatorname{sinc}\left[\frac{2N v_o v^2 v T^2}{c \Delta y R_o}\left(\hat{y} - y_o \frac{\Delta v}{vT} + N\Delta y\right)\right] \cos\left(4\pi u_o \hat{x} + \phi\right)$$

$$+ \text{bias terms} \qquad (11.62)$$

where ϕ is a constant phase term. $Q(x, y)$ is the final output of the processor; its form demonstrates the imaging capability of the SAR processor. In the x direction, $Q(x, y)$ is a sinc with width $\pi v_s/b\tau$ and centered at $x = 2R_o v_s/c$. In other words, we obtain the image of the point scatterer located at $x = x_o$ on the ground, focused in the x direction. The carrier at spatial frequency $2u_o$ arises from the inclusion of the reference signal in the AO device, which allows the recording of the phase of the detected signals on the CCD. This is essential because the Doppler information used to focus the return signal in the azimuth (y) direction is encoded in the phase of the range-compressed signal $Q(x, y, n)$. In addition, since the output forms on a carrier, it can easily be separated from the bias terms in Equation (11.62).

In the y direction, $Q(x, y)$ is also a sinc function whose position is proportional to y_o, the location of the point scatterer on the ground in the azimuth direction. NvT is the distance that the vehicle carrying the radar travels during the time interval NT and is equal to the synthetic aperture of the system. The entire pattern is shifted in the y dimension by $N\Delta y$. For a CCD with N pixels in the horizontal direction, this implies that after the signal is integrated on the CCD for N pulses to produce a focused image, it arrives at the edge of the device (at $y = N\Delta y$) where a separate CCD stage transfers an entire line of the data (a slice of the image for each azimuth position) to the output amplifier of the CCD. Azimuth slices are produced continuously as long as the flight continues, producing an image of a long strip on the ground parallel to the direction of flight.

In most cases, parameters of the radar/target geometry, such as velocity and altitude of the SAR platform and the direction in which the antenna is pointed, change dynamically. The real-time SAR processor must be able to adapt rapidly to such changes in order to provide a well-focused image continuously. To accomplish this with the system of Fig. 11.16, the mask could be replaced by a real-time 2-D SLM on which the proper reference function is formed and updated as needed. Examination of the required 2-D reference function reveals that when overall range to the target is large compared to the swath width, the reference function is approximated well by a 1-D base-band linear FM signal in the azimuth direction, whose scale varies linearly in range to account for the range/azimuth coupling of the geometry. This suggests that an AO device can be used to input the reference function to the processor and that range/azimuth coupling can be introduced by an appropriate anamorphic lens.

An optical system that generates the 2-D azimuth reference function using a 1-D AO device is shown schematically in Fig. 11.17. An electronic chirp signal is applied to an AO device along with a sinusoidal reference signal. After the acoustic signals have propagated to the center of the AO device, the laser diode is pulsed in the same manner as the radar signal is sampled in the AO device of the SAR processor described previously. An anamorphic lens arrangement spreads diffracted light uniformly in the x direction at the detector plane. Undiffracted light (not shown in Fig. 11.17) is blocked and not used. In the y direction, diffracted light from a sinusoidal reference signal in the AO device is a plane wave that is transformed into a cylindrical wave at the detector plane. The light diffracted by the chirp signal is a cylindrical wave that is transformed into a cylindrical wave of different curvature at the detector plane. Waves from the chirp and reference signals have different radii of curvature, and therefore the resulting interference pattern on the detector will be approximately equal to a linear FM signal in the y direction. The scaling that is required in the x direction to account for the range/azimuth coupling is accomplished by tilting a cylindrical lens away from the y axis, as shown in Fig. 11.17.

A programmable real-time SAR-processing architecture is formed by integrating the interferometric technique for SAR azimuth filtering described in Fig. 11.17 with the range-processing technique described previously. One approach is a two-arm interferometer in which the light that is modulated by the range-focusing AO device follows a different path to the output CCD detector than does the light that forms the azimuth filter. Alternatively, it is possible to cascade the two systems so that the interfering light beams follow the same path through the optical system, thereby forming a one-arm interferometer that minimizes the effect of mechanical vibrations. The cascaded system is shown in Fig. 11.18. As in the original SAR processor, the radar signal is range-focused, and its phase is detected interferometrically at the detector. Since the light is now also diffracted by the second AO device (containing the azimuth reference signal), it is effectively multiplied by the interferometrically generated reference function. The programmable architecture gives the optical SAR processor flexibility to adapt to changes in the radar/target geometry by electronically changing the starting frequency and

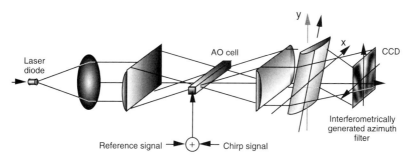

Figure 11.17 Optical architecture for generating 2-D azimuth reference function for optical SAR processor.

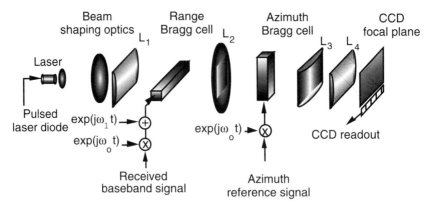

Figure 11.18 Real-time programmable SAR image formation architecture using two orthogonal Bragg cells for range and azimuth focusing.

chirp rate of the azimuth reference signal. Equally significant is the fact that this architecture allows the azimuth correlation to be performed by shifting the reference function in the AO device rather than shifting and adding the charge on the CCD. This capability is useful for the spotlight mode SAR, when it is desired to form the image in a frame-by-frame format rather than in a long strip.

For this architecture the transmitted signal is

$$s(t) = \sum_{n=1}^{N-1} \text{rect}\left(\frac{t-nT}{\tau}\right) \exp\left[jb(t-nT)^2\right] exp(j\omega t) \quad (11.63)$$

and the received signal from a single point scatterer is

$$s\left[t - \frac{2r(t)}{c}\right] = \sum_{n=1}^{N-1} \text{rect}\left(\frac{t - 2r_0/c - nT}{\tau}\right) \exp\left[jb\left(t - \frac{2r_0}{c} - nT\right)^2\right]$$

$$\exp(j\omega t) \exp\left[\frac{j\omega_0(vnT - y_0)^2}{r_0}\right] \quad (11.64)$$

The received signal can be reinterpreted as a 2-D asymmetric zone plate defined by

$$s(\tau', t') = \text{rect}\left(\frac{\tau' - 2r_0/c}{\tau}\right) \exp\left[jb\left(\frac{\tau' - 2r_0}{c}\right)^2\right]$$

$$\exp\left[\frac{j\omega(vt' - y_0)^2}{r_0}\right] \exp[j\omega t] \quad (11.65)$$

where $\tau' = t - nT$ and $t' = nT$. As a result of the SAR imaging process as described in Fig. 11.18, the signal is focused (or spatially integrated) in range by the first Bragg cell, phase-detected interferometrically in azimuth, and focused in azimuth (t') by temporal correlation with a reference function defined as

$$I_{\text{mask}}(x, y) = 1 + \cos\left(\frac{\omega v'^2 t'^2}{r_0}\right) \quad (11.66)$$

which is generated in the second Bragg cell (or defined over a 2-D mask in front of the CCD).
As a result, the final output image is

$$I(x, y) = 1 + \text{sinc}^2 \left[\frac{(\hat{x} - 2r_o(v/c)) \tau b}{\pi v_a^2} \right]$$

$$+ 2 \text{sinc} \left[(x - 2r_o(v/c)) \tau b / \pi v_a^2 \right] \text{sinc} \left[\frac{(y - y_o) b_2 y}{\pi} \right] \quad (11.67)$$

where v_a is the Bragg cell acoustic velocity, b is the chirp rate, $b_2 = \omega(v^2/v'^2)r_o$, and Y is the integration limit. Note the presence of the first two terms yields a bias, and the third term is the desired image of the original finite source.

PROBLEM EXERCISES

1. For an arbitrary function $f(x)$ with Fourier transform $F(k)$ show that

$$\int_{-\infty}^{\infty} x f(x) \, dx = \frac{-1}{2\pi j} \frac{dF}{dk} \bigg|_{k=0}$$

and

$$\int_{-\infty}^{\infty} x^2 f(x) \, dx = \frac{-1}{4\pi^2} \frac{d^2 F}{dk^2} \bigg|_{k=0}$$

2. If $f(x, y)$ has a Fourier transform given by $F(u, v)$, show that

$$\frac{\int_{-\infty}^{\infty} \int_{-\infty}^{\infty} f(x, y) \, dx dy}{f(0, 0)} = \frac{4\pi^2 F(0, 0)}{\int_{-\infty}^{\infty} \int_{-\infty}^{\infty} F(u, v) \, dudv}$$

where the expression on the left-hand side is the equivalent area of the function $f(x, y)$, and the expression on the right-hand side is the equivalent area of its corresponding Fourier transform $F(u, v)$, denoted by Δ_{xy} and Δ_{uv}, respectively. What is this result a statement of?

3. Consider a short symmetric pulse $f(x)$ with pulse width Δx that has a maximum value at $x = 0$ so that $f(0) \geq f(x)$ and that has a Fourier transform $F(k)$. Show that the uncertainty principle holds, namely, $\Delta x \Delta k \geq 2\pi$, where Δk is the nominal spatial frequency bandwidth of the pulse.

4. Work through the intermediate steps of Example 11.2 and explicitly derive the final result for the ambiguity function. Then plot it versus delay and Doppler separately. Show that cuts along other multiples of T are given by

$$|\chi(pT, f)| = \frac{1}{N} \left| \frac{\sin(\pi f \tau_p)}{\pi f \tau_p} \frac{\sin(\pi f (N - |p|) T)}{\pi f T} \right|, \quad |p| \leq N - 1$$

In addition to the analytical derivation, use MATLAB to plot your results.

5. A chirp radar transmitter generates a waveform given by

$$s(t) = \text{rect}\left(\frac{t}{\tau}\right) \cos\left(\omega t + \frac{1}{2}\alpha t^2\right)$$

which is reflected from a stationary target at range r with cross-section $\sigma(r)$.

(a) Describe an optical architecture that would perform a matched-filter operation over a time interval of T seconds following pulse transmission.

(b) Write an expression for the received signal and the corresponding matched filter.

(c) Assuming that the received signal and matched filter are implemented as multiplication masks in the Fourier domain and that the intermediate plane is film-scanned at velocity v_s so that $x = v_s t$, determine their product and Fourier transform.

6. Given the simplified architecture for ambiguity function processing as shown in Fig. 11.6, define the lens elements and their arrangement and spacing required to implement it. (*Hint*: A total of four lenses will be required.)

7. Defining the laser (source) modulation signal as $f_1(t) = f(t) \exp(-j\pi\alpha t^2)$, the first Bragg cell modulation as $f_2(t) = g(t)$, and the second Bragg cell modulation as $f_3(t) = \exp(j\pi\alpha t^2)$ and assuming that the corresponding intensity modulation terms (I_1, I_2, and I_3) are the product of the source modulation and the AO modulations in each case, do the following:

(a) Show that

$$I_o(t) = A_o \left\{ 1 + \sqrt{2} m_o \operatorname{Re}\left[f_1(t) \exp(j\omega_o t) \right] \right\}$$

$$I_1(t) = A_1 \left\{ 1 + +2m_1^2 |f_2(t)|^2 + 2\sqrt{2} m_1 \operatorname{Re}\left[f_2(t) \exp(j\omega_1 t) \right] \right\},$$

and

$$I_2(t) = A_2 \left\{ 1 + +2m_2^2 |f_3(t)|^2 + 2\sqrt{2} m_2 \operatorname{Re}\left[f_3(t) \exp(-j\omega_2 t) \right] \right\}$$

where ω_0, ω_1, and ω_2 are frequency differences between the reference and the double-sideband supressed carrier modulations at $\omega_c + \omega_0$, $\omega_c + \omega_1$, and $\omega_c + \omega_2$, respectively.

(b) Show that the integrated intensity at the detector plane is a triple product,

$$R(\tau_1, \tau_2) = \int_{-\infty}^{?\infty} I_o(t) I_1(t - \tau_1) I_2(t - \tau_2) \, dt$$

and that this expression contains the desired ambiguity function

$$\operatorname{Re}\left\{ A_{fg}(\tau_1, \alpha\tau_2) \exp\left[-j2\pi \left(f_1\tau_1 - f_2\tau_2 + \frac{1}{2}\alpha\tau_2^2 \right) \right] \right\}$$

where τ and ω_D are the delay and Doppler, respectively.

8. For the SAR architecture of Fig. 11.18, use the processes of range focusing, interferometric detection, and azimuth temporal correlation to obtain the result of Equation (11.67), where the matched filter for range compression is $\exp[-jb(x + x')^2 / v_a^2]$, where v_a is the first Bragg cell acoustic velocity, and the interfering signal for interferometric detection is $\exp(j\omega t')$.

9. In SAR image generation the received signal located in the Fourier domain is given by $S(u, v)$. The corresponding SAR image in the spatial domain, $s(x, y)$, is defined as

$$s(x, y) = \int_{-\infty}^{\infty} \int_{-\infty}^{\infty} S(u, v) \exp[j2\pi(ux + vy)] \, du \, dv$$

Show that the Radon transform or projection of the SAR image $s(x, y)$ along one axis as defined by

$$l(x) = \int_{-\infty}^{\infty} s(x, y) \, dy$$

can be expressed as

$$l(x) = \int_{-\infty}^{\infty} S(u, 0) \exp[j2\pi ux] \, du$$

thus avoiding the full 2-D Fourier transform in order to get useful projection slices of the SAR image.

10. For the linear FM chirp pulse with rectangular envelope produced by a high-range-resolution radar, the waveform is

$$s(t) = \frac{1}{\sqrt{T}} \text{rect}\left(\frac{|\tau|}{T}\right) \exp\left(\frac{j2\pi\alpha t^2}{2}\right)$$

Calculate the ambiguity function and show that it is given by

$$|\chi(t, f_D)| = \text{tri}\left(1 - \frac{|\tau|}{T}\right) \frac{\sin[\pi(k\tau + f_D)(T - |\tau|)]}{[\pi(k\tau + f_D)(T - |\tau|)]} \text{rect}\left(\frac{t}{2T}\right)$$

where τ is the delay and f_D is the Doppler frequency. Plot the result on each axis, label, and denote the functional form of each slice in τ and f_D. Use MATLAB. Can you plot the full functional form in isometric view?

11. Calculate the Wigner distribution given by

$$W(x, f) = \int u(x + x'/2) u(x - x'/2) \exp(-j2\pi f x') \, dx'$$

for the cases of (a) ideal monochromatic plane wave, (b) infinitely narrow pulse, and (c) infintely long linear FM signal. Use MATLAB to plot your results.

12. Consider a real signal defined by

$$s(t) = \cos(\omega_o t + \pi \alpha t^2) \text{rect}\left(\frac{t - T/2}{T}\right)$$

where $\omega_o = 2\pi f_o$ is the lower frequency, $\alpha = B/T$ is the FM slope, and B is the signal bandwidth.
(a) Write the analytic form of this signal.
(b) Then derive the form of $W(t, f)$ explicitly.
(c) Plot the result in time and frequency separately. Use MATLAB to plot your results.

13. Calculate the Wigner distribution for the following signals
(a) $f_1(t) = 1 + \cos(2\pi f_o^2 t^2)$
(b) $f_2(t) = 1 + \sin[A \sin(\omega_m t) + \omega_o t]$

and plot the results. Then take the 2-D Fourier transform of each to obtain the corresponding ambiguity function and plot these results with MATLAB. Interpret your results in both domains.

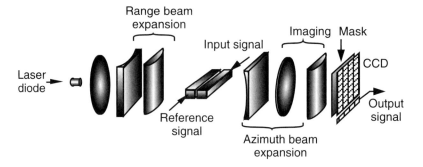

14. The joint collinear interferometric correlator architecture illustrated was proposed by Montgomery to generate SAR images. A coherent horizontally expanded laser beam illuminates two Bragg cells in sequence, and the diffracted beams are imaged in range and expanded in azimuth onto the detector plane.

    (a) Write expressions for the diffracted beams given that the received signal is

    $$s(t) = s_o(t) \exp(j2\pi f_1 t)$$

    and the reference signal is

    $$h(t) = h_o \exp\left[j2\pi \left(f_1 t + \frac{1}{2}\alpha t^2\right)\right]$$

    (b) Calculate the charge $Q(nT, x)$ produced at a single photodiode (without the mask) due to the nth pulse using the expression

    $$Q(nT, x) = K \int_\tau |s(t) + h(t)|^2 \, dt$$

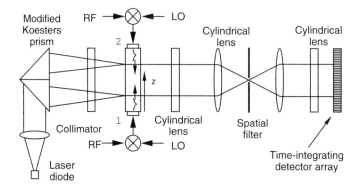

15. Consider the architecture shown in the figure. This is a coherent time-integrating correlator in which two counterpropagating signals are input to a Bragg cell. The beam-forming optics (including the two orthogonal cylindrical lenses) and spatial filter comprise the remainder of the architecture. The input signal is $x_1(t) = A(t) \cos \omega_s t$, where $A(t)$ is broadband modulation on the carrier of frequency ω_s. A plane wave incident on the Bragg cell at angle $2\theta_B$ can be written in complex form as $\exp(j\omega_L t) \exp(jz \sin 2\theta_B / \lambda)$, where ω_L and λ are the light frequency and wavelength, respectively, and z is the distance from the origin (at position 1). The acoustic traveling wave is

given by $A(t - z/v) \exp[j\omega_s(t - z/v)]$ for port 1 of the Bragg cell. What is the expression for the output light distribution? Express it in a real form, label it as $y_1(t, z)$, and incorporate the incident laser beam shape factor $a(z)$. Write down the equivalent expression for the other port (port 2), where the counterpropagating term is $B(t)\cos\omega_s t$. The result of imaging these two functions through the architecture and detecting them with a time-integrating square-law detector array is

$$V(z) = \int_0^T I(t, z)\, dt$$

where

$$I(t, z) \propto [y_1(t, z) + y_2(t, z)]^2 = y_1^2(t, z) + y_2^2(t, z) + 2y_1(t, z)y_2(t, z)$$

Expand this expression for $I(t, z)$ explicitly and incorporate it into the expression for $V(z)$. Simplify if possible.

16. For the previous problem assume that the time integrating detector array low passband limit excludes any terms with frequencies above ω_L. Given that

$$\sin(2\theta_B) = \left(\frac{\omega_o}{c}\right)^{-1}\left(\frac{\omega_L}{v}\right)$$

where ω_o is the design center frequency of the Bragg cell, determine $V(z)$ and show that it consists of two (DC) bias terms and one cross-term. Identify the cross-correlation term.

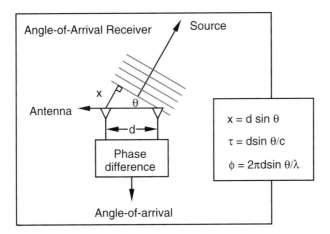

17. The architecture (Problem 15) can be used to estimate angle-of-arrival (AOA) as shown in the figure above, where outputs of the receiver go to ports 1 and 2 in the architecture. An AOA measurement first requires a time delay measurement. (Phase can also be used to refine the measurement, but it is limited to one cycle by ambiguity considerations.) The time delay measurement is commonly set by processing considerations to be $\Delta t \sim 1/10\Delta f$, where Δf is the Bragg cell bandwidth. For antenna element spacing of 10 m, what is the time-delay resolution? What is the corresponding angle accuracy $\Delta\theta$ for small angles? If the pattern can be improved by 1/50th of a cycle, what is the angle of arrival accuracy in degrees? If the correlation is imaged onto a 512-element, 25-μm pixel size array and the maximum possible delay is 30 nsec, what is the maximum possible spatial

extent of the correlation pattern given that the material used is lithium niobate? What must the lens magnification be to distribute the time delays from 0 to 30 nsec across the array?

18. (a) For the sum of two exponentials defined by

$$s_1(t) = [\exp(j\omega_1 t) + \exp(j\omega_2 t)]$$

determine the Wigner distribution and plot your results by hand.

(b) For the sum of two Gaussian signals defined by

$$s(t) = \left(\frac{\alpha}{\pi}\right)^{1/4} \left\{ \exp\left[-\frac{\alpha}{2}(t-t_1)^2 + j\omega_1 t\right] + \exp\left[-\frac{\alpha}{2}(t-t_2)^2 + j\omega_2 t\right] \right\}$$

calculate the Wigner distribution and plot your results using MATLAB. Your results should contain three terms, one of which oscillates, is bipolar, and is located between two otherwise identical positive Gaussian terms (except for their relative locations in the time-frequency domain). What happens to the middle term as the separation between the two Gaussian terms increases?

BIBLIOGRAPHY

I. Abramovitz et al., "Coherent Time-Integration Processors," in *Acousto-Optic Signal Processing–Theory and Implementation*, N. J. Berg and J. N. Lee, eds., Marcel Dekker, New York (1983).

H. O. Bartlett, K.-H. Brenner, and A. W. Lohmann, "The Wigner Distribution Function and Its Optical Rotation," *Optics Commun.* **32**, 32 (1980).

B. Boashash, O. P. Kenny, and H. J. Whitehouse, "Radar Imaging Using the Wigner–Ville Distribution," pp. 282–294, Proc. SPIE, Vol. 1154, *Real-Time Signal Processing XII* (1989).

W. S. Burdic, *Radar Signal Analysis*, Prentice-Hall, Englewood Cliffs, NJ (1968).

D. Casasent and M. Carletto, "Multidimensional Adaptive Radar Processing Using an Iterative Optical Matrix–Vector Processor," *Opt. Eng.* **21**, 814 (September/October 1982).

D. Casasent and D. Psaltis, "General Formulation for Optical Signal Processing Architectures," *Opt. Eng.* **19**, 962–974 (1984).

L. Cohen, "Time-Frequency Distributions—A Review," *Proc. IEEE* **77** (7), pp. 941–981 (1989).

C. E. Cook and M. Bernfeld, *Radar Signals, An Introduction to the Theory and Application*, Academic Press, New York (1967).

C. D. Daniel, "Concepts and Techniques for Real-Time Optical Synthetic Aperture Radar Data Processing," *IEE Proc.* **133**, Pt. J, No. 1, 7 (1986).

K. J. DeVos, "System Identification and Signal Detection Using the Wigner Distribution," M.S. Thesis, Carnegie–Mellon University (October 1985).

C. Elachi, "Spaceborne Synthetic-Aperture Imaging Radars: Applications, Techniques and Technology," *Proc. IEEE* **70**, 1191 (1982).

J. L. Horner, ed., *Optical Signal Processing*, Academic Press, San Diego, CA (1987).

J. N. Lee, "Optical and Acousto-optical Techniques in Radar and Sonar," Proc. SPIE, Vol. 456, *Optical Computing*, pp. 96–104 (1984).

P. A. Molley and K. T. Stalker, "Acousto-Optic Signal Processing for Real-Time Image Recognition," *Opt. Eng.* **29**, 1073 (1990).

R. M. Montgomery, "Acousto-optical Signal Processing System," U.S. Patent #3,634,749, January 1972.

D. Psaltis and M. Haney, "Acousto-optic Synthetic Aperture Radar Processors," in *Optical Signal Processing*, J. L. Horner, ed., Academic Press, New York, pp. 191–241 (1987).

D. Psaltis and K. V. Wagner, "Real-Time Optical Synthetic Aperture Radar (SAR) Processor," *Opt. Eng.* **21**, 822 (1982).

A. W. Rihacek, *Principles of High Resolution Radar*, McGraw-Hill, New York (1969).

R. J. Sadler and M. R. Buttinger, "Acousto-optic Ambiguity Function Processor," *IEE Proc.* **133**, Pt. J, No. 1, 35 (1986).

H. H. Szu, "Two-Dimensional Optical Processing of One-Dimensional Acoustic Data," *Opt. Eng.* **21**, 804 (1982).

S. Qian and D. Chen, *Joint-Time Frequency Analysis: Methods and Applications*, Prentice-Hall PTR, Upper Saddle River, NJ (1996).

D. R. Wehner, *High Resolution Radar*, Artech House, Norwood, MA (1987).

Chapter 12
PATTERN RECOGNITION APPLICATIONS

12.1 OVERVIEW

Optical pattern recognition is another equally important area of application of optical processing. Here optical correlation, spectrum analysis, or other "feature extraction" techniques may be used to determine key features of objects. Typically digital sequential computing devices would follow the feature extraction step to perform the object classification function, which is often based on using these key features for pattern recognition. This process is suggested by Fig. 12.1. However, parallel optical computing architectures have also recently been developed, which employ neural network concepts to integrate the feature extraction and object classification functions into one. These multilayer network architectures are amenable to optoelectronic solutions, where optical free-space coupling and nonlinear optoelectronic devices are married. Micro-optical arrays and 2-D matrix SLMs can be employed in building such architectures.

Specific examples of optical architectures will be discussed for calculating Mellin transforms, moments of a 2-D image, Walsh–Hadamard coefficients, angular correlation features, and Radon transforms. Finally, the use of optical neural networks will be considered as a means to perform object classification.

The general problem of pattern recognition can be described in terms of a process summarized in Fig. 12.2. Although this is not the only way to look at pattern recognition,

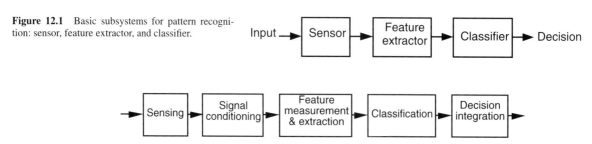

Figure 12.1 Basic subsystems for pattern recognition: sensor, feature extractor, and classifier.

Figure 12.2 Generic pattern recognition functions: sensing, signal conditioning, feature measurement and extraction, classification, and decision integration.

it is a common approach that delineates the subsystem functions and potential roles for optical processors. The sensor measures an object in at least two dimensions using various potential parameters, all of which are represented as complex or real signals—that is, amplitude or intensity. After sensing, the image may or may not be in a recognizable form. For instance, in SAR the received signal is a 2-D linear superposition of Fresnel zone responses. As we have already seen, in the case of SAR imaging, optical processing can be used to create an image of the object. From a pattern recognition point of view, this could be termed "signal conditioning," although signal conditioning usually implies a whole host of things, such as high- or low-pass filtering, A/D conversion, and so on. After sensing and signal conditioning, features are measured and extracted from the image ostensibly to do two things:

1. Reduce the dimensionality or space–bandwidth product of the sensed data to hopefully enhance throughput or reduce data storage requirements for subsequent processing.
2. Reduce the sensitivity of the data to variations in position, scale, and rotation to hopefully make the subsequent processing invariant to such geometric image distortions or other errors.

If the extraction of the features is accomplished effectively, then the classification stage of the process will be made easier. In other words, key discriminants are features extracted that easily separate distinct classes of objects. Feature extraction is also important to reduce the computational bottleneck in the subsequent classification process. The classification step can be accomplished by correlation of sensed data with reference data in a sequential manner, similar in principle to classical matched filtering or correlation. More sophisticated approaches, such as associative processing using parallel computational tools or neural networks, can also be employed. After the object is correctly identified, multiple looks may enhance decision confidence or output signal-to-noise ratio in the decision integration step.

One of the main applications of optical signal processing is feature extraction. There are a host of possible transformations that have been investigated, including Cartesian or polar (wedge-ring detection) Fourier transforms, the Radon (or Hough), Mellin, and Walsh–Hadamard transforms, and moments. In addition, correlation can be used directly on the input data to obtain features suitable for classification or to perform the matching directly on raw data.

12.2 FEATURE EXTRACTION

We will first discuss several examples of feature extraction, a process which usually involves projection of the measured feature vector (or its transform) by a vector inner product operation.

The simplest and most common optical feature extraction technique is the Fourier transform, illustrated in Fig. 12.3 and given (in polar form) by

$$F(\rho, \phi) = \int_0^{-\infty} \int_0^{-2\pi} f(r, \theta) \exp[-j2\pi r\rho \cos(\theta - \phi)] r \, dr \, d\theta \quad (12.1)$$

A special output plane detector, called the *wedge-ring detector*, can be used to sample the output. Knowing that the output is symmetric about the origin means that the radial samples can be made on one half-plane and that the angular samples can be made on the other half-plane. By

PATTERN RECOGNITION APPLICATIONS

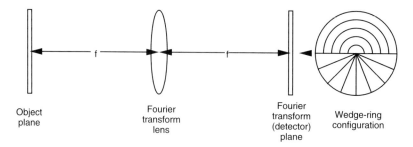

Figure 12.3 Fourier transform architecture employing a wedge-ring detector.

virtue of the wedge-ring sampling of the Fourier space, the wedge samples are scale-invariant and the ring samples are rotationally invariant. Since the detectors are square-law, the output is equivalent to the magnitude of the Fourier transform and hence translation-invariant.

Another transform that has, at least, scale invariance, is the Mellin transform, which was discussed in Chapter 10 with regard to matched filter/correlators. The 2-D Mellin transform is

$$F_M(s_x, s_y) = \int_{-\infty}^{\infty} \int_{-\infty}^{\infty} f(x, y) x^{s_x-1} y^{s_y-1} \, dx \, dy \qquad (12.2)$$

where $(s_x, s_y) = 1, 2, 3, \ldots$ yield the zeroth, first, second, and so on, moments of $f(x, y)$ along each axis.

Rather than show a particular Mellin transform example (since we have already covered that in Chapter 3), let us move on to the direct calculation of moments and show how it is done optically. Moments for a 2-D image are calculated according to the equation

$$m_{pq} = \int_{-\infty}^{\infty} \int_{-\infty}^{\infty} f(x, y) x^p y^q \, dx \, dy \qquad (12.3)$$

where $p = s_x - 1$ and $q = s_y - 1$. Actually Equation (12.3) is derivable from a more general equation given by

$$g(x', y') = \int_{-\infty}^{\infty} \int_{-\infty}^{\infty} f(x'y) h(x + x', y + y') \, dx \, dy \qquad (12.4)$$

which looks like a cross-correlation. Here we let $x' = 0$ and $y' = 0$ and $h(x, y) = m(x, y) = x^p y^q$ ($p, q = 0, 1, 2, \ldots$) so that $M_{pq} = g(0, 0)$. The quantities M_{pq} are the moments, and $m(x, y)$ is the mask that, when properly encoded with a superposition of cosines at multiple frequencies, yields moments up to the desired order in a coherent optical parallel architecture. Higher moments characterize location, shape, symmetry, and so on, of an object function—for example, mean, variance, skewness, kurtosis, and so on. To accomplish this optically in the architecture shown in Fig. 12.4, the monomials $x^p y^q$ are recorded on different spatial frequency carriers on the mask at plane P_2. Thus different moments appear at different detectors in the output plane P_3 as determined by the different spatial frequency carriers. Postdetection digital electronics must be used to compute the so-called invariant moments because those calculated here by the optical architecture are simple moments.

The values of simple moments are sensitive to position, scale, rotation, and contrast of an image. To make them invariant to these factors appropriate nonlinear combinations of m_{pq}

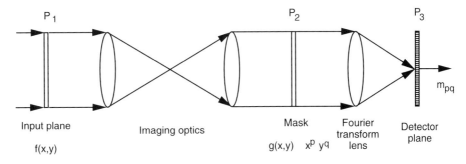

Figure 12.4 Optical architecture for calculating moments of an input image.

can be constructed. For instance, central moments (μ_{pq}) are first constructed that are invariant to position. They are defined by

$$\mu_{pq} = \int_{-\infty}^{\infty} \int_{-\infty}^{\infty} f(x, y)(x - \bar{x})^p (y - \bar{y})^q \, dx dy \tag{12.5}$$

Rotation- and scale-invariant moments of order (p, q) can defined in terms of central moments as described by Casasent, but are not readily computable by optical means.

As a particular design example, consider the case of a space-integrating architecture for moment calculations. In this architecture shown in Fig. 12.5 the arrangement is similar to the triple product processor. This arrangement of orthogonal AO cells is possible by virtue of the fact that moments are separable in Cartesian coordinates. If the pulse width τ of the laser is much less than the inverse bandwidth of the signals input into the AO cells, then the illumination will act as a delta function in time and freeze an image of the moment generating function when it slides to the center of the AO cell in each case (x and y). Thus the signals t_p and t_q must be synchronously applied to the AO cells as amplitude modulation on an RF carrier at the AO cell center frequency; that is,

$$s_{1p}(t) = t^p \exp(j\omega_0 t) \tag{12.6}$$

and

$$s_{2q}(t) = t^q \exp(j\omega_0 t) \tag{12.7}$$

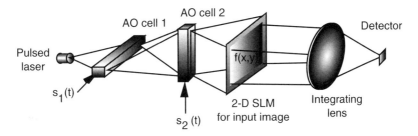

Figure 12.5 Space integrating architecture for moment calculations based on triple product concept.

For odd powers there is a change of sign at $t = 0$ that is represented by a 180° phase shift. This implies that destructive interference occurs at the integrated output, which can be interferometrically detected. When these signals simultaneously slide to the centers of the AO cells, the laser diode is pulsed, collimated, and focused into a horizontal strip that enters the first AO cell at the Bragg angle. Diffracted light is collimated in y and Fourier transformed in x to produce a vertical strip that illuminates the second AO cell at its Bragg angle. Both undiffracted components are filtered, and doubly diffracted light is expanded in x to a square so that the first AO cell is imaged onto $f(x, y)$ in the $-x$ direction and the second AO cell is imaged onto $f(x, y)$ in the $-y$ direction. The light amplitude incident on the image plane is similar to a vector outer product and is modulated by

$$E_i(x, y) = \delta(t) s_{1p}(x + v_s t) s_{2q}(y + v_s t) \tag{12.8}$$

or

$$E_i(x, y) = \delta(t) s_{1p}(x) s_{2q}(y) \tag{12.9}$$

This result is multiplied by the image transmissivity $f(x, y)$, and their product is Fourier-transformed by a spherical lens. This transform can be sampled at its center or integrated over its entire extent to yield an interferometrically detected signal proportional to the desired moment. Thus

$$m_{pq} = \int_{-\infty}^{\infty} \int_{-\infty}^{\infty} s_{1p}(x) s_{2q}(y) f(x, y) \, dx \, dy \tag{12.10}$$

or

$$m_{pq} = \int_{-\infty}^{\infty} \int_{-\infty}^{\infty} x^p y^q f(x, y) \, dx \, dy \tag{12.11}$$

All moments can be calculated very rapidly with high accuracy and dynamic range on a single detector pair. For instance, for an aperture time of 10 msec, moments through the 10th can be calculated in 500 μsec. Therefore, many windowed sets can be calculated during each frame time of a 2-D SLM. Alternatively, moments can be calculated simultaneously with lower dynamic range in only a few microseconds using temporal frequency multiplexing of moment generating functions. In that case the signals in the AO cells are

$$s_1(t) = \sum_{p=0}^{P} t^p \exp[j(\omega_o + p\Delta\omega)t] \tag{12.12}$$

and

$$s_2(t) = \sum_{q=0}^{Q} t^q \exp[j(\omega_o + q\Delta\omega)t] \tag{12.13}$$

where $\Delta f (= \Delta\omega/2\pi)$ is the frequency separation of the frequency-multiplexed moment-generating functions. The architecture is now modified to look like that in Fig. 12.6. For an image bandwidth B (cy/mm) and acoustic velocity v_s, different moments would be spatially

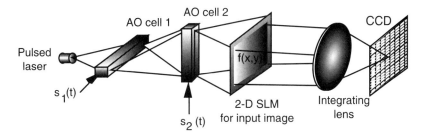

Figure 12.6 Modified optical architecture for moment calculations, which uses temporal frequency multiplexing.

separated in the output plane if $\Delta f/v_s > B$, and thus could be sampled by the detector array.

To obtain the central moments, we must first incorporate the centroid location in each dimension by finding the zeroth- and first-order moments and using $\bar{x} = m_{10}/m_{00}$ and $\bar{y} = m_{01}/m_{00}$ or by operating the source continuously and locating the zero crossing of the detector's temporal output as the generating functions for m_{10} or m_{01} slide through the AO cells. Under these circumstances the input moment-generating functions to the AO cells have relative delay given by

$$\tau_D = \tau_1 - \tau_2 = \bar{x}/v_s - \bar{y}/v_s \tag{12.14}$$

and the laser diode is pulsed at $t = \tau_1$. This aligns the origin of the moment-generating functions with the image centroid and spatially integrated outputs so as to compute the central moments, m_{pq}. Windowing the functions $s_1(t)$ and $s_2(t)$ is equivalent to windowing the region of $f(x, y)$, where μ_{pq} are calculated.

Another useful transform that was discussed in Chapter 3 is the Radon transform, given by

$$F_R(x', \phi) = \int_{-\infty}^{\infty} \int_{-\infty}^{\infty} f(r)\delta(x' - \mathbf{r} \cdot \hat{\mathbf{n}}) \, d^2r \tag{12.15}$$

where $\mathbf{x}' = \mathbf{R}\mathbf{x}$, as described previously. One of the simplest architectures to perform this operation is shown in Fig. 12.7. A mechanically driven image rotator (e.g., Dove prism) is used to perform the linear transformation $\mathbf{R}$ on object space coordinates $\mathbf{x}$ (at plane P_1) to yield coordinates $\mathbf{x}'$ at the intermediate plane (P_2). Then the cylindrical lens L_1 focuses the light in the vertical (y') axis, and the cylindrical lens L_2 images the light in the horizontal (x') axis such that the image at (x', y') is anamorphically relayed to the 1-D detector array at plane P_3. In other words, the lateral magnification in x' is unity ($m_x = 1$) and in y' is much less than unity ($m_y \ll 1$). For a given rotation angle ϕ_k the Radon slice $F_R(x_k'', \phi_k)$ is read-out of the 1-D detector array.

There are several alternative schemes to calculate the Radon transform, among which are two techniques that do not require moving parts. One technique involves the use of a video feedback architecture discussed in Chapter 10. In this architecture the output processor consists of two orthogonal cylindrical lenses as discussed previously (see Problem 14, Chapter 4). The linear self-scanned array reads-out successive Radon slices, which can be digitized for further

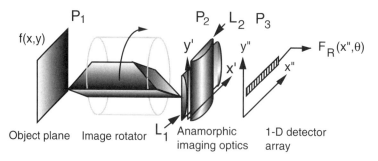

Figure 12.7 Radon transform architecture employing orthogonally oriented cylindrical imaging lenses and a mechanical image rotator (e.g., Dove prism; see Fig. 4.15).

computations. As an example of the use of this Radon architecture concept, consider the theoretical and experimental calculation of the Radon transform of a simple object.

EXAMPLE 12.1 Radon Transform of a Simple Object Figure 12.8 contains the geometrical description of an object which consists of an elliptical pillbox with three unit impulse functions on it. The unit impulse functions are actually approximated with finite-width cylinders in order to more easily render their Radon transform plot. The corresponding Radon slices of this surface (from 0° to 90°, every 30° increments) are shown in Fig. 12.9. These are the experimental results of optically computing the Radon transform. These results were obtained from an oscilloscope display of the readout from a linear self-scanned array in which a computer generated the image rotation for playback on a miniature television display, emulating a mechanical or video feedback architecture for image rotation.

To speed-up the process of computing the Radon transform, we must avoid both the mechanical and video feedback architectures. A multiaperture array can enable the computation of the Radon transform for discrete increments of angle ϕ_k by employing a Dove prism array (see Fig. 4.15). This scheme is shown is Fig. 12.10. In this architecture the initial lenslet array (L_1) consists of an $N \times N$ array ($N = 4$) of spherical lenses, which replicate the input display image. Each lenslet must produce a collimated beam. These multiple parallel beams pass through a Dove prism array (image rotator). Each Dove prism is rotated by a fixed amount through angles spanning 0 to π radians in increments of $\phi_k = \pi/15$, that is, $\pi/(N^2 - 1)$, where $N^2 - 1$ is the number of unique angle samples in 2π. Subsequently each replicated/rotated image is rotated by $2\pi/15$ and then subjected to a pair of orthogonal cylindrical lenses to yield

Figure 12.8 Elliptical pillbox with three cylinders on top.

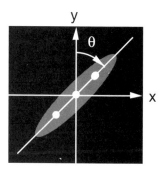

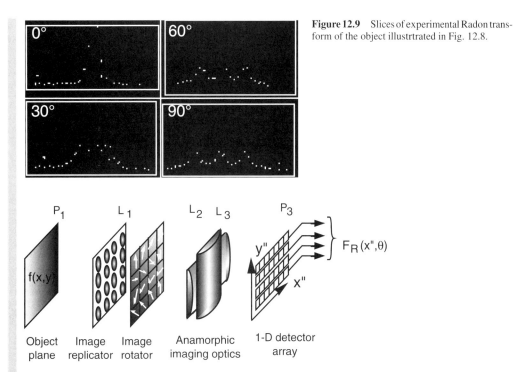

Figure 12.9 Slices of experimental Radon transform of the object illustrtrated in Fig. 12.8.

Figure 12.10 Multiaperture lenslet/Dove prism array processor concept for calculating the Radon transform. An actual Dove prism array was pictured in Fig. 4.15.

a matrix of Radon transforms at the output detector plane. The output detector plane must be constructed of N linear detector arrays, each with sufficient pixels to adequately sample each Radon transform slice.

Another example of a feature extractor that employs multiaperture lenslet arrays is the Walsh–Hadamard transform processor. This processor, first proposed by Glaser, employs a lenslet array to replicate the input image as before. Here, however, the lenslet array outputs are passed through a binary mask placed directly in front of the lenslet array and in front of a fully parallel readout focal plane array, as shown in Fig. 12.11. The mask corresponding to the Walsh–Hadamard transform is shown in Fig. 12.12, where the elements of the mask are unipolar binary representations of what actually is a bipolar set of basis functions.

After each detector in the parallel readout array collects light, the result is an array of (zero-lag) cross-correlations whose relative magnitudes correspond to the relative weights of the Walsh–Hadamard coefficients of a particular input image. What's unique about this approach is that, contrary to the usual sequence shown in Fig. 12.1, the roles of feature extractor and sensor are reversed. Instead of a high space–bandwidth product sensor connected to a feature extractor that may often perform intensive hardwired data compression to support a slower sequential classifier, we now have considerable space–bandwidth compression occurring before the sensor. Thus the sensor can be a much-reduced space–bandwidth product focal plane array, but with parallel readout. A parallel classifier, however, can be directly coupled to

Figure 12.11 Lenslet array processor for calculating Walsh–Hadamard transform.

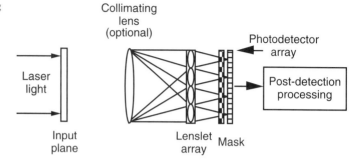

Figure 12.12 Walsh–Hadamard transform basis function mask for an 8 × 8-sized basis set. The light regions represent +1, and the dark regions represent −1.

it. The parallel classifier could, in fact, be a multilayer neural network classifier. We will discuss this later, but first consider the following example of a Walsh–Hadamard feature extractor.

EXAMPLE 12.2 Walsh-Hadamard Optical Feature Extractor Given the image shown in Fig. 12.13(a) we can calculate its Walsh–Hadamard transform as shown in Fig. 12.13(d). If, however, we anamorphically scale and then rotate this simple image as shown in Figs. 12.13(b) and 12.13(c), then the transform changes as shown in Figs. 12.13(e) and 12.13(f), respectively. Thus, the architecture discussed previously is not invariant with respect to elementary image transformations. This is a consequence of using simple correlation, which is at the heart of this architecture. Somehow, the classifier must be able to accommodate multiple aspects of the

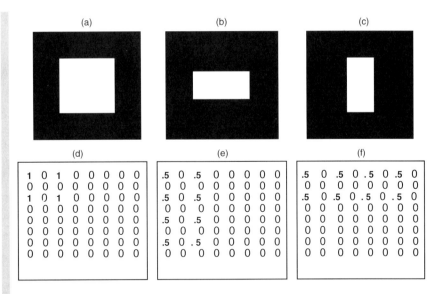

Figure 12.13 (a) Simple square input image and corresponding Walsh–Hadamard transform (d). (b) Rectangular shape and (c) its rotated version, and their corresponding transforms (e) and (f).

object in its reference memory, or the feature extractor must be made invariant with respect to these distortions. Finally, it is also obvious that the Walsh–Hadamard transform mask used in this architecture is unipolar but should actually represent bipolar basis functions. To accomplish this, either the mask must be doubled in size to get the complementary (black–white reversed) mask in parallel or it must be time-multiplexed at a 50% duty cycle between the unipolar mask and its complement, which takes twice as long.

It is clear that this approach has its limitations, but in general this architecture could calculate any linear transformation of the form

$$F(u, v) = \sum_{x=0}^{N-1} \sum_{y=0}^{N-1} f(x, y) K(x, y; u, v) \qquad (12.16)$$

where the transformation kernel K for the Walsh–Hadamard transform is

$$K(x, y; u, v) = (-1)^{p(x,y,u,v)} \qquad (12.17)$$

where

$$p(x, y, u, v) = \sum_{i=1}^{N} (u_i x_i + v_i y_i) \text{ modulo } 2 \qquad (12.18)$$

and u_i, v_i, x_i, and y_i are binary representations of u, v, x, and y, e.g.: $(u)_{\text{decimal}} = (u_{n-1}, u_{n-2}, \ldots, u_1, u_0)_{\text{binary}}$, where $u_i = 0$ or 1. Equation (12.16) can also be written more compactly as a matrix–matrix product, namely,

$$[F(u, v)] = [H(u.v)][f(x, y)][H(u, v)]^T \qquad (12.19)$$

where $[H(u, v)]$ is the Hadamard transform matrix, and $[H(u, v)]^T$ is its transpose.

As a final example of feature extraction using optics, let us reconsider the angular correlator described in Chapter 10. Recall that an architecture using a simple video feedback loop was coupled to a single photodetector using a spherical lens to calculate angular correlation. Once the angular correlation values are obtained, key features of the object can be calculated. To see this, consider the following example.

EXAMPLE 12.3 Angular Cross-Correlation of Simple Binary Objects The angular correlation function can yield object boundary estimates if we just cross-correlate a rectangular slit with the input image. In this case the angular cross-correlation function for a binarized image is given by

$$R_{n'}(\phi) = \int_0^{2\pi} \int_0^\infty \text{rect}\left[\frac{r}{r(\phi')}\right] \text{rect}\left[\frac{r}{r_{\text{rect}}(\phi' + \phi)}\right] r\, dr\, d\phi'$$

where the second term in the integrand, $\text{rect}[r_{\text{rect}}/r(\phi' + \phi)]$, is a formal functional description of a rectangular slit of uniform intensity rotated by an angle ϕ. Optical angular cross-correlation can be used to determine object boundary estimates directly. It is clear that if the rectangular slit approaches a long (greater than the field of view) and infinitesimally narrow slit defined as $\delta[r \sin(\phi' - \phi)]$, we will recover exactly the boundary of the object defined as

$$R_{rr'}(\ ) = r(\phi) + r(\phi + \phi')$$

Of course this is physically impractical, since the signal-to-noise level would go to zero and no boundary could be recovered. A compromise is possible by trading-off slit-width (resolution) versus signal-to-noise ratio. An expedient choice of reference function (i.e., a narrow rectangular slit) will enable the recovery of the object boundary to some degree of accuracy dependent upon the width of the slit relative to the proper boundary (Nyquist) sample rate. Such a reference image also enables one to determine the orientation angle that an oblong object makes with respect to the reference image and hence direction within the field of view, a result which is useful for object recognition; that is, it enables partitioning of the total recognition task into several simpler recognition tasks based on orientation.

Consider the shapes shown in Fig. 12.14 and their corresponding angular correlation functions shown in Fig. 12.15. The angular correlation values are the boundaries of each object. These results were obtained by a digital simulation of the angular correlator optical architecture.

A multiaperture lenslet array version of an angular correlator is shown in Fig. 12.16. In this architecture a multiple lenslet array replicates exactly as described earlier for the Radon transform, except the Dove prism array is deleted and the slits are rotated by the appropriate increments, as shown in the inset. The superpositions of the replicated object space image and each rotated slit pattern are then focused in-parallel onto a 2-D parallel-readout array. This processor is much faster than the feedback-based scheme.

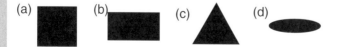

Figure 12.14 Basic geometric shapes for calculating angular correlation: (a) square, (b) 2:1 rectangle, (c) equilateral triangle, (d) 4:1 ellipse.

12.2 FEATURE EXTRACTION   323

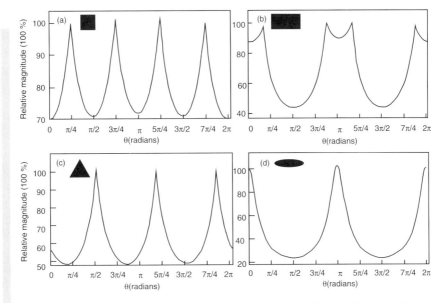

Figure 12.15 Angular correlation functions corresponding to the diameter functions of the shapes shown in Fig. 12.14: (a) square, (b) 2:1 rectangle, (c) equilateral triangle, (d) 4:1 ellipse. Note that the equilateral triangle has an arbitrary starting phase, which indicates the orientation of the object with respect to a reference axis, in addition to the fact that the period of each pattern is a distinguishing feature for most shapes.

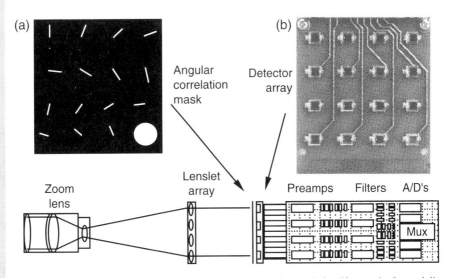

Figure 12.16 Multiple aperture lenslet array processor for angular correlation. Note mask of rotated slits and parallel readout detector array in the insets (a) and (b).

12.3 MATRIX–VECTOR MULTIPLICATION

Before proceeding to address the classification function in pattern recognition, it is important to understand the operation of a fundamental optical processor architecture that performs vector–matrix multiplication. The optical matrix–vector multiplier is an important architecture in connection with neural networks as well as linear algebraic processing. The basic architecture is shown in Fig. 12.17. This particular architecture looks rather simple because the anamorphic optics required are not shown in order to simplify understanding the matrix–vector multiplication operation. Light from the LED sources is spread horizontally either by cylindrical lenses, optical fibers, or planar light guides to illuminate the 2-D mask or SLM representing matrix **H**. Light leaving plane P_1 is the vector **f**, which, if looked at as a continuous operation, can represent the function $f(x)$. Light leaving plane P_2 is the product of $f(x)$ and the impulse function $h(x', y')$ or equivalently **Hf**, where **H** is the matrix version of $h(x', y')$. This light is focused (spatially integrated) in x and imaged in y onto P_3 to yield $g(y')$, which is equivalent to the discrete vector **g**. Hence $\mathbf{g} = \mathbf{Hf}$.

A more detailed account of the optics is shown in Fig. 12.18. Here the intensity of light emitted by the LED array is denoted by $f(n)$. The spherical lens L_1 collimates light from each LED in the horizontal direction. The combination of L_1 and L_2 produce an image of the LED array at P_1 in the vertical direction. Consequently the 2-D mask at P_2 is illuminated uniformly in the horizontal direction such that each row of the mask or SLM is illuminated by a different LED. The intensity of the mask is denoted by $H(n, m)$, which signifies that a pixel of the mask with area Δx by Δy located at $x = m\Delta x$ and $y = n\Delta y$ has transmittance $H(n, m)$. This mask is imaged in both directions onto the output plane P_3. In the vertical direction, L_3 and L_4 produce a demagnified image of the mask that is smaller than the detector height. Therefore the photogenerated signal at the mth detector is proportional to the total intensity transmitted through the mth column of the mask, namely,

$$g(m) = \sum_{n=0}^{N-1} H(n,m) f(n) \tag{12.20}$$

where N is the number of LEDs. Vector–matrix multiplication is equivalent to a 1-D linear

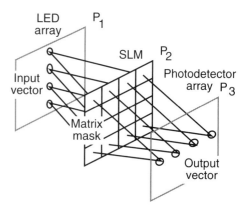

Figure 12.17 Original Stanford optical vector–matrix multiplier design.

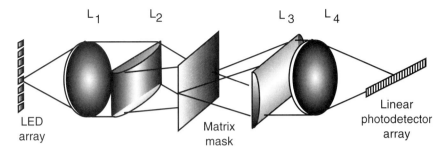

Figure 12.18 Detailed optical layout for the optical vector–matrix multiplier.

operation with a space-variant kernel. For continuous signals the analogous mathematical operation is

$$g(y') = \int f(x) h(x; y') \, dx \qquad (12.21)$$

No functional relationship exists between x and y', which also precludes a scanning (or shift invariant) relationship between them—that is, a relationship of the form $h(x; y') = h(x; vt') = h(x - vt')$.

For 10^2 LEDs and 10^2 detectors, 10^4 analog multiplications can be performed. For a 100-nsec dwell time typical of high-resolution TV-compatible processing, the processing rate is 10^4 multiplications/10^{-7} sec or 10^{11} multiplications/sec (which is not trivial). Since light intensity is used with LEDs, transmitted and detected light signals must always be non-negative. Special coding techniques, which have been treated as described in the technical literature, must be used if both positive and negative or complex numbers are to be accommodated.

There are several different applications of matrix–vector multiplication. One of these is the discrete Fourier transform, which is used for spectral estimation and can be applied to a number of problems, including feature extraction for pattern recognition. Later we will see that this architecture is also basic to optical implementation of neural networks.

EXAMPLE 12.4 Discrete Fourier Transform Implemented as a Vector–Matrix Operation
The discrete Fourier transform of a signal sequence $f(n)$ is usually written

$$g(k) = \frac{1}{N} \sum_{n=0}^{N-1} f(n) \exp(-j2\pi nk/N)$$

where $n = 0, 1, 2, \ldots, N-1$. This can be rewritten as a vector–matrix operation:

$$g = f \exp \left\{ -j \frac{2\pi}{N} \begin{bmatrix} 0 & 0 & 0 \ldots & 0 \\ 0 & 1 & 2 \ldots & N-1 \\ 0 & 2 & 4 \ldots & 2(N-1) \\ 0 & 3 & 6 \ldots & 3(N-1) \\ \vdots & & \vdots & \vdots \\ 0 & N-1 & 2(N-1) \ldots & (N-1)^2 \end{bmatrix} \right\}$$

Table 12.1
Linear Signal Processing Operations Implemented by Optical Matrix–Vector Multiplication

Operation	Kernel
Convolution	$H(n-k)$
Cross-correlation	$H(k-n)$
Autocorrelation	$f(k-n)$
Fourier transform	$\exp(-j2\pi nk/N)$
Cosine transform	$\cos(2\pi nk/N)$
Sine transform	$\sin(2\pi nk/N)$
Laplace transform	$\exp(-2\pi nk/N)$
Hankel transform	$2\pi n J_0(2\pi nk)$
Mellin transform	n^{k-1}
Abel transform	$2n(n^2-k^2)^{-1/2}$
Hilbert transform	$[\pi(n-k)]^{-1}$
Hartley transform	$\cos(2\pi nk/N)+\sin(2\pi nk/N)$

or, equivalently, **g** = **Hf** or

$$g(m) = \sum_{n=0}^{N-1} H(n,m) f(n)$$

In this case we simply apply the optical vector–matrix multiplier in a straightforward fashion. In fact, many useful linear operations can be cast in this form. Table 12.1 summarizes most of these cases for $H(n,k)$.

EXAMPLE 12.5 Incoherent Vector Matrix Optical Processor   These operations can also be implemented in a different architecture illustrated in Fig. 12.19. This incoherent electro-optical processor developed by Bromley, Bocker, and co-workers uses a single high-speed temporally modulated LED to address the architecture with the input function $f(n)$. It is also attractive by virtue of being compact, since it requires no image transfer optics if the SLM and CCD can be matched in size. The optical matrix mask contains a pattern of apertures of varying transmissivity equivalent to $H(n,k)$. A CCD imager is then used to form the output sequence:

$$g(m) = \sum_{n=0}^{N-1} H(n,m) f(n)$$

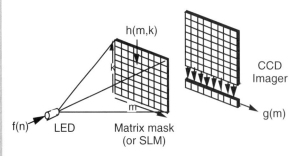

Figure 12.19 Compact incoherent optical vector–matrix processor developed by for processing signals using fixed masks.

where $m = 1, 2, 3, \ldots, M$. The sum is performed by the CCD by appropriate clocking waveforms that induce charge transfer between pixels, which accumulate (or add) transferred charges by the shift-and-add (or TDI) mode. For operating the CCD we require a vertical shift rate equal to the horizontal shift rate divided by M; that is, $v_V = v_H/M$. (Normally the input rate would equal the vertical shift rate; that is, $v_{in} = v_V$.) The input signal $f(n)$ must be non-negative. Bipolar signals can be accommodated if two processors are used or if a bias is employed. The mask transmittance must also be non-negative.

For the case where $N = M = 512$ and where the sample rate for $f(n)$ is v_s, then $v_{in} = v_V = v_s$ and $v_H = Mv_s$. For a modest value of $v_s = 32 \times 10^3$ samples/sec the horizontal shift rate is $v_H = 1.64 \times 10^7$ shifts/sec. The corresponding multiplication rate is $MNv_s \approx 8.4 \times 10^9$ multiplications/sec.

Figure 12.20 shows an example of a Fourier transform mask used in the above-described architecture. On the left side of the mask the transmittance is proportional to $1 + \cos(2\pi mn)$. On the right side it is proportional to $1 + \sin(2\pi mn)$. Thus the corresponding values of $g(m)$ represent real and imaginary parts of the Fourier coefficients. Thus this mask-encoded processor computes 64 simultaneous 128-point discrete Fourier transform coefficients.

Other applications of this approach include multichannel cross-correlation and multichannel finite impulse response (FIR) filtering. Multichannel correlation is calculated between the input waveform denoted by $f(n+n')$ and M signals comprising simultaneously M columns of the mask $H(m, n)$. That is, all time shifts (n') of the input are compared in parallel with the M reference signals comprising $H(m, n)$. For instance, in range-Doppler radar the locations (m, n') of the peaks on the CCD plane give range (proportional to n') and velocity (proportional to m) of the possible targets. In multichannel filtering for a given value of m we implement the equation

$$g(m, n') = \sum_{n=0}^{M-1} f(n+n')H(m, n) \qquad (m = 1, 2, \ldots, M; n' = \ldots, -2, -1, 0, 1, 2, \ldots)$$

which describes a shift-invariant FIR filter. In this case each of the M columns of the mask can be a different impulse response. Hence each column of the processor can have different characteristics—for example, low-pass, high-pass, bandpass, or different center or

Figure 12.20 Fourier transform mask used in vector–matrix processor of Fig. 12.19.

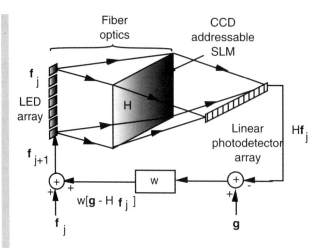

Figure 12.21 Optical vector–matrix multiplier architecture with feedback that implements modern spectrum estimation.

cutoff frequencies.

In addition to the simple FIR filtering, correlation, and discrete transform operations described above, we can also employ the optical vector–matrix multiplier to solve for the matrix inversion problem—that is, $\mathbf{f} = \mathbf{H}^{-1}\mathbf{g}$. This is the implicit solution of the equation $\mathbf{y} = \mathbf{H}\mathbf{f}$, which requires iteration and feedback, as shown in Fig. 12.21. Iterative processing schemes circumvent digital interfacing and low-speed input/output and are expressed by the following difference equation:

$$\mathbf{f}_{j+1} = (\mathbf{I} - \mathbf{H})\mathbf{f}_j + \mathbf{y} \quad (12.22)$$

where $\mathbf{I}$ is the identy matrix. In fact, Equation (12.22) is the implicit solution to the matrix equation $\mathbf{g} = \mathbf{H}\mathbf{f}$.

Consider the case of autoregressive spectral estimation. We will briefly highlight this example here.

EXAMPLE 12.6 Autoregressive (Maximum Entropy) Spectral Estimation   Modern methods of spectral estimation are based on autoregressive or maximum entropy techniques. The autoregressive (or all-pole) modeling technique attempts to fit a polynomial in z^{-1} of order N given by

$$H(z) = 1 - a_1 z^{-1} - a_2 z^{-2} - \cdots - a_N z^{-N}$$

to the input sensor data such that, when evaluated on the unit circle $[z = \exp(j\omega)]$. The power spectrum is given by

$$S_{\text{in}}(\omega) = \frac{S_{\text{out}}(\omega)}{|H(\omega)|^2}$$

where $S_{\text{out}}(\omega)$ is a white spectrum. The coefficients $(a_1, a_2, \ldots a_N)$ that whiten $S_{\text{out}}(\omega)$ are the solution to the fundamental Yule–Walker equation:

$$\begin{bmatrix} R(0) & R(1) & \ldots & R(N-1) \\ R(1) & R(0) & \ldots & R(N-2) \\ \vdots & & & \\ R(N-1) & \ldots & \ldots & R(0) \end{bmatrix} \begin{bmatrix} a_1 \\ a_2 \\ \vdots \\ a_N \end{bmatrix} = \begin{bmatrix} R(1) \\ R(2) \\ \vdots \\ R(N) \end{bmatrix}$$

where the matrix elements, $R(n)$, are

$$R(n) = \langle x(k) x(k+n) \rangle$$

where $x(k)$ is the input data sequence (or vector). It is therefore possible to solve for $S_{\text{in}}(\omega)$ using the architecture shown in Fig. 12.21 since the solution is iterative and requires a matrix inversion similar to that described in the discussion leading to Equation (12.21).

There are several limitations to the matrix–vector multiplier. First, the accuracy that can be achieved is limited by the accuracy of control of the input light source and the readout of the output intensities. Accuracy is, in turn, limited by the dynamic range of the source and/or detector a ways. Finally, rapid updating of the matrix M (or A) requires a high quality 2-D read–write SLM. Developments to improve accuracy have been addressed, including making analog/optical into digital/optical architectures and adopting systolic-array processing approaches.

12.4 OPTICAL NEURAL NETWORKS

In this section we will describe how optics can best be used to perform the last major step in the pattern recognition process: classification. In the classical approach to pattern recognition outlined earlier (Fig. 12.1) there are three major steps: sensing, feature measurement/extraction, and classification. Typically, hybrid systems have been conceived where optics is used to reduce the space–bandwidth product of the sensed image to an optimal or suboptimal set of features. Subsequently, the reduced feature set is sensed by a detector (or detector array), thus converting it to an analog electronic signal, which can be subsequently digitized. Further computation for classification can therefore be accomplished by a particular algorithm using a serial(or parallel) digital processor.

The process of developing a classifier is usually an iterative design procedure involving training and testing of the classification algorithm. Statistical measures of effectiveness, such as minimum mean-square error or maximum likelihood, are established by an analysis of the ensemble of feature vectors for a particular classification experiment. The classifier algorithm is often evaluated in terms of the probability of correct classification (as well as probability of incorrect classification) and an associated confidence level.

One of the concerns in classification is the ability of the classifier to verify the correct classification of an object when the sensed object (test signal) is distorted by noise or other systematic distortions, such as translation, scale, and rotation. This is why we were concerned about these earlier when discussing matched-filter correlators, since they can be used for classification, in addition to their typical role of detection (or location estimation). The degree to which classifiers are able to correctly recognize objects which have been distorted or, for that matter, to correctly classify objects not in the same classes trained-for (but only slightly different) is termed loosely: robustness. We desire classifiers to be robust. Much interest has been expressed in using so-called neural networks to address the robustness problem. Furthermore, the massive parallelism and interconnectivity of the neural network approach suggests that optics would be a natural technology to implement it.

Neural networks derive some of their basic characteristics from studies of the brain. The brain processes information in a collective manner using a vast network of densely interconnected nonlinear decision-making elements, known as neurons. A "soft" decision is made by an individual neuron by combining weighted inputs from other neurons. This simplified paradigm is illustrated in Fig. 12.22. The important point here is that neural networks emphasize the high connectivity or communication between computational units with parallel architectures in contrast to traditional electronic computers with von Neuman (serial) architectures. Optics is particularly appropriate to connectionist architectures characteristic of neural networks because of the ability of light beams to pass through one another without interference and because of the natural tendency for light to fan-out from diffractive spatially modulating structures. Neural networks also possess an attribute that is very desirable in pattern recognition: fault tolerance—that is, tolerance to individual component failure as well as tolerance for errors embedded in the data being processed. Here "error" means noise, clutter, and loss of information in the scene.

Neural networks are very much related to associative processing or associative memories. Before describing optical neural networks for classification, let us summarize the basic concepts of associative memories. An associative memory is a mapping from data to data, and it usually is nonlinear. An associative memory is a mechanism for storing signals or patterns represented by the vectors **x** and **y** so that at a later time the sensing of one pattern, **x**, will result in the recall of the other pattern, **y**. (In other words, associative memory is virtually the same as pattern recognition.) The vectors can be discrete binary numbers that represent images, character strings from text, spectrum samples, state variables of a control system, features extracted from a signal or image, or other sensor outputs. Some associative processing schemes offer the prospect of fault tolerance or distortion immunity greater than that achievable with classic correlators. Hence associative processors serve as the basis of robust pattern discrimination. In fact, so-called novelty filters are associative memories that respond only to patterns not previously encountered. Using feedback of output patterns and nonlinearities (thresholds), enhanced performance may be achieved for sensed patterns that are distorted.

The performance of an associative memory is basically determined by three things: its

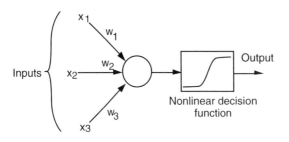

Figure 12.22 Simplified paradigm for a neural network: the processing node between interconnects, where weighted sums are subjected to a "soft" threshold decision processing element.

ability to store a large number of associations (i.e., its capacity), the ease with which the parameters of the memory can be determined (i.e., learning), and how it responds to test cases not in the original training set [i.e., generalization ("robustness")].

A linear associative memory, such as a matrix $\mathbf{M}$, is the simplest type of associative memory, where matrix $\mathbf{M}$ associates an output pattern or feature vector $\mathbf{y}_k$ with the input pattern $\mathbf{x}_k$. (The matrix $\mathbf{M}$ can also be thought of as a system impulse response.) Thus associative memories can best be understood using linear algebra. Associative recall of $\mathbf{y}_k$ given $\mathbf{x}_k$ can therefore be expressed as a simple matrix–vector product, namely, $c_k \mathbf{y}_k = \mathbf{M} \mathbf{x}_k$, where the vectors $\mathbf{x}$ and y can be defined as partitioned matrices:

$$\mathbf{X} = (\mathbf{x}_1, \mathbf{x}_2, \ldots, \mathbf{x}_M) \tag{12.23}$$

and

$$\mathbf{Y} = (\mathbf{y}_1, \mathbf{y}_2, \ldots, \mathbf{y}_M) \tag{12.24}$$

and c_k are a set of constants. $\mathbf{M}$ is the memory addressed by the vectors $\mathbf{x}_k$, which produces the associated stored vectors $\mathbf{y}_k$ with weights c_k. For all $c_k = 1$ we have an autoassociative memory, which ideally reproduces an input that exactly matches one of the stored vectors. This match occurs if one of the stored vectors is close enough to the input, even if the input is incomplete or distorted. In this case the equation $c_k \mathbf{x}_k = \mathbf{M} \mathbf{x}_k$ is the eigenvalue equation for the transformation given by $\mathbf{M}$, and the vectors are stored as eigenfunctions of $\mathbf{M}$. However, even though the input vector $\mathbf{x}$ is one of the stored vectors, there is no guarantee that the output will be $\mathbf{x}$. The output is actually a linear combination of stored vectors so that the vector that correlates best with the input is amplified the most.

Although it is desired that the associative memory produce an output that most closely matches with the input, it does so only approximately, with the desired vector that is stored representing the signal and the weighted contributions from the other stored vectors representing the noise. By virtue of appropriate binarization of the input and stored vectors and thresholding of the output, it is possible to recover the exact binary vector stored in memory that correlates maximally with the input. Even if $\mathbf{x}$ is not exactly equal to one of the binary stored vectors, it is a close enough approximation with a high correlation. By iteratively using the output as the new input vector, eventually the output can be made to converge to the correct stored vector. When the input vector differs from the stored vector by some amount such that the output stabilizes to the correct association, we have arrived at a minimum measure called the *Hamming distance*. The number of locations in the vector space where these two vectors can differ and still be within a radius that ensures attraction to the correct stable association is, in fact, the Hamming distance. (The Hamming distance between two binary vectors of the same dimension is the number of bits in disagreement between them.) It is also a measure equally important to the storage requirements of the associative memory. The smaller the radius of attraction (or Hamming distance), the larger the storage capacity. The radius of attraction can vary between zero and N for the cases where $M = N$ (all stored vectors are linearly independent) and $M = 1$ (all input vectors converge to the stored vector or its complement). The tradeoff between radius of attraction and storage capacity is an important performance tradeoff, which is expressed as the product of the radius of attraction $(+1)$ and the storage capacity.

Now let's consider the following example of an associative memory used for pattern recognition.

EXAMPLE 12.7 Associative Memory Used for Pattern Matching An associative memory can be used to implement pattern matching (or correlation) in the following manner. Assume that the reference patterns to be matched against a sensed pattern are stored as rows of a matrix **P** and that they are stored as bipolar binary values (± 1's). To match the stored patterns against the input pattern, the inner product of the transpose of the input vector and the reference pattern matrix is calculated. For example, the pattern matrix might be

$$\mathbf{P} = \begin{bmatrix} -1 & 1 & -1 & 1 \\ -1 & -1 & 1 & 1 \\ 1 & -1 & 1 & -1 \\ -1 & 1 & 1 & -1 \end{bmatrix}$$

where each row is a different pattern. If the input pattern vector is

$$\mathbf{I} = \begin{bmatrix} -1 \\ -1 \\ 1 \\ 1 \end{bmatrix}$$

then the inner product yields

$$\mathbf{P} \cdot \mathbf{I} = \begin{bmatrix} -1 & 1 & -1 & 1 \\ -1 & -1 & 1 & 1 \\ 1 & -1 & 1 & -1 \\ -1 & 1 & 1 & -1 \end{bmatrix} \begin{bmatrix} -1 \\ -1 \\ 1 \\ 1 \end{bmatrix} = \begin{bmatrix} 0 \\ 4 \\ 0 \\ 0 \end{bmatrix}$$

This result indicates that the input pattern matches the second row of the pattern matrix.

For such an associative memory the correlation matrix **R** is formed by taking the inner product of the pattern matrix and its transpose, namely, $\mathbf{R} = \mathbf{P} \cdot \mathbf{P}^T$, which is given here by

$$\mathbf{R} = \begin{bmatrix} -1 & 1 & -1 & 1 \\ -1 & -1 & 1 & 1 \\ 1 & -1 & 1 & -1 \\ -1 & 1 & 1 & -1 \end{bmatrix} \begin{bmatrix} -1 & 1 & -1 & 1 \\ -1 & -1 & 1 & 1 \\ 1 & -1 & 1 & -1 \\ -1 & 1 & 1 & -1 \end{bmatrix} = \begin{bmatrix} 4 & 0 & -4 & 0 \\ 0 & 4 & 0 & 0 \\ -4 & 0 & 4 & 0 \\ 0 & 0 & 0 & 4 \end{bmatrix}$$

The correlation matrix shows that each input pattern matches itself exactly and is easily distinguishable from other patterns within the matrix **P**. In general, correlation matrices of similar patterns contain numbers close to the maximum along the diagonal, similar to "confusion" matrices used in traditional pattern recognition.

Now that a little background has been given on associative memories, we can focus on an optical implementation of neural networks. For the sake of simplicity we will consider the simple perceptron first (as a neural network designed to handle 1-D inputs). The simple perceptron consists of a nonlinear transformation $F(\ldots)$ of the weighted sum of an input vector x_i that yields an output vector y_j, namely,

$$y_j = F\left(\sum_i W_{ij} x_i\right) \tag{12.25}$$

where W_{ik} is the weight matrix and F is often the sigmoidal or sgn function. This was schematically described in Fig. 12.22 for a sigmoidal-like nonlinearity. A matrix–vector multiplier can be used in which the nonlinearity can be implemented in the detector plane. The simplest form of this network consists of a single layer.

Training of this algorithm can be carried out by the Widrow–Hoff algorithm in which the squared error between the desired and actual output vector is minimized. The rule for updating the weight matrix is

$$\Delta W_{ij} = \gamma (t_i - y_i) x_j \tag{12.26}$$

where t_i is the (desired) target vector, x_j is the input vector, y_j is the output vector, and γ is the learning rate. That is, the weight matrix at time $t+1$ is the weight matrix at time t plus the update matrix ΔW_{ij} at time t. This learning rule is described in block diagram form in Fig. 12.23. Implementation of this procedure is often carried out on a digital processor/controller connected to an SLM to which the matrix W_{ij} is downloaded in the matrix–vector multiplier.

Studies show that a two-layer network is better than a single-layer network—that is, has more capacity and more error tolerance, where the second layer decodes the encoded first layer output. Two-layer perceptrons can thus be implemented for increased pattern recognition capacity. Several options are possible to obtain the second layer, including feeding-back the first layer (SLM) for a second pass, cascading a second SLM for a second layer, or using electronics for the second layer (matrix–vector multiplier).

A hidden layer is often employed between the input and output layers. Figure 12.24 illustrates the signal flow architecture to do this. Such higher-order networks increase complexity over a single-layer perceptron by introducing higher-order interconnections between the input and output units. For second-order networks the input–output relation is

$$y_j = F\left(\sum_i \sum_k W_{ijk} x_j x_k \right) \tag{12.27}$$

which has the advantage of being invariant with respect to translation of the input vector.

In other terms, translation invariance is expressed as

$$F\left(\sum_j \sum_k W_{ijk} x_j x_k \right) = F\left(\sum_j \sum_k W_{ijk} x_{j+m} x_{k+m} \right) \tag{12.28}$$

The product of two input vectors may be implemented by two passes through an SLM. The training algorithm is now

$$\Delta W_{ijk} = \gamma (t_i - y_i) x_j x_k \tag{12.29}$$

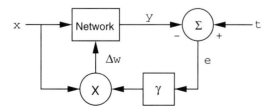

Figure 12.23 Block diagram of delta learning rule (Widrow–Hoff algorithm) for neural network training.

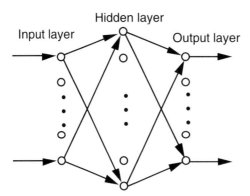

Figure 12.24 Schematic of multiple-layer neural network, consisting of input and output layers and a hidden layer.

which is similar to the original Widrow–Hoff technique. By virtue of the translation invariance described above, the 3-D weight matrix W_{ijk} can be reduced to a 2-D matrix, enabling it to be implemented with a 2-D SLM. By making $m = j$, which we are free to do because m is arbitrary, then $W_{ijk} \rightarrow W_{ik-j}$. Then Equation (12.27) becomes

$$y_j = F\left(\sum_n W_{in} x_j x_{j+n}\right) \qquad (12.30)$$

where $n = k - j$.

To understand the basic operation of a neural network, consider the following numerical example of the Hopfield net.

EXAMPLE 12.8 Associative Recall Using the Hopfield Net The Hopfield net performs associative recall between two vectors (or patterns). It also employs feedback to iterate the associative recall process in order to converge to a correct association. In order to do this, a memory matrix is formed in terms of the superposition of all possible vector products of the bipolar binary representations of the original pattern vectors. For instance, two patterns to be distinguished are given in binary form as

$$\mathbf{A}_1 = (1\,0\,1\,0\,1\,0) \quad \text{and} \quad \mathbf{A}_2 = (1\,1\,1\,0\,0\,0)$$

They are each associated with vectors $\mathbf{B}_1 = (1\,1\,0\,0)$ and $\mathbf{B}_2 = (1\,0\,1\,0)$. To form the associative memory $\mathbf{M}$, we first make the above vectors bipolar: $\mathbf{A}_1 \rightarrow \mathbf{X}_1 = (1\,-1\,1\,-1\,1\,-1)$ and $\mathbf{A}_2 \rightarrow \mathbf{X}_2 = (1\,1\,1\,-1\,-1\,-1)$, and $\mathbf{B}_1 \rightarrow \mathbf{Y}_1 = (1\,1\,-1\,-1)$ and $\mathbf{B}_2 \rightarrow \mathbf{Y}_2 = (1\,-1\,1\,-1)$. Then we form the matrix $\mathbf{M}$ as

$$\mathbf{M} = \begin{bmatrix} 2 & 0 & 0 & -2 \\ 0 & -2 & 2 & 0 \\ 2 & 0 & 0 & -2 \\ -2 & 0 & 0 & 2 \\ 0 & 2 & -2 & 0 \\ -2 & 0 & 0 & 2 \end{bmatrix}$$

which is constructed by calculating $\mathbf{M} = \mathbf{X}_1^T \mathbf{Y}_1 + \mathbf{X}_2^T \mathbf{Y}_2$. For example, if $\mathbf{A}_1$ is presented to the memory matrix $\mathbf{M}$, the association $(\mathbf{A}_1, \mathbf{B}_1)$ should be recalled and the response $\mathbf{B}_1$ is obtained. Since we are dealing with binary inputs and outputs, we will apply a threshold at zero to the output of the matrix–vector product $\mathbf{A}_k \mathbf{M}$ in order to recover the binary version of $\mathbf{B}$. Thus

$$\mathbf{A}_1 \mathbf{M} = \begin{bmatrix} 1 \\ 0 \\ 1 \\ 0 \\ 1 \\ 0 \end{bmatrix} \begin{bmatrix} 2 & 0 & 0 & -2 \\ 0 & -2 & 2 & 0 \\ 2 & 0 & 0 & -2 \\ -2 & 0 & 0 & 2 \\ 0 & 2 & -2 & 0 \\ -2 & 0 & 0 & 2 \end{bmatrix} = \begin{bmatrix} 4 \\ 2 \\ -2 \\ -4 \end{bmatrix} \rightarrow \begin{bmatrix} 1 \\ 1 \\ 0 \\ 0 \end{bmatrix} = \mathbf{B}_1$$

which is, indeed, $\mathbf{B}_1$. Thus $\mathbf{A}_1$ evokes $\mathbf{B}_1$. Then, to check for convergence we calculate

$$\tilde{\mathbf{B}}_1 \mathbf{M}^T = \begin{bmatrix} 4 \\ 2 \\ -2 \\ -4 \end{bmatrix} \begin{bmatrix} 2 & 0 & 2 & -2 & 0 & -2 \\ 0 & -2 & 0 & 0 & 2 & 0 \\ 0 & 2 & 0 & 0 & -2 & 0 \\ -2 & 0 & -2 & 2 & 0 & 2 \end{bmatrix}$$

$$= \begin{bmatrix} 16 \\ -8 \\ 16 \\ -16 \\ 8 \\ -16 \end{bmatrix} \begin{bmatrix} 1 \\ 0 \\ 1 \\ 0 \\ 1 \\ 0 \end{bmatrix} = \mathbf{A}_1$$

where $\tilde{\mathbf{B}}_1$ signifies un-normalized bipolar version of $\mathbf{B}_1$. Thus $\mathbf{B}_1$ evokes $\mathbf{A}_1$ and so forth, and the pattern ($\mathbf{A}_1$, $\mathbf{B}_1$) reverberates across the network.

Although there are many and varied optical architectures for neural networks, we will focus on only a couple. The first is based on a simplified version of the optical vector–matrix (OVM) multiplier architecture originally shown in Fig. 12.18 but with electronic feedback similar to that shown in Fig. 12.21. Now we have the architecture shown in Fig. 12.25. Each layer of a hypothetical multilayer neural network can be treated as one iteration through this architecture. In other words, one pass through the optical vector–matrix multiplier is equivalent to Equation (12.30).

Using the architecture of Fig. 12.25, assume that an eight-element input vector is multiplied by an 8×8 matrix. An anamorphic (essentially cylindrical) lens is used to fan-beam expand the eight elements of the input LED array. After passing through the SLM with transmissive weights $\{W_{ij}\}$ representing the 8×8 weight matrix, the transmitted fan beams are then focused in an orthogonal direction onto a photodiode array again using anamorphic optics. As a result the light incident on the first photodetector is given by the sum of the products of the light intensities and transmittances of the SLM for column 1. In general, the first pass-through yields

$$y_j = \text{sgn}\left(\sum_{i=1}^{N} W_{ij} x_i\right) \tag{12.31}$$

where y_j is the network output of neuron j (corresponding to the output of photodetector j), W_{ij} is the weight from neuron i to neuron j (corresponding to the weight mask at row

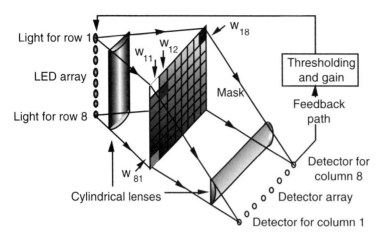

Figure 12.25 Optical vector–matrix multiplier architecture for implementing a neural network for pattern recognition.

i and column j), and x_i is the input vector. The equation above is essentially the same as Equation (12.30) except that we have specified F to be a thresholding (or sgn) function. This network can operate at very high clock rates whereby the entire matrix–vector multiplication can occur in a few nanoseconds and which is essentially independent of matrix size. To use this system adaptively requires an update of the weight matrix W_{ij}. This can be accomplished by reprogramming the SLM after a minimum number required to represent the several layers of a multilayer network paradigm.

The network described above is really a feedforward network in the sense that outputs from a given layer are fedback to the inputs simply to represent subsequent layers of a multiple-layer neural network. This algorithm was first proposed by Hopfield. A one-dimensional system for implementing a Hopfield associative memory using geometric-type interconnects was developed by Farhat, Psaltis, and co-workers. It utilizes a standard incoherent vector–matrix multiplier scheme to model the synaptic connections as shown in Fig. 12.25, where the nonlinear activation function and feedback are performed electronically.

If we now feedback the photodetector outputs of the network to drive the LEDs, we can create the so-called Hopfield net. A threshold (and gain) are introduced after each photodetector equivalent to the "sgn" function. An additional requirement on the weight matrix W_{ij} is that all $W_{ij} = 0$ for $i = j$. This ensures stability. This architecture can also be used to implement an associative memory.

The interconnection weights (connecting the ith-neuron to the jth neuron) for a Hopfield network are given by

$$W_{ij} = \sum_{m=1}^{M} x_{im} x_{jm} \tag{12.32}$$

where the M vectors are those to be stored in this memory. Thus the weight matrix can be seen to be the sum of the outer product of each vector with itself for the M stored vectors. This produces a memory that is autoassociative, as discussed earlier. When an input that is one of the stored vectors is applied, the input to each neurons activation function is

12.4 OPTICAL NEURAL NETWORKS

$$\text{Net} = \sum_{j=1}^{N} \sum_{m=1}^{M} x_{im} \, x_{jm} \, x_{jk} \tag{12.33}$$

$$= (N-1)x_{i,k} + \sum_{m=1}^{M} a_m x_{im} \tag{12.34}$$

where

$$a_m = \sum_{j=1}^{N} x_{jm} \, x_{jk} \tag{12.35}$$

The network input to the neuron is a multiple of the applied vector x_k plus crosstalk terms a_m. The value of the crosstalk coefficient, a_m, is $\sqrt{(N-1)}$ on the average. Thus thresholding [applying a function $f(x) = \text{sgn}(x)$ to the net values] will cause the network outputs to be the same as the inputs. Then

$$y_{i,k} = \text{sgn} \left\{ \sum_{j=1}^{N} W_{ij} \, x_{jk} \right\} \tag{12.36}$$

Since this output is then used as the input for the next iteration, the output vector will in general converge to the stored vector.

This architecture has two main features of interest. The most important feature is that when initialized with a vector that doesn't correspond to any of the stored vectors, it will converge to the stored vector that is the minimum Hamming distance from the input vector. This produces a system that can recognize distorted or incomplete inputs. Another convenient feature is that it can be trained by the straightforward calculation as shown above. As it turns out, this eliminates the need for lengthy training sessions typical of statistical classifiers.

Another very convenient feature of the Hopfield net implemented here is that convergence generally occurs even when the synaptic weights are clipped so that negative elements become -1 and positive elements become $+1$. This demonstrates the error tolerance of the system. The synaptic weights $\{-1, 0, 1\}$ are implemented by spatial encoding of the weighting mask. Then two rows of the synaptic mask and two photodetectors are associated with each LED. The diodes provide currents of opposite sign when illuminated so that ± 1 can be realized. (Since this system is based on a free space optical propagation for interconnection scheme, the neuron and interconnection density are limited by diffraction.)

The architecture presented above can be extended so that it accepts two-dimensional inputs. This is implemented optically as shown in Fig. 12.26. Although the functioning is similar to the 1-D system, it is obvious from Fig. 12.26 that the $N^2 \times N^2$ weight matrix presents difficulties in actual implementation.

In the 2-D case the input is a 2-D matrix X_{ij}, and the associative memory being addressed is a four-dimensional matrix W_{ijkl}. Hence we have a 2-D matrix retrieved from memory by the operation

$$\hat{X}_{ij} = \sum_{k,l} W_{ijkl} x_{kl} \quad (i, j, k, l = 1, 2, \ldots, N) \tag{12.37}$$

where the procedure for obtaining the memory matrix W_{ijkl} is to form the outer product

$$W_{ijkl} = \sum_{m=1}^{M} X_{ijm} X_{kl} \tag{12.38}$$

for the set of m bipolar binary patterns y_{ij}^m. The trick is to decompose W_{ijkl} into a partitioned memory mask of 2-D submatrices. Hence W_{ijkl} is set equal to

$$\begin{aligned} W_{ijkl} =\, & W_{11kl} + W_{12kl} + \ldots + W_{1jkl} \\ & + W_{21kl} + W_{22kl} + \ldots + W_{2jkl} \\ & + \ldots \\ & + W_{i1kl} + W_{i2kl} + \ldots + W_{ijkl} \end{aligned} \tag{12.39}$$

and is arranged as shown in Fig. 12.26. In order to address this partitioned matrix a 2-D lenslet array may be used.

Many of the problems in implementing the two-dimensional system discussed above can be overcome by using holographic interconnections between the neurons. When a hologram of an object is recorded and subsequently addressed by a reference beam (similar to the one used to generate it in the first place) the object is reconstructed. Conversely, the reference beam can be recreated if the object beam is used. Hence these two beams are associated with each other, just like two patterns in a associative memory. Implementations of a Hopfield net using planar holograms have been presented by Psaltis and Abu-Moustafa and by Psaltis and Farhat. This system, shown in Fig. 12.27, is basically an image correlator with feedback.

If we think in terms of correlators with their stored reference images, these images can

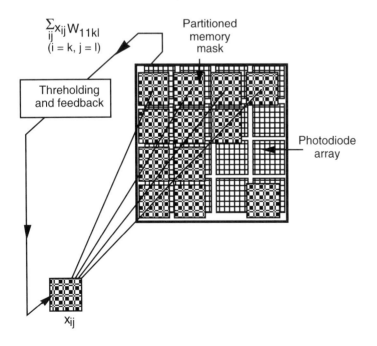

Figure 12.26 Two-dimensional implementation of a Hopfield optical neural network.

be recovered using thin or volume holograms that are coherently illuminated and iteratively addressed in an optical feedback loop. An image is introduced into the system and optically correlated with all the stored reference images simultaneously. When these correlations are fed back they are thresholded so that the strongest correlation reinforces and enhances the input image, even if it was initially incomplete or noisy. The image continues to pass around the loop and approaches the desired image, which is in fact the pattern to be recognized. In this regard the holographic correlator obviates the feature extraction step and simply recovers the desired image (or object in a scene), thus classifying it. This approach can also be regarded as an associative memory.

Let us analyze in more detail the architecture developed by Psaltis et al. as shown in Fig. 12.27. As we have just discussed, for associative memories that store 2-D patterns, a four-dimensional linear transformation matrix W_{ijkl} has to be constructed, but to implement it in planar (2-D) optics would require partitioning it into multiple 2-D masks addressed via multiple aperture lenslet arrays. In the case of holographic memories treated here, however, we must replace the set of m discrete binary input patterns x_{ij}^m with the set of continuous 2-D functions $f_m(x, y)$. Hence we must form the four-dimensional function analogous to W_{ijkl} discussed earlier [see Equation (12.39)], namely,

$$\mathbf{W}(x, y, x', y') = \sum_{m=1}^{M} f_m(x, y) f_m(x', y') \qquad (12.40)$$

In a manner analogous to that leading to Equation (12.33) we obtain the output produced by the (outer-product) associative memory when it is addressed by the input image $f(x, y)$, that is,

$$\hat{f}(x', y') = \int\int \mathbf{W}(x, y, x', y',) f(x, y)\, dx\, dy \qquad (12.41)$$

$$= \sum_{m=1}^{M} \left[\int\int f_m(x, y) f(x, y)\, dx\, dy \right] f_m(x', y') \qquad (12.42)$$

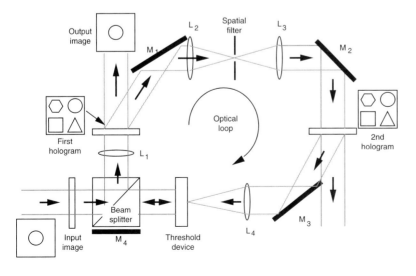

Figure 12.27 Holographic implementation of a neural network model of an associative memory (after Farhat and Psaltis).

where the 2-D integration inside the brackets here corresponds directly to a 2-D correlation of $f(x, y)$ with $f_m(x, y)$, evaluated at the origin (zero lag). This suggests the use of a coherent optical correlator architecture like that developed by VanderLugt.

In this system an input image containing a circle enters the architecture via a beamsplitter. A spatial light modulator located at plane P_1 is used to detect the input image and yields a thresholded version of it on the other side. In the other path, lens L_1 Fourier transforms this pattern (since it is reflected back via mirror M_4) and places it at plane P_2 where a hologram is located (before propagating on to mirror M_1). This Fourier transform hologram at P_2 contains all the stored references and is created by a procedure similar to the way a matched filter was created, as discussed in Chapter 10. A second Fourier transform lens L_2 will thus yield the correlation of the composite hologram and the thresholded input image. The correlation peaks associated with each of the images that make up the composite and the input are located at different positions according to their original positions. Since only inner products (not the full correlations) need be recovered for continuing the feedback process, we can use pinholes as shown in plane P_3. These pinholes are arrayed to merely sample each correlation at its zero-lag value, so that the amplitude transmitted through each hole is

$$\hat{f}_m(0, 0) = \int\int f_m(x - x', y - y') f(x, y) \, dx dy \,|_{x'=0,\, y'=0} \quad (12.43)$$

$$= \int\int f_m(x, y) f(x, y) \, dx dy \quad (12.44)$$

The resulting light is recollimated by lens L_3 after which it illuminates a second hologram at plane P_4. This hologram is identical to the first. Then the final lens L_4 (after mirror M_3) yields the Fourier transform of the light transmitted through the hologram at P_4 back at plane P_1. Hence this part of the architecture from plane P_4 to P_1 is also a correlator. Looked at ideally, the light transmitted through each pinhole (in plane P_3) reconstructs the entire composite image stored in the hologram at plane P_2. Light transmitted through a pinhole on which the mth inner product is formed is located in the pinhole array at the position of the mth image in the original composite. Thus the reconstruction obtained in plane P_1 coming from this particular pinhole is shifted so that the mth image appears centered on the optical axis. Thus the light incident on plane P_1 is a superposition of all stored images each weighted as described by Equation (12.44). The SLM thresholds this result, which becomes the next input for the optical loop. Since the brightest image reaching the back of the threshold device represents the best match to the input image, it is essentially the one that is transmitted through for the second pass. Successive passes around the loop continue to enhance the best match, which is finally retrieved as an output image leaving the architecture through the beamsplitter (output image with circle).

Brady and Psaltis have shown that this type of system can also be implemented using volume holograms, which have interconnection densities estimated at 10^8 to 10^9. This still has limitations for neural nets. The degrees of freedom for interconnection is only $0(N^3)$, whereas full interconnection of all neurons requires $0(N^4)$. This degeneracy can give rise to multiple couplings in the medium. It appears that systems involving holographic interconnections are probably the most logical development path for optical neural nets. This is mainly due to the large potential interconnectivity and neuron density expected by using holographic elements. These systems should also have the advantage of possessing the capability to correlate incomplete or distorted input data.

Another type of feedback can be employed in neural networks using the optical path directly. Here we feedback the photodetector outputs optically in a direction opposite to the input, as shown in Fig. 12.28, and arrive at a bidirectional associative memory developed by Kosko. Notice that at each end (input and output) there is a detector–source pair. This means that there will be counterpropagating beams from each set of sources. Thus light from the left transmits to the right and vice versa. After detection the received signal may be thresholded. Each detector on the right sees an entire row of the SLM (weight) pattern. Stability is ensured even if the main diagonal does not equal zero.

An important feature of optical neural networks, when using incoherent light (i.e., bipolarity), can be accommodated by doubling the number of rows (or columns) and encoding the positive values in one subrow and the negative values in the adjacent subrow.

PROBLEM EXERCISES

1. Show that the orientation θ of an object $f(x, y)$ in the field of view of an image can be determined from the expression

$$\theta = \frac{1}{2}\tan^{-1}\left(\frac{m_{10}}{m_{20} - m_{10}}\right)$$

where m_{pq} are the ordinary image moments:

$$m_{pq} = \int_{-\infty}^{\infty} \int_{-\infty}^{\infty} x^p y^q f(x, y)\, dx\, dy$$

2. Given the Radon transform $F_R(x', \theta)$ of an object where x' is related to x by the rotation matrix $\mathbf{R}$, derive an expression for the moments of $F_R(x', \theta)$ given by

$$m_n^\phi = \sum_{i=0}^{n} \binom{n}{i} \cos^{n-1}\phi \sin\phi\, m_{n-i,i}$$

where $m_{n-i,i}$ are the Cartesian moments of the original function $f(x, y)$. Invert this result to express the Cartesian moments in terms of the Radon transform moments and show

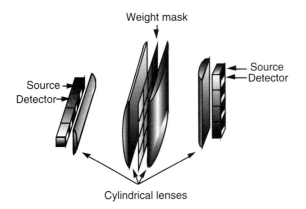

Figure 12.28 Bidirectional associative memory implemented as an optical vector–matrix multiplier used bidirectionally by virtue of interleaved detector/transmitter pairs in each focal plane (after Kosko).

$$m_{00} = m_0^\phi \quad (\text{all}\phi)$$

$$m_{j0} = m_j^0 \quad (j = 1, 2, 3)$$

$$m_{j0} = m_j^{\pi/2} \quad (j = 1, 2, 3)$$

$$m_{11} = m_2^{\pi/4} - \frac{1}{2}\left(m_2^0 + m_2^{\pi/2}\right)$$

$$m_{21} = \frac{1}{3}\left[\sqrt{2}\left(m_3^{\pi/4} + m_3^{3\pi/4}\right) - \pi/4\right] + m_3^{\pi/2}$$

$$m_{12} = \frac{1}{3}\left[\sqrt{2}\left(m_3^{\pi/4} - m_3^{3\pi/4}\right) - \pi/4\right] - m_3^{\pi/2}$$

From these results it is clear that only four projections are needed (at $\phi = 0, \pi/4, \pi2, 3\pi/4$) to compute the Cartesian moments. Discuss the relative computational requirements for calculating Cartesian moments directly versus by Radon moments.

3. Derive the expression for the angular autocorrelation of a rectangular pillbox and show that it is given by

$$R_{rr}(\phi) = \begin{cases} w^2 \csc\phi, & \phi > 2\tan^{-1}(w/l) \\ lw - \frac{\sec\phi \tan(\phi/2)}{2}[l^2 + w^2 + 2lw\sin(\phi/2)], & \phi \leq 2\tan^{-1}(w/l) \end{cases}$$

where w is width, l is length, and $x' = \mathbf{R}x$. What happens at $l = w$? Plot this function for $l = 4w, 2w,$ and w. Show how the values $R(0)$ and $R(\pi/2)$ can be used to calculate the area, aspect ratio, length, and width of the rectangle.

4. Express the boundary of an elliptical pillbox with semi-major axis a and semi-minor axis b in polar coordinates. Then show that the angular correlation of the ellipse with a narrow slit of width w_s that satisfies the Nyquist criterion for the boundary function is given by

$$w_s = \frac{2ab\sin(\theta_s/2)}{[a^2 + (b^2 - a^2)\cos^2(\theta_s/2)]^{1/2}}$$

where θ_s is the required Nyquist angle sampling rate derived from the boundary Fourier spectrum.

5. Calculate the Fourier transform of an elliptical pillbox using Equation (12.1) and the result of Problem 4. Then sample and integrate over finite intervals in angle and radius to simulate the wedge-ring detector. Pick suitable intervals to recover the elliptical shape if the inverse Fourier transform is applied. Compare the resulting reconstruction to the original ellipse using a mean-square error criterion.

6. Using the architecture of Fig. 12.5, calculate the moments of a rectangular box and a Gaussian spot. Incorporate acoustic attenuation in the AO cells and Gaussian beam taper in the laser. How do those device models affect the resulting moments compared to the ideal moments for these shapes?

7. Consider the architecture of Fig. 12.7 for calculating the Radon transform. The angular resolution of this scheme is limited by the time required to get data from the linear detector readout array and the rotational speed of the prism. For an input focal plane array frame (or display) time of 30 msec

(with no interlace), what must the rotation rate of the Dove prism be and the sampling rate of the linear detector readout array be not to limit the frame time? If the input dynamic range is 8 bits (256), what does the dynamic range of the linear array have to be to accommodate the maximum possible variations in the output signal?

8. Explicitly derive the results of Example 12.2 for the Walsh–Hadamard transform for an 8×8 mask size and an image resolution of 8×8 pixels.

9. For the three images shown below, calculate the ordinary moments m_{pq} and the central moments μ_{pq} up to third order.

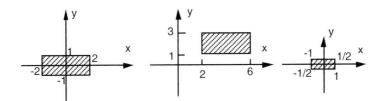

10. Calculate the moments up to fourth order for the following probability density functions: (a) Gaussian, (b) Poisson, (c) Rayleigh, (d) uniform, (e) gamma, which is given by

$$p(x) = \begin{cases} \dfrac{1}{\beta^{\alpha+1}\alpha!} x^{\alpha} \exp(-x/\beta), & x > 0, \alpha > 0, \beta > 0] \\ 0, & \text{elsewhere} \end{cases}$$

(f) Plot the first versus second moments, and the third versus fourth moments. What relation appears to hold between these moments in each plot? Using these functions as models of time-domain pulse shapes from a radar, add a little noise and plot the results again for 50 trials of a random number generator for the noise. Be sure to keep the noise variance low initially to clearly see the sensitivity of your plots to noise.

11. Calculate the Radon transform of an ellipse with three delta functions on it like that shown in Fig. 12.8. Compare your analytic results to that shown in Fig. 12.9 for the four cases shown.

12. For the architecture of Fig. 12.19 calculate the cosine transform of an input signal defined as

$$f(n) = \delta(n) + \delta(n - n_0)$$

for $N = 128$ using MATLAB. Model $H(n, m)$ as having exact weights and as having uneven weights Gaussian white-noise distributed with known variance. Incorporate a Gaussian laser beam shape of known variance into the mask to compare its effect on the accuracy of the resulting Fourier transform versus mask noise alone. Assess the significance of your results.

13. Given the systolic processing architecture shown in the figure below for a matrix–vector multiplier, carry out a vector–matrix multiplication of the form

$$\begin{bmatrix} y_1 \\ y_2 \end{bmatrix} = \begin{bmatrix} a_{11} & a_{12} \\ a_{21} & a_{22} \end{bmatrix} \begin{bmatrix} x_1 \\ x_2 \end{bmatrix}$$

and explain how this operation is performed by labeling the inputs and outputs at each clock cycle and showing the corresponding ray trace. Assume that the region of support for each component of the input vector is less than half the Bragg cell extent. Show how this architecture could be extended to a 3 × 3 matrix operation.

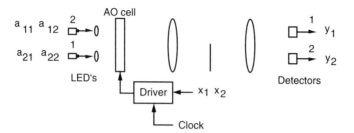

14. In the matrix–vector processor described in Fig. 12.18 the general form of the input–output relationship is given by Equation (12.20), namely,

$$g(m) = \sum_{n=0}^{N-1} f(n) H(n, m)$$

Write down the form of $H(n, m)$ for (a) the cosine transform, (b) the sine transform, (c) the Hilbert transform, and (d) the chirp-z transform. In each case write out explicitly the matrix for the 4 × 4 case.

15. Model a two-tone signal with white Gaussian noise to second order with the autoregressive model of Example 12.6 and calculate its power spectrum. Use a random noise generator from MATLAB. Explicitly write out the equations to carry out this estimation procedure and map out the relationship to the architecture of Fig. 12.21.

16. Consider a Hopfield net with two exemplar pattern vectors given by

$$\mathbf{A}_1 = (1\,0\,0\,1\,0\,0)$$

$$\mathbf{A}_2 = (0\,0\,1\,1\,0\,0)$$

Convert these to bipolar binary form and construct a weight matrix W via an outer (vector) product. This matrix represents symbolically the synaptic weights of the net. Now consider the input vector given by

$$\mathbf{B} = (1\,1\,0\,1\,0\,0)$$

Calculate the vector–matrix product and iterate until the output vector stabilizes. How is the output vector related to one of the exemplars? In this case the Hamming distance is large, and it is indicative of a limited dimensionality of the network.

17. Determine the associative memory correlation matrix for the case where the pattern matrix is

$$\begin{bmatrix} -1 & 1 & 1 & -1 & 1 \\ 1 & -1 & 1 & 1 & -1 \\ -1 & 1 & -1 & 1 & 1 \\ -1 & 1 & 1 & -1 & 1 \\ 1 & -1 & -1 & 1 & -1 \end{bmatrix}$$

and the input pattern is

$$\begin{bmatrix} -1 \\ 1 \\ -1 \\ 1 \\ 1 \end{bmatrix}$$

18. For the weight matrix M of Example 12.8 and the input vectors

$$\mathbf{A}_1 = (1\ 1\ 0\ 1\ 1\ 0)$$

and

$$\mathbf{A}_2 = (1\ 0\ 1\ 1\ 0\ 1)$$

find the associated output vectors $\mathbf{B}_1$ and $\mathbf{B}_2$. Construct an architecture like that shown in Fig. 12.28 to implement this. Check for convergence with and without additive white Gaussian noise. Account for the required source nonuniformity at the weight mask plane using the cosine fourth law in the corresponding axis for the left propagating and right propagating light. How many iterations are tolerable before uncorrected nonuniformities dominate the Gaussian noise contribution from the detectors? Assume that the detectors are linear and that the weight mask is bipolar. Does it help to incorporate a threshold in the feedback between source and detector?

19. For the simple neural network (Perceptron) shown in the figure below with an output function y given by

$$y(t) = \text{sgn}[w_1 x_1(t) + w_2 x_2(t) - \theta]$$

where $w_1 = 0.5$ and $w_2 = 0.2$ initially and where θ is the threshold value ($= 1$), we want to distinguish two classes of patterns whose two inputs are each either below 1/3 or above 2/3 (i.e., "low" versus "high" inputs), but where intermediate inputs are not involved, such as (0.5, 0.5). Assume that the learning constant is $\gamma = 0.1$ in the equation

$$w_i(t+1) = w_i(t) + \gamma[d(t) - y(t)]x_i(t)$$

where $d(t)$ is the desired (target) vector. Initiate this network with inputs (0, 0) and check to see if the correct output is achieved. Do the same for initial inputs (1, 1). When the incorrect output is thus evoked, update the weights with the appropriate correction factor and iterate this process until the weight updates are minimized and stable. This simple perceptron, however, cannot resolve mixed pairs (one high and one low), which you can verify by the same procedure; that is, it cannot resolve an exclusive-OR logic operation, which is a type of correlation used in digital correlators. How would such a procedure be implemented optically?

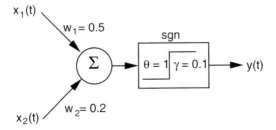

BIBLIOGRAPHY

Y. S. Abu-Mostafa and D. Psaltis, "Optical Neural Computers," *Sci. Am.* **256** (3), 88–95 (March 1987).

M. Agu, A. Akiba, and S. Kamermaru, "Multimatched Filtering System as a Model of Biological Visual System," Proc. SPIE, Vol. 1014, *Micro-Optics*, p. 144 (1988).

R. P. Bocker, K. Bromley, and S. R. Clayton, "A Digital Optical Architecture for Performing Matrix Algebra," *Real-Time Signal Processing VI*, K. Bromley, ed, Proc. SPIE 431, 194 (1983).

B. G. Boone, O. B. Shukla, and D. H. Terry, "Extraction of Features from Images Using Video Feedback," SPIE Conference Proceedings on Automatic Object Recognition, Orlando, FL, April 1–5, 1991.

D. Brady and D. Psaltis, "Perceptron Learning in Optical Neural Computers," *Proceedings, Scottish Universities Summer Program on Optical Computing*, pp. 251–64 (1989).

K. Bromley, A. C. H. Louie, R. D. Martin, J. J. Symanski, T. E. Keenan, and M. A. Monhan, "Electro-Optical Signal Processing Module," Proc. SPIE, Vol. 180, *Real-Time Signal Processing II*, p. 107 (1979).

D. Casasent, "Coherent Optical Pattern Recognition: A Review," *Opt. Eng.* **24**, 26 (1985).

D. Casasent and M. Carletto, "Multidimensional Adaptive Radar Processing Using an Iterative Optical Matrix–Vector Processor," *Opt. Eng.* **21**, 814 (September/October 1982).

D. Casasent and C. Neumann, "Iterative Optical Vector–Matrix Processors," Proceedings, NASA Conference on Applications of Optical Processing to Aerospace Needs, August 1981.

N. H. Farhat, D. Psaltis, A. Prata, and E. Paek, "Optical Implementation of the Hopfield Model," *Appl. Opt.* **24**, 1469 (1985).

N. George, J. Thomasson, and A. Spindel, "Photodetector for Real-Time Pattern Recognition," U.S. Patent # 3,689,772 (1970).

G. R. Gindi and A. F. Gmitro, "Optical Feature Extraction via the Radon Transform," *Opt. Eng.* **23**, 499 (1984).

I. Glaser, "Lenslet Array Processors," *Appl. Opt.* **21**, 1271 (1982).

J. J. Hopfield, in *Proceedings of National Academy of Science USA*, No. 79, pp. 2554–2558 (1982).

J. L. Horner, ed., *Optical Signal Processing*, Academic Press, San Diego, CA (1987).

B. K. Jenkins and A. R. Tanguay, Jr., "Photonic Implementations of Neural Networks" in *Neural Networks for Signal Processing*, B. Kosko, ed., Prentice-Hall, Englewood Cliffs, NJ, pp. 287–382 (1992).

B. Kosko, "Optical Bidirectional Associative Memories," *SPIE Proc. Image Understanding*, Vol. 758 (1987).

B. Kosko, "Bidirectional Associative Memories," *IEEE Trans. Syst. Man, and Cybernetics* **18**, 49 (1988).

R. P. Lippman, "An Introduction to Computing with Neural Nets," *IEEE ASSP* magazine, pp. 4–22 (April 1987).

T. Lu, S. Wu, X. Xu, and F. T. S. Yu, "Two-Dimensional Programmable Optical Neural Network," *Appl. Opt.* **28** 4908 (1989).

J. Ohta, J. Sharpe, and K. Johnson, "An Optoelectronic Feature Extractor for Classification of Fingerprints," *SPIE Proceedings Conference on Optoelectronic Neural Networks*, Vol. 2026 San Diego, CA (1993), pp. 394–402.

D. Psaltis and N. H. Farhat, "Optical Information Processing Based on an Associative Memory Model of Neural Nets with Thresholding and Feedback," *Opt. Lett.* **10**, N2 (February 1985).

W. T. Rhodes, "Acousto-optic Algebraic Processing Architectures," *Proc. IEEE* **72**, 820–830 (1984).

W. T. Rhodes, "Optical Matrix–Vector Processors: Basic Concepts," Proc. SPIE, Vol. 614, *Highly Parallel Signal Processing Architectures*, p. 146 (1986).

M. G. Robinson, L. Zhang, and K. M. Johnson, "Optical Neurocomputer Architectures Using Spatial Light Modulators," Proc. SPIE, Vol. 1469, *Applications of Artifical Neural Models II*, pp. 240–249 (1991).

F. Rosenblatt, "The Perceptron: A Probabilistic Model for Information Storage and Organization in the Brain," *Psychol. Rev.* **65**, 386 (1958).

Y. Sheng, C. Lejeune, and H. H. Arsenault, "Frequency-Domain Fourier–Mellin Descriptors for Invariant Pattern Recognition," *Opt. Eng.* **27**, 354 (1988).

O. B. Shukla and B. G. Boone, "Optical Feature Extraction Using the Radon Transform and Angular Correlation," *Proceedings, SPIE Conference on Optical Information Processing Systems and Architectures IV*, Vol. 1772, pp. 142–156 (1992).

P. D. Wasserman, *Neural Computing: Theory and Practice*, Van Nostrand Reinhold, New York (1989).

B. Widrow and M. Hoff, "Adaptive Switching Circuits," *IRE WESCON Convention Record* (1960).

R. C. D. Young and C. R. Chatwin, "Computation of the Forward and Inverse Radon Transform Via the Central Slice Theorem Employing a Non-Scanning Optical Technique," *SPIE Proc.*, Vol. 2752, 306 (1996).

F. T. S. Yu, "Optical Neural Networks: Architecture, Design and Models," in *Progress in Optics*, Vol. XXXIII, E. Wolf, ed., Elsevier Science Publishers, pp. 61–144, (1993).

Appendix A
MATHEMATICAL TABLES

Table 1.1
Properties of Dirac Delta Function

$\delta(-x, -y) = \delta(x, y)$

$\delta(\pm ax, \pm by) = \dfrac{1}{|ab|}\delta(x, y)$

$\delta(x^2 - a^2) = \dfrac{1}{2a}[\delta(x - a) + \delta(x + a)]$

$\displaystyle\int_{-\infty}^{\infty} f(x) \dfrac{d}{dx}\delta(x - a)\, dx = -\left.\dfrac{df}{dx}\right|_{x=a}$

$\displaystyle\int_{x_1}^{x_2} \delta(x' - x)\delta(x' - x_0)\, dx' = \delta(x - x_0), \qquad x_1 < x_0 < x_2$

$\displaystyle\int_{-\infty}^{\infty} \delta[f(x)]\, dx = \sum_k \dfrac{1}{|f'(x_k)|}$

where $f'(x_k) \neq 0$, $f(x_k) = 0$, and $f'(x) \equiv \dfrac{df(x)}{dx}$

Table 1.2
Models for the Dirac Delta Function

$\displaystyle\lim_{a \to 0} \dfrac{1}{a}\text{rect}\left(\dfrac{x}{a}\right), \qquad a \geq 0$

$\displaystyle\lim_{a \to \infty} \dfrac{a}{2}\exp(-a|x|), \qquad a > 0$

$\displaystyle\lim_{a \to 0} \dfrac{1}{\sqrt{2\pi}\, a}\exp\left(\dfrac{-x^2}{2a^2}\right), \qquad a \geq 0$

$\displaystyle\lim_{a \to \infty} \dfrac{\sin(max)}{\pi x}, \qquad \begin{array}{l} a > 0 \\ m > 0 \end{array}$

Table 1.3
General Properties of Correlation

Property	Mathematical Expression
Commutative	$f \otimes g = g \otimes f$
Associative	$f \otimes (g \otimes h) = (f \otimes g) \otimes h$
Distributive	$f \otimes (g + h) = f \otimes g + f \otimes h$

Table 1.4
Properties of Two-Dimensional Fourier Transforms

Property	Space Domain	Fourier Domain
Linearity	$af(x, y) + bg(x, y)$	$aF(u, v) + bG(u, v)$
Sign change	$f(-x, -y)$	$F(-u, -v)$
Scale change	$f(x/a, y/b)$	$\|ab\|F(au, bv)$
Shift	$f(x - a, y - b)$	$F(u, v)\exp[-j2\pi(au + bv)]$
Frequency shift	$f(x, y)\exp[j2\pi(ax + by)]$	$F(u - a, v - b)$
Conjugation	$f^*(x, y)$	$F^*(-u, -v)$
Space	$\partial f(x, y)/\partial x$	$j2\pi u F(u, v)$
Differentiation	$\partial f(x, y)/\partial y$	$j2\pi v F(u, v)$
Frequency	$-j2\pi x f(x, y)$	$\partial F(u, v)/\partial u$
Differentiation	$-j2\pi y f(x, y)$	$\partial F(u, v)/\partial v$
Convolution theorem	$f(x, y) * g(x, y)$	$F(u, v)G(u, v)$
Product theorem	$f(x, y)g(x, y)$	$F(u, v) * G(u, v)$
Parseval's theorem	$\int\int_{-\infty}^{\infty} f^*(x, y)g(x, y)\,dxdy$	$\int\int_{-\infty}^{\infty} F^*(u, v)G(u, v)\,dudv$

Table 1.5
Transform Pairs of Separable Functions in Rectangular Coordinates

Function	Transform
$\delta(x)\delta(y)$	1
1	$\delta(u)\delta(v)$
$\text{rect}(x/x_0)\,\text{rect}(y/y_0)$	$\text{sinc}(\pi x_0 u)\,\text{sinc}(\pi y_0 v)$
$\text{tri}(x/x_0)\,\text{tri}(y/y_0)$	$\text{sinc}^2(\pi x_0 u)\,\text{sinc}^2(\pi y_0 v)$
$\exp[-\pi(x^2 + y^2)]$	$\exp[-4\pi(u^2 + v^2)]$
$\exp[j2\pi(x_0 + y_0)]$	$\delta(u - u_0)\delta(v - v_0)$
$\text{sgn}(x)\,\text{sgn}(y)$	$(1/j\pi u)(1/j\pi v)$
$\text{comb}(x/x_0)\,\text{comb}(y/y_0)$	$x_0 y_0 \text{comb}(x_0 u)\,\text{comb}(y_0 v)$

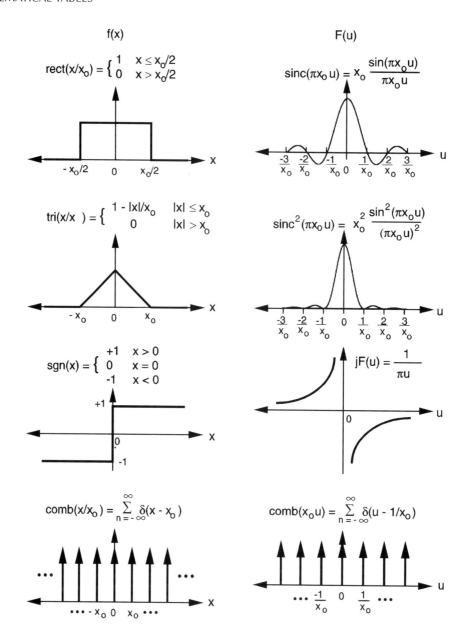

Figure A.1 Definition of functional notation and corresponding Fourier transforms.

Table 1.6.
Properties of Bessel Functions

$J_{-n}(z) = (-1)^n J_n(z)$

$\dfrac{dJ_0(z)}{dz} = -J_1(z)$

$\dfrac{d}{dz} z^n J_0(z) = z^n J_{n+1}(z)$

$\displaystyle\int J_1(ax)\,dx = \dfrac{-1}{a} J_0(x)$

$\displaystyle\int x J_0(ax)\,dx = \dfrac{x}{a^2} J_1(ax)$

$\displaystyle\int_0^\infty J_n(ax)\,dx = \dfrac{1}{a} \quad \text{Re}(n) > -1$

$\displaystyle\int_0^a x^{n+1} J_n(x)\,dx = a^{n+1} J_{n+1}(x)$

$\displaystyle\int_0^a J_1(x)\,dx = 1 - J_0(a)$

Table 1.7
Hankel Transform Pairs

Function	Hankel Transform
$J_0(2\pi ar)$	$\delta(r-a)/2\pi a$
$\sin(2\pi ar)/r$	$\operatorname{rect}(r/2a)/(a^2 - \rho^2)^{1/2}$
$\operatorname{jinc}(r) = J_1(\pi r)/2r$	$\operatorname{rect}(\rho)$
$\delta(r-a)$	$2\pi a J_0(2\pi a\rho)$
$\exp(-\pi r^2)$	$\exp(-\pi \rho^2)$
$1/r$	$1/\rho$
$\exp(j\pi r^2)$	$-j\exp(-j\pi\rho^2)$
$\exp(-ar)$	$2\pi a/(4\pi^2\rho^2 + a^2)^{3/2}$
$\delta(r)/\pi r$	1
$\exp(-ar^2)$	$\exp(-\rho^2/4a)/2a$

Table 1.8
Properties of the Hankel Transform

Property	Space Domain	Hankel Domain
Linearity	$af(r) + bg(r)$	$aF(\rho) + bG(\rho)$
Shift	\multicolumn{2}{c}{Shift of origin destroys circular symmetry}	
Scale change (Similarity theorem)	$g(ar)$	$a^{-2} G(\rho/a)$
Convolution theorem	$\displaystyle\int_{-\infty}^{\infty} f(r')g(R) r'\,dr'\,d\theta$	$F(r)G(r)$
where	$R^2 = r^2 + r'^2 - 2rr'\cos\theta$	
Parseval's theorem	$\displaystyle\int_{-\infty}^{\infty} f(r)g^*(r) r\,dr$	$\displaystyle\int_{-\infty}^{\infty} F(\rho)G^*(\rho)\rho\,d\rho$

Table 3.1
Properties of Hilbert Transform

Property	Signal Domain	Hilbert Domain
Similarity	$f(ax)$	$F_H(ax)$
Superposition	$af(x) + bg(x)$	$aF_H(ax) + bG_H(ax)$
Shift	$f(x-a)$	$F_H(x-a)$
Parseval's Theorem	$\int_{-\infty}^{\infty} |f(x)|^2 \, dx$	$\int_{-\infty}^{\infty} |F_H(x)|^2 \, dx$
Autocorrelation	$\int_{-\infty}^{\infty} f^*(x)f(x-x')\,dx$	$\int_{-\infty}^{\infty} F_H^*(x)F_H(x-x')\,dx$
Convolution	$f(x) * g(x)$	$-F_H(x) * G_H(x)$

Table 3.2
Hilbert Transform Pairs

Function	Hilbert Transform
$\cos x$	$-\sin x$
$\sin x$	$\cos x$
$\sin x / x$	$(\cos x - 1)/x$
$\text{rect}(x)$	$\ln\left|(x-\tfrac{1}{2})/(x+\tfrac{1}{2})\right|/\pi$
$1/(1+x^2)$	$-x/(1+x^2)$
$\delta(x)$	$-1/\pi x$

Table 3.3
Mellin Transform Pairs

Function	Mellin Transform
$\delta(x-a)$	a^{s-1}
$\exp(-ax)$	$a^s \Gamma(s)$ (Re $a > 0$, Re $s > 0$)
$\exp(-x^2)$	$\Gamma(s/2)/2$
$\sin x$	$\Gamma(s)\sin(\pi s/2)$
$\cos x$	$\Gamma(s)\cos(\pi s/2)$

Note: $\Gamma(x)$ is the Gamma function defined by

$$\Gamma(x) = \int_0^{\infty} t^{x-1} e^{-t} \, dt$$

for integer and noninteger values.

Appendix B
ANNOTATED BIBLIOGRAPHY

A listing with comments of most well-known and recent books on optical signal processing and Fourier optics is provided in this appendix.

Fourier Optics, J. W. Goodman, McGraw–Hill, New York (1968)

This is probably the most famous text in the field of Fourier optics. It is a seminal text by a nationally and internationally known expert in the field. It emphasizes the basics of physical optics, Fourier transform properties of lenses, traditional coherent optical processing techniques, and holography. Many challenging problem exercises are provided for the student. This text is very appropriate as a prerequisite for more advanced courses in optics including advanced optical signal processing.

Optical Information Processing, Optical Signal Processing, Fourier Optics, F. T. S. Yu, John Wiley & Sons, New York (1983)

This is another book that covers much of Fourier optics with applications to signal processing, again by a well-known expert in the field. There is extensive coverage of holography, especially polychromatic and noncoherent techniques, as well as a considerable number of problems.

Optical Processing, Computing, and Neural Networks, F. T. S. Yu, and S. Jutamulia, John Wiley & Sons, New York (1992)

This is among several recent textbooks on optical signal processing and computing. It is similar to Yu's previous work, but goes further by including optical computing and neural networks, as well as more current topics. Extensive homework problems are included.

Principles of Optical Engineering, F. T. S. Yu, John Wiley & Sons, New York (1990)

This is a more basic text on optical engineering that embodies a significant portion devoted to optical signal processing. The examples and problems are more accessible to the engineering student who hasn't had much previous exposure to optics.

The Fourier Transform and Its Applications, R. N. Bracewell, McGraw–Hill, New York (1978)

This textbook is a more general text on the Fourier transform and its applications in various fields. It also covers other, related mathematical transforms, such as the Hankel, Abel, and Fourier–Bessel transforms. Many physically intuitive graphics and interpretations are provided to gain insight into the mathematical foundations.

Optical Signal Processing, J. L. Horner, ed., Academic Press, Orlando, FL (1987)

This edited work by J. L. Horner is a excellent compedium of papers written especially for this volume by a number of experts in the field. For the working scientist or engineer, this collection is probably one of the most useful, especially for transitioning from formal theory to practical applications.

Optical Computing, D. G. Feitelson, MIT Press, Cambridge, MA (1988)

This book is a survey for computer scientists, which emphasizes technology and computing with a minimum of mathematics. This volume is of a survey nature, and it offers the serious investigator a wealth of information and an extensive bibliography.

Optical Computer Architectures, A. D. McAulay, John Wiley & Sons, New York (1991)

This book is a first-year graduate text and covers optical *computing* extensively, *not* signal processing. It emphasizes more of the hardware and implementation aspects of optical computing than the theory or algorithms.

Acousto-Optic Signal Processing, N. J. Berg, and J. N. Lee, eds., Marcel Dekker, New York (1983)

This book focuses almost entirely on acousto-optics at a detailed technical level. It is essentially a collection of well-prepared pedagogically-oriented papers. There are no problem exercises, but applications are covered extensively.

Optical Signal Processing, A. VanderLugt, John Wiley & Sons, New York (1992)

This recent text by one of the leaders in the field covers optical signal processing extensively, with many excellent examples, applications, and homework problems, especially using acousto-optic devices.

Fundamentals of Photonics, B. E. A. Saleh and M. C. Teich, John Wiley & Sons, New York (1991)

This recent work provides an extensive coverage of all aspects of photonics, which includes acousto-optic and electro-optic devices, as well as many other technologies and their physics of operation.

Electro-Optics, J. L. Pinson, John Wiley & Sons, New York (1985)

This textbook, although currently out-of-print, is a nice, compact summary of all the basic aspects of electro-optics, including a chapter on optical signal processing.

Applications of Optical Fourier Transforms, H. Stark, ed., Academic Press, Orlando, FL (1982)

This is an edited volume of tutorial papers on the theory and application of optical Fourier transforms by many experts in the field. It has extensive references but no problems, since it is intended more as a survey of the field.

Systems and Transforms with Applications to Optics, A. Papoulis, McGraw–Hill, New York (1968)

Although this volume also covers, like the previous one, applications of Fourier transforms, it is considerably more pedagogical in style and written with much more emphasis on the mathematical aspects of the subject. It also contains numerous examples and problems.

Appendix C
SOFTWARE FOR MODELING AND VISUALIZATION*

METHODOLOGY FOR MODELING AND SIMULATION

In this appendix we summarize a modeling and simulation methodology for synthesizing and analyzing optical processing systems, including devices and architectures, as well practical and useful code for implementing the models with MATLAB from MathWorks, Inc. Since MATLAB is a very popular and accessible software language, most of the tools used to enable the student to apply the principles described herein are provided as M-files. In some cases some of the code is in "C", accessible as a (.mex) file from MATLAB. Of course, the most complete and up-to-date information sources on MATLAB software are the reference and users' manuals, as well as handy references by K. Sigmon and by Biran and Breiner. Most of the code included with the accompanying disk was developed from the various sources listed. Additional sources of information are available from the MathWorks, Inc., makers of MATLAB software at their website at http://www.mathworks.com or at their anonymous ftp site at ftp://ftp.mathworks.com. The following website (http://www.jhuapl.edu/authorbook/boone) provides updated information and allows user feedback. Most of the M-files described in the text were originally created using MATLAB version 4.2a for the Macintosh, but should readily work in DOS or UNIX environments. In addition to the following technical description of the application software tools, we provide, when it is essential, programming tips to expedite efficient use of the code.

It should be mentioned that there are other sources of software for modeling optical diffraction and Fourier transform-based optical processing, which were used to generate some of the graphical results in this book, especially in Chapter 1. These include DIFFRAC and FFT Vision.

It might seem ironic to resort to what is essentially digital signal processing to understand optical processing, but of course, digital computers are often used to model and simulate many nondigital systems under development in order to understand all aspects of a system design prior to commiting large sums of money to actually building it. However, our goal is, after all, to understand the signal processing principles and to assess the impact of the unique properties of

---

*The author is grateful for the contribution of O. B. Shukla to the software described in this section.

optical signal propagation through free space as well as optical signal generation, modulation, and detection via the appropriate electro-optical devices.

The overall modeling methodology is depicted in Fig. C.1. Part (a) shows the generic model of a spatial light modulator (SLM). Part (b) shows the core of the generic model of an optical processor. This may be one stage of a multistage architecture as suggested by Fig. C.1(c). It consists of an input plane which may represent an input light source and transparency, or a laser or laser diode array trans-illuminating an SLM. The intermediate plane would typically be an SLM, mask, or lens, and the output plane might be a photodetector or focal plane array (FPA), or possibly another SLM.

The input and output propagators could be one of several, depending on the exact configuration. A particular arrangement is depicted in Fig. C.2, where the intermediate plane is a spherical lens. For this case the possible configurations are (a) Fourier transform, (b) finite conjugate ratio imaging, (c) cascaded telecentric imaging, or (d) any combination of the above in either axis (x or y) normal to the optical (z) axis. The general expressions for the corresponding propagators in the specific case of Fig. C.2 are:

$$u_2(x_2, y_2) = C_1 u_1(x_1, y_1) t(x_1, y_1) * \exp[-jk(x_1^2 + y_1^2)/2d_1]$$

for the input propagator (plane 1 to plane 2) and

$$u_3(x_3, y_3) = C_2 u_2(x_2, y_2) P(x_2, y_2) \exp[-jk(x_2^2 + y_2^2)/2f] * \exp[jk(x_2^2 + y_2^2)/2d_2]$$

for the output propagator (plane 2 to plane 3), where the asterisk denotes convolution.

The functional block diagram corresponding to this case is illustrated in Fig. C.3 with the added input and output device planes—that is, SLM and FPA, also shown. The software is implemented **slm_arch.mfile**. The constants C_1 and C_2 are $\exp(jkd_1)/j\lambda d_1$ and $\exp(jkd_2)/j\lambda d_2$, respectively. Examples of input light distribution functions are

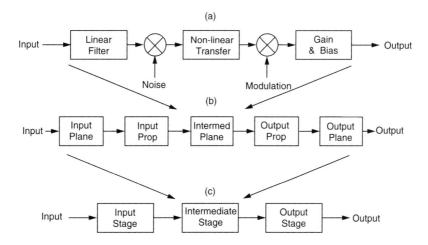

Figure C.1 Modelling methodology for optical signal processing devices and architectures. Part (a) is a generic SLM model, part (b) is the generic single stage model (e.g., that shown below in Fig. C.2), and part (c) is the representation of a cascaded system architecture.

Figure C.2 Example of the simplest optical architecture with input, intermediate, and output planes (or pupils), and object and image spaces where the input and output free-space propagators operate.

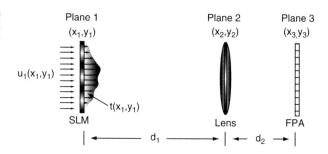

1. *Plane wave:*

$$u(x_1, y_1) = \exp(jkx_1 \sin\theta)$$

where the angle θ is measured in the x–z plane for propagation along the z axis, and the entrance aperture plane and other pupils are defined in x–y coordinates.

2. *Diverging spherical wave:*

$$\frac{1}{j\lambda\rho} \exp[jk(x_1^2 + y_1^2)/2\rho]$$

3. *Converging spherical wave:*

$$\frac{1}{j\lambda\rho} \exp[-jk(x_1^2 + y_1^2)/2\rho]$$

where we have ignored the time-dependent phasor $\exp(j\omega t)$, a fixed phase offset $\exp(j\phi)$, and the exact amplitudes.

The generic model of an SLM or FPA is shown in Fig. C.4. The components include input and output scale factors (i.e., gains, k_1 and k_2) for unit conversions, additive shot noise based on the input signal, and additive thermal noise, which is assumed Gaussian (with a bias). In addition, nonuniform responsivity or transmission noise can be included in the nonlinear signal transfer characteristic (which can thus vary with pixel location). Finally, a linear filter is included to represent the point spread function of the device. This is put after the nonlinear transfer characteristic to allow additional smoothing required to compensate for any high spatial frequencies created by it, particularly for the case where it is a binary threshold operation (see pp. 51–52 in Chapter 2, especially Fig. 2.13).

The parameters used in the model are defined below.

$$s(x, y) = \text{input signal}$$
$$s'(x, y) = \text{output signal}$$
$$h(x, y) = \text{impulse response}$$
$$n_s(x, y) = \text{signal (shot) noise}$$
$$n_G(x, y) = \text{Gaussian (thermal) noise}$$
$$n_R(x, y) = \text{responsivity noise}$$
$$f(s) = \text{nonlinear transfer characteristic}$$

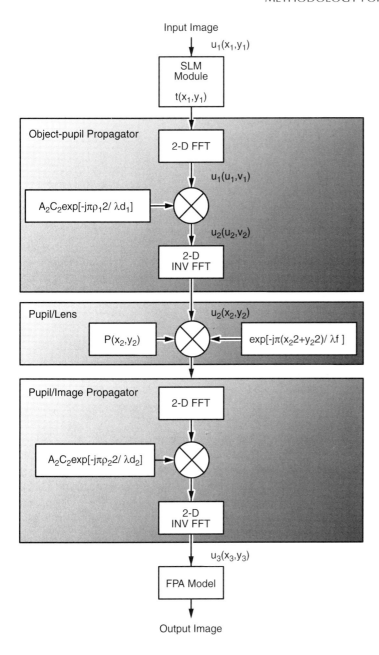

Figure C.3 Flowchart of the basic processes representing the example architecture for modelling and simulation shown in Figure C.2.

The input and output scale factors (k_1 and k_2), as well as the shot noise and Gaussian noise scale factors (k_s and k_G), can be added by the user if desired for unit conversion or to explicitly represent an actual gain factor or variable SNR.

In addition to the generic 2-D SLM model, we can model the 1-D acousto-optic Bragg cell as depicted in Fig. C.5. Here we have illustrated all the basic features described in Chapter 8,

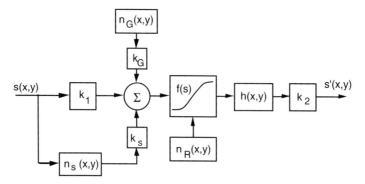

Figure C.4 Generic electro-optical interface model representing either a spatial light modulator or focal plane array.

including input Gaussian beam intensity, Bragg cell aperture, acoustic attenuation, acoustic bandshaping, nonlinear response, heterodyne modulation and filtering, time-to-space scaling, frequency-to-angle scaling, and, of course, diffraction efficiency between the input and first-order beams. However, we will model only the essential spatial modulation effects below.

In setting up a model of an optical processing system, the actual architecture and several components must be specified, including the input plane, intermediate plane, output plane, and type of propagator or transformation desired. The modeler should consider the type of architecture to be modeled, i.e., coherent or noncoherent; time-integrating, space-integrating, or time–space-integrating; as well as the complexity, i.e., single stage (as suggested in Fig. C.2) or cascaded, the number of stages, and the spatial alignment and temporal synchronization requirements.

The input plane specification could, in general, include geometric factors (aperture size, shape, aspect ratio, orientation, obliquity, and apodization) as well as input wavefront characteristics [plane wave or spherical wave, angle of incidence, spatial shape, temporal modulation (duty cycle, pulse width, and pulse repetition frequency), beam width (or waist), polarization (linear, circular, etc.), spectral characteristics (wavelength and bandwidth), and degree of coherence].

The intermediate plane specification could, in general, include the form of addressing (optical or electrical); transducer characteristics (efficiency for amplitude, phase, polarization, diffraction, etc. conversion); input or controlling function (analytical function or user defined look-up table); dimensions (or aperture size, shape, aspect ratio, orientation, obliquity, and apodization); number of pixels and their size, spacing, and dead space; spatial resolution, either as the point spread function or modulation transfer function or as the diffraction limit or contrast measure/criterion; the point transfer characteristic, including its input/output range, dynamic range, functional form (linear or nonlinear), spectral dependence; temporal response; noise characteristics (amplitude, phase, spatial or fixed pattern, temporal, etc.) and associated parameters (mean or bias, variance, and SNR); statistical distribution (Gaussian or non-Gaussian); and frequency content (white or colored).

The output plane specification could include the type of detector (single photodetector, linear self-scanned array, CCD, or CID) or parallel readout 2-D array; the sensing mode (square-law or linear); readout characteristics (frame rate, dwell time, scan regimen—for example,

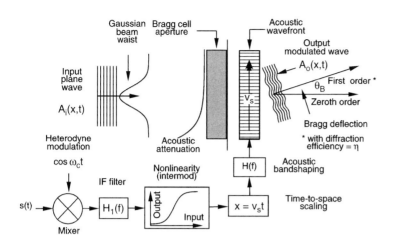

Figure C.5 Diagram of a basic model of the acousto-optic Bragg cell.

TDI mode, frame or line transfer, etc.); as well as all the other characteristics similar to the intermediate plane specifications (dimensions, pixel characteristics, spatial resolution, point transfer characteristic, and noise characteristics).

The propagator specifications should indicate (a) the distance between input and output planes and (b) the optical elements [i.e., type of lenses (circular, cylindrical, conical), their aperture characteristics (size, shape, aspect ratio, orientation, obliquity, apodization, and optical density), and refractive power (including possibly their radii of curvature, index of refraction, focal length, and aberrations)] and nonlenticular elements [e.g., prisms (porro, penta, dove, etc.), beam splitters (polarizing, nonpolarizing), mirrors (reflecting, partially reflecting, flat, curved), and filters (spectral, spatial, and/or neutral density)].

The input test images should be carefully prepared. Are they analytical or user defined? Examples of analytical forms are: rectangular and circular apertures, grating functions, and simple apodization functions. The actual amplitude, phase, or just intensity transmission distribution could be sine or cosine waves, square waves (bar pattern), Gaussian, or random noise patterns (Gaussian or uniform). The actual data could be real or complex, unipolar or bipolar. Therefore the architecture may have to be coherent (amplitude) or incoherent (intensity) with spatial or temporal multiplexing (or some particular coding) to handle complex or bipolar signals with only intensity. Finally: Is it sufficient to represent the incident wave with a given number of pixels (e.g., 512 × 512) and bits (e.g., 8 bits)?

With these as preliminaries, we will now summarize the code included in the disk and describe the function of each component.

LISTINGS OF MATLAB CODE FOR MODELING AND SIMULATION

Often the most effective way to convey a facility with software tools such as MATLAB is simply to expose the student to many examples of programs that perform a basic set of operations. These can then be extended or modified for other situations. Herewith we cover several topics briefly before explaining the basic integrated C-code that enables the student to use the model described in the first part of this appendix. These preliminary topics are:

1. Simple plot routines
2. Common functional forms and their Fourier transforms
3. Fourier transform and convolution
4. Correlation
5. Image filtering
6. Linear transforms
7. Vector–matrix and matrix–matrix multiplication

PLOT ROUTINES

The following plot routines are intended to familiarize the user with various options for plotting analytical functions. Many problems in the text require this capability. Some one-dimensional plots, each demonstrating a different plot style, are:

1. *Chirp waveform:*
   ```
   % Line plot of a chirp
   % Defines a vector of values ranging from 0 to 10 in 0.01 steps
   x = 0:0.01:10;
   y = sin(x.^2);
   plot(x,y)
   grid
   ```

2. *Gaussian plot:*
   ```
   % Bar plot of a Gaussian curve
   x = -2.9:0.2:2.9;
   bar(x,exp(-x.*x));
   ```

3. *Sine wave:*
   ```
   % Stairstep plot of sine wave
   x = 0:0.25:10;
   stairs(x,sin(x));
   grid
   ```

4. *Polar plot:*
   ```
   % Polar plot
   t = 0:.01:2*pi;
   polar(t,abs(sin(2*t).*cos(2*t)));
   ```

5. *Damped chirp waveform:*
   ```
   % Stem plot of exponentially damped chirp
   x = eps 0:0.1:4;
   y = sin(x.^2).*exp(-x);
   stem(x,y)
   ```

6. *Sinc intensity pattern (array factor)*
   ```
   % Sinc-squared array factor function
   N = 6;
   x = 0:0.01:10;
   y = (sin(N*x).*sin(N*x))./(sin(x).*sin(x));
   ```

```
plot(x,y)
grid
```

COMMON FUNCTIONAL FORMS AND THEIR FOURIER TRANSFORMS

Below some more complex (2-D) functional forms are plotted and their corresponding Fourier transforms are calculated and plotted.

1. *Square aperture and Fourier transform*
```
% Square aperture and Fourier transform
sq = zeros(128);     % define a 128 x 128 array filled with zeros
sq(60:67,60:67) = ones(8);   % fill the center with ones to create
an
aperture
figure(1)
mesh(sq)    % display as a 3-D plot
colormap(gray(256)), imagesc(sq)    % can alternatively display as
an image

% extract the real portion of the Fourier transform
sqreft = real(fftshift(fft2(fftshift(sq))));
figure(2)
mesh(sqreft)    % display

% compute the magnitude of the Fourier transform
sqmdft = abs(fftshift(fft2(fftshift(sq))));figure(3)
% display the magnitude of the Fourier transform
mesh(sqmdft)
```
2. *Circular aperture and Fourier transform*
```
% Circular aperture and Fourier transform
% variables for defining a circular aperture
rows = 128;
cols = 128;
r_cen = 64;
c_cen = 64;
radius = 8;
ap = zeros(rows,cols);    % intialize an array filled with zeros

for i = 0:radius,    % for the given radius, fill in a circle positioned
for ang = 0:0.01:2*pi,    % at the center of the array
x = i*cos(ang);
y = i*sin(ang);
ap(r_cen - y,c_cen + x) = 1;
end
end
```

```
figure(1)    % display the aperture
mesh(ap)
colormap(gray(256)), imagesc(ap)
% can alternatively view the aperture in an image format

apreft = real(fftshift(fft2(fftshift(ap))));    % extract the real
portion of
the Fourier transform
figure(2)
colormap(gray(256)), imagesc(apreft)    % display

% compute the magnitude of the Fourier transform
apmdft = abs(fftshift(fft2(fftshift(ap))));
figure(3)
colormap(gray(256)), imagesc(apmdft)    % display the results
colormap(gray(128)), imagesc(apmdft)
```

FOURIER TRANSFORM AND CONVOLUTION

In this section, convolution and fast convolution are employed to generate the point spread function and optical transfer function of simple optical apertures. In addition, the Fresnel transform is introduced using the Fourier transform to demonstrate its applicability to calculating the near-field irradiance from an aperture.

1. *Point spread function of a square aperture by fast convolution:*
```
% Point spread function
% Define a 128 x 128 array filled with zeros
sq = zeros(128);
% Fill the center (8 x 8 square) with ones
sq(60:67,60:67) = ones(8);
figure (1)
colormap(gray(256))
imagesc(sq)

% Compute the Fourier transform
sqft = fftshift(fft2(fftshift(sq)));
colormap(gray(256))
imagesc(real(sqft))    % display the results
figure (2)

% compute the magnitude
psf = abs(sqft);
colormap(gray(256))
imagesc(psf)    % Display the magnitude
figure(3)
sqftph = angle(sqft);    % extract the phase
figure(4)
```

```
     colormap(gray(256))
     imagesc(sqftph)      % display the phase
```
2. *OTF of a square aperture by fast convolution:*
```
   % OTF of square aperture by fast convolution
   % define a 128 x 128 array of zeros
   % define a square aperture in the center of the array
   sq = zeros(128);
   sq(55:75,55:75) = ones(21);
   % display a plane with a clear aperture
   figure(1)
   mesh(sq)

   % can alternatively view the aperture as an image
   colormap (gray(256)), imagesc(sq)

   % compute the 2D Fourier transform
   sqft = fftshift(fft2(fftshift(sq)));
   % compute the magnitude
   psf = (real(sqft)).^2 + (imag(sqft)).^2;
   % display the magnitude
   figure(2)
   mesh(psf)

   % compute the inverse Fourier transform
   otf = fftshift(ifft2(fftshift(psf)));
   % compute the magnitude
   otf = (real(otf)).^2 + ( imag(otf)).^2;

   % display the OTF
   figure(3)
   mesh(otf)
```
3. *OTF of a square aperture by convolution:*
```
   % OTF of square aperture by convolution
   sq = zeros(128);
   sq(49:80,49:80) = ones(32);
   otf = conv2(ones(32),ones(32));
   figure(1)
   mesh (otf)
   colormap(gray(128)), imagesc(oft)
```
4. *Fresnel irradiance using Fourier transform by fast convolution:*
```
   % Fresnel irradiance by Fourier transform
   i=sqrt(-1);    % Make sure that the variable i is imaginary;
   [x,y] = meshdom(-1:1/16:15/16,-1:1/16:15/16);   % Define the
   coordinate space
   chirp2d = (x.^2 + y.^2).    % Compute the argument of the
   exponent
   ap = zeros(128);    % Define a 128 x 128 array of zeroes
   mesh(ap)    % Plot the aperture
   % Apply phase factor to the square aperture
```

```
ap(49:80,49:80) = ones(32).*(exp(i.*2.*pi.*chirp2d));
ft = fftshift(fft2(fftshift(ap)));     % Compute the FFT
fresirr = abs(ft);    % Compute the irradiance
    % Plot the results
colormap(gray(128)), images(fresirr)
```

CORRELATION

In this section an example is given of Cartesian correlation, including full gray-scale, binary correlation, and phase-only correlation. In addition, angular (or polar) correlation is included to support feature extraction.

The following MATLAB code implements the gray-scale, phase-only, and binary phase-only (BPOF) matched filter operation between the letter "G" and the letter "O".

```
% Full gray scale, binary and phase-only correlation
% Define the letter 'O' as input function to match
lettero = zeros(64);
lettero(50:79,55:74) = ones(30,20);
lettero(55:74,60:74) = zeros(20,15);
figure(1)
colormap(gray(256))
imagesc(lettero)

% define the letter 'G' as matching function
letterg = zeros(128);
letterg(50:79,55:74) = ones(30,20);
letterg(55:74,60:74) = zeros(20,15);
letterg(55:59,70:74) = zeros(5,5);
letterg(65:74,70:74) = ones(10,5);
letterg(64:68,66:74) = ones(5,9);
figure(2)
colormap(gray(256))
imagesc(letterg)

% Perform 2-D FFT of the letter ''G'' to obtain filter functions
transx0y0 = letterg;
fieldx1y1 = fftshift(fft2((fftshift(transx0y0))));
maxfield = max(max(abs(fieldx1y1)));
phase = angle(fieldx1y1);
figure(3)
colormap(gray(256))
imagesc(fieldx1y1)
figure(4)
colormap(gray(256))
imagesc(phase)

% Convert to binary phase
binaryphase = pi.*(phase<0);
figure(5)
colormap(gray(256))
```

```
imagesc(binaryphase)

% Using letter ''G'' as input, compute filter functions
pof = exp(j*phase);
bpof = exp(j*binaryphase);
matched_g = fieldx1y1./maxfield;
%matched_po = pof./maxfield;
%matched_bpo = bpof./maxfield;
figure(6)
colormap(gray(256))
imagesc(matched_g)
%figure(7)
%colormap(gray(256))
%imagesc(matched_po)
%figure(8)
%colormap(gray(256))
%imagesc(matched_bpo)

% Choose either letter ''O'' or ''G'' as input to the optical
correlator
% Compute incident power.
% Noise may be specified by the SNR
signoise = 1e1;
% Add Gaussian noise
inputx0y0 = lettero + (1/signoise).*randn(128);
inputpower = sum(sum(inputx0y0.^2))
figure(9)
colormap(gray(256))
imagesc(inputx0y0)

% Using chosen input object, e.g., letter ''O'' or letter ''G'',
% compute output from correlator
% with one of the filters inserted in the filter plane
fieldx1y1 = fftshift(fft2(inputx0y0));
transx1y1 = conj(fieldx1y1.*conj(matched_g));
%transx1y1 = conj(fieldx1y1.*conj(matched_po));
%transx1y1 = conj(fieldx1y1.*conj(matched_bpo));
fieldx2y2 = fftshift(fft2(transx1y1));
amp = abs(fieldx2y2);
corrpower = sum(sum(amp.^2))
intensity = amp.^2;
corrpeak = intensity(65,65)

% Plot correlation plane intensity.
s = [1,1,1];
m = [45,25];
colormap(gray(256))
figure(10)
imagesc(intensity)
% mesh(intensity,m,s)
```

To implement angular auto- and cross-correlation, which are C-code routines that are called as mex files, use the following operations:

```
output = ang_auto_correl(input_image,ang_stepsize)
```
and
```
output = ang_cross_correl(image1,image2,ang_stepsize)
```
where angular step size must be chosen to achieve the user's desired object feature extraction resolution. The second image in the angular cross-correlation operation can be specified to be a slit function in order to recover the object boundary.

IMAGE FILTERING

Image filtering is accomplished simply by using the matched spatial filter correlator in which the intermediate Fourier plane is used to place a particular filter (or mask) to filter the input image, yielding a desired output image. Linear filters such as wavelet transforms can be implemented this way, as well as the effects of nonlinear filters, such as binarization, square-law detection, and sigmoidal transfer characteristics.

The following MATLAB code will generate a simple matched filter process where the intermediate Fourier plane is used for object frequency domain spatial filtering. In this case, a "notch" filter in the y-direction serves to enhance the object function edges in the corresponding output plane. An all-pass filter in the x-direction essentially leaves the other (orthogonal) edges alone.

```
sq = zeros(128);
sq(60:67,56:71) = ones(8,16);
figure(1)
colormap(gray(256)),imagesc(sq)
notch = ones(128);
notch (56:71,1:128) = zeros(16,128);
figure(2)
colormap(gray(256)),imagesc(notch)
sqft = fftshift(fft2(fftshift(sq)));
figure(3)
sqreft = real(sqft);
colormap(gray(256)),imagesc(sqreft)
prod = notch.*conj(sqft);
figure(4)
colormap(gray(256)),imagesc(prod)
filt = real(fftshift(fft2(fftshift(prod))));
figure(5)
colormap(gray(256)),imagesc(filt)
```

Binarization is simply implemented by the following script:
```
thresh = 128;
binary_image = (image> thresh)
```

The effect of binarizing the Fourier transform magnitude can be visualized by the following MATLAB script, which essentially implements the matched filter (inverse Fourier transform) on the binarized magnitude of the 2-D sinc function. The effect will be to suppress the output contrast and the generate aliases by virtue of the fact that binarization actually adds additional high frequency content not in the original Fourier spectrum.

```
sq = zeros(128);
sq(60:67,56:71) = ones(8,16);
figure(1)
colormap(gray(256)),imagesc(sq)
sqft = fftshift(fft2(fftshift(sq)));
figure(2)
sqreft = real(sqft);
colormap(gray(256)),imagesc(sqreft)
binary = (sqreft>0);
figure(3)
colormap(gray(256)),imagesc(binary)
filt = real(fftshift(fft2(fftshift(binary))));
figure(4)
colormap(gray(256)),imagesc(filt)
```
A square-law detection of an entire image field is simply implemented by
```
output = image.*image
```
where image is defined in the same way as the aperture functions described in several of the programs above. After squaring, the output can be normalized as follows:
```
norm_image = output./max(max(output))
```
The nonlinear sigmoidal transfer characteristic defined in Chapter 2, Example 2.3 is implemented as follows:
```
alpha = #
beta = #
gamma = #
xo = #
sigmoid = alpha./(1 + exp(gamma.*(image - xo))) + beta
```
where the user specifies the desired parameters **alpha, beta, gamma,** and **xo**. Or the user can simply write the following single-line instruction:
```
output = sigmoid(input_image, alpha, beta, gamma, xo)
```
  We can systematically explore the signal distorting (or enhancing) effects of the sigmoidal transfer characteristic used in the Fourier domain on an intermediate plane (spatial light modulating) signal by looking at the difference in the modulation signal in both the spatial and spatial frequency domains and the sigmoid's effect on the reconstructed signal. The following script is a start on such an analysis.
```
[x,y] = meshgrid(-16:1:16,-16:1:16);
mod = 1.*cos(x./4).*cos(y./4);
alpha = 1;
beta = 0;
gamma = 1;
xo = 0;
output = alpha./(1+exp(gamma.*(mod - xo))) + beta;
figure(1)
colormap(gray(256)),imagesc(mod)
figure(2)
colormap(gray(256)),imagesc(output)
difference = abs(output- mod);
figure(3)
colormap(gray(256)),imagesc(difference)
sqreft1 = real(fftshift(fft2(fftshift(mod))));
```

```
figure(4)
colormap(gray(256)),imagesc(sqreft1)
sqreft2 = real(fftshift(fft2(fftshift(output))));
figure(5)
colormap(gray(256)),imagesc(sqreft2)
sqreft1log = log(sqreft1);
sqreft2log = log(sqreft2);
recon = real(fftshift(fft2(fftshift(sqreft2))));
figure(6)
colormap(gray(256)),imagesc(recon)
```

To implement a wavelet transform to place it in the Fourier plane of a matched filter architecture consider the 2-D Gabor filter in the spatial frequency domain as defined in Equation (3.87). In MATLAB script it is

```
i = sqrt(-1);
pi = 3.14159;
xo = 32;
yo = 32;
uo = 1;
vo = .5;
alpha = 16;
beta = 8;
[x,y] = meshgrid(0:1:64,0:1:64);
arg1 = -pi*(((x - xo).^2)/alpha.^2 + ((y - yo).^2)/beta.^2);
factor1 = exp(arg1);
factor2 = (cos(uo.*(x - xo))).*(cos(vo.*(y - xo)));
gabor = factor1.*factor2;
figure(1)
colormap(gray(256)),imagesc(factor1)
figure(2)
colormap(gray(256)),imagesc(factor2)
figure(3)
colormap(gray(256)),imagesc(gabor)
```

This result can then be input into the simple BPOF matched filter model described above, or the OSP architecture simulation described below. The filter can be repeatedly applied as the parameters **jo, ko, xo, yo, alpha,** and **beta** are varied (corresponding to the parameters of Equation (3.87)). A sequence of outputs is required to fully represent a 2-D wavelet transform because it yields a 4-D output. Simpler 1-D wavelet filters that yield 2-D output functions can also be defined.

LINEAR TRANSFORMS

In this section examples are given of code to implement the Radon, Hough, and Mellin transforms for feature extraction. The Radon transform is implemented by the M-file defined as

```
% Create square object field
object=0.0*ones(64,64);
% Create two squares of ones
object(32--8:32+8,16--8:16+8)=1*ones(17,17);
object(32--8:32+8,48--8:48+8)=1*ones(17,17);
```

```
% Plot object
colormap(jet(64)),imagesc(abs(object));colorbar
title('Test Object');
pause
% Compute Radon transform
theta=0:1:360;
P=RADON(object,theta);
colormap(jet(64)),imagesc(abs(P));colorbar
pause
```

where the angle to compute the projection of the input image is all that is required to be specified other than the input image (or object). The Radon transform can then be plotted in an image or mesh format. The Hough transform is the Radon transform of a binarized function and is implemented in the same way, namely,

```
binary_image = (input_image>0)
output = radon(input_image, angle)
```

The Mellin transform can be used to calculate the moments of a function. One of the simplest examples of moments is the 1-D moment of a Gaussian function given by the following MATLAB script.

```
i = [0:1:4]';
for j = 0:4;
x = [-8:.5:8]';
integrand = (exp(-x.^2)).*(x.^j);
y(j+1) = trapz(x,integrand);
end;
stem(i,y)
```

VECTOR–MATRIX AND MATRIX–MATRIX MULTIPLICATION

A simple inner (scalar) product operation is defined by **z** = **x.*y**, where **x** and **y** are vectors defined in row and column formats, respectively. If **x** and **y** are matrices, say 10×10 and 10×20, then the matrix–matrix operation is also given by **z** = **x*y**, and is a 10×20 matrix. If we define **x'** as the transpose of **x**, then the outer product is **z** = **x*y'**. Finally, the vector–matrix operation is **z** = **x.*y**. These are very simple MATLAB operations because it is naturally based on matrices.

An example of interest to optical signal processing is the Walsh–Hadamard transform, which can be modeled with the following code. In MATLAB the command **hadamard (N)** creates the bipolar Hadamard matrix of size $N \times N$. Using this, the full set Walsh–Hadamard weight matricies for an 8×8 input image size can be created with the following m-file:

```
h = hadamard(8);
for k = 1:8
for j = 1:8
A = h(:,j)*h(:,k)';
end
end
C = [h(:,1)*h(:,1)'  h(:,1)*h(:,2)'  h(:,1)*h(:,3)'  h(:,1)*h(:,4)'
h(:,1)*h(:,5)'  h(:,1)*h(:,6)'  h(:,1)*h(:,7)'  h(:,1)*h(:,8)';
h(:,2)*h(:,1)'  h(:,2)*h(:,2)'  h(:,2)*h(:,3)'  h(:,2)*h(:,4)'
h(:,2)*h(:,5)'  h(:,2)*h(:,6)'  h(:,2)*h(:,7)'  h(:,2)*h(:,8)';
```

```
                h(:,3)*h(:,1)'  h(:,3)*h(:,2)'  h(:,3)*h(:,3)'  h(:,3)*h(:,4)'
                h(:,3)*h(:,5)'  h(:,3)*h(:,6)'  h(:,3)*h(:,7)'  h(:,3)*h(:,8)';
                h(:,4)*h(:,1)'  h(:,4)*h(:,2)'  h(:,4)*h(:,3)'  h(:,4)*h(:,4)'
                h(:,4)*h(:,5)'  h(:,4)*h(:,6)'  h(:,4)*h(:,7)'  h(:,4)*h(:,8)';
                h(:,5)*h(:,1)'  h(:,5)*h(:,2)'  h(:,5)*h(:,3)'  h(:,5)*h(:,4)'
                h(:,5)*h(:,5)'  h(:,5)*h(:,6)'  h(:,5)*h(:,7)'  h(:,5)*h(:,8)';
                h(:,6)*h(:,1)'  h(:,6)*h(:,2)'  h(:,6)*h(:,3)'  h(:,6)*h(:,4)'
                h(:,6)*h(:,5)'  h(:,6)*h(:,6)'  h(:,6)*h(:,7)'  h(:,6)*h(:,8)';
                h(:,7)*h(:,1)'  h(:,7)*h(:,2)'  h(:,7)*h(:,3)'  h(:,7)*h(:,4)'
                h(:,7)*h(:,5)'  h(:,7)*h(:,6)'  h(:,7)*h(:,7)'  h(:,7)*h(:,8)';
                h(:,8)*h(:,1)'  h(:,8)*h(:,2)'  h(:,8)*h(:,3)'  h(:,8)*h(:,4)'
                h(:,8)*h(:,5)'  h(:,8)*h(:,6)'  h(:,8)*h(:,7)'  h(:,8)*h(:,8)'];
                D = (C > 0);
                colormap(gray(128)),imagesc(D);
```

The student, however, will want to generate a sequence-ordered mask to easily compare with the example in Chapter 12. Subsequently, the actual weights to represent a given input image or function must be generated. This will be left as an exercise for the student.

OPTICAL PROCESSING ARCHITECTURE SIMULATION

The optical processor simulation employs several basic components that can be called repeatedly to implement cascaded operations. These components include a free-space propagator and a spatial light modulator (SLM), which incorporates an option to represent a square-law detection operation to represent a focal plane array. A separate one-dimensional model is used to represent an AO Bragg cell.

PROPAGATOR SPECIFICATIONS

The propagator function described in Fig. C.3 is defined as part of the MATLAB M-file given by
   output = output_field(input_field,d,lam,xmin,xmax,ymin,ymax)
where the **input_field** is that matrix defining the incident field on the input plane (and specified by the user as empirical data, or an analytic form such as a plane wave that would be generated by a laser and collimator lens). The parameter **d** is the propagation distance. The parameter **lam** is the wavelength, and the x, y values (**xmin,xmax,ymin,ymax**) correspond to the extremal coordinates of the input plane. The output is a matrix that defines the output field for subsequent processing. This propagator can be repeatedly invoked for each free-space propagation step, with intermediate plane phase and/or amplitude modulations imposed by lenses or spatial light modulators (SLM's).

SLM DEFINITION

The function defined by
   output = slm_model(input_field,input_mod_fcn,noise_prod,noise_add)
specifies the basic elements of the SLM, including the input light field, the electrical modulating signal, multiplicative noise representing the non-uniformity of response, and additive noise.

The input modulation function actually represents a generic, potentially nonlinear function of the input signal in which the multiplicative and additive noise are imposed after the nonlinearity is applied to the input field. The noise models available include uniform, Gaussian, and Poisson noise fields with user specified means and standard deviations. The nonlinearities include simple binarization, square-law detection, and a sigmoidal transfer characteristic.

An example of a sequence of instructions to implement the architecture of Fig. C.2 is as follows:

```
% Make sure that the variable i is imaginary;
i = sqrt(-1);
% Define a 128 x 128 array filled with zeros
sq = zeros(128)
% Define square aperture with ones
sq(60:67,60:67) = ones(8);
% Calculates SLM output given input irradiance field
% Specify input modulation function, product noise and additive noise
slm_out = slm_model(input_field,input_mod_fcn,noise_prod,noise_add)
output = output_field(slm_out,d1,lam,xmin,xmax,ymin,ymax)

% Define the coordinate space
[x,y] = meshgrid(-64:1:63,-64:1:63);
% Compute phase delay of the lens
phase = (x.^2 + y.^2);

% Define lens pupil by first intializing an array filled with zeros
rows = 128;
cols = 128;
r_cen = 64;
c_cen = 64;
radius = 8;
pupil = zeros(rows,cols);

% For a given radus, fill in a circle of ones positioned at the center of the array
for i= 0:radius,
for ang = 0:0.01:2*pi,
x = i*cos(ang);
y = i*sin(ang);
ap(r_cen - y,c_cen + x) = 1;
end
end

% display the aperture
figure(1)
colormap(gray(256))
imagesc(ap)
% Compute lens phase factor
lens = pupil.*exp(i.2.pi.phase);
```

```
input_field = lens.*output;
amp = output_field(input_field,d2,lam,xmin,xmax,ymin,ymax);
mx_amp = max(max(amp));

% Plot output amplitude image
colormap(gray(256))
imagesc(amp)

% Calculate detected (focal plane) image including Gaussian noise
% Use k factor to scale noise relative to maximum amplitude at
focal plane
fpa = amp.^2 + (k.*mx_amp).^2.*randn(size(amp));
norm_image = fpa./max(max(fpa))

% Plot normalized focal plane output
colormap(gray(256))
imagesc(norm_image)
```

ACOUSTO-OPTIC BRAGG CELL MODEL

The Bragg cell model is a 1-D model that is implemented with the following command:
```
ao_out = aocell(gauss_beam, aperture, mod_signal,
ao_atten,diff_eff)
```
where **mod_signal** is the spatial equivalent of the time-domain modulating signal that is applied to the transducer port. It will in reality be a heterodyned signal and will be affected by mixer and acoustic nonlinearities, transducer insertion loss due to mismatch, and bandpass filtering effects of the electrical prefilters and the input transducer, but we specify here only the spatial functions and parameters that directly affect the spatial shape of the diffraction efficiency into the first-order beam, that is, Gaussian beam shape of the input laser field, acoustic attenuation effect, and the possible truncation of the Bragg cell aperture.

Therefore we must define **gauss_beam** in terms of a suitable mean and variance relative to the Bragg cell aperture, **aperture** as the actual Bragg cell length, **ao-atten** as the acoustic attenuation constant of the AO cell, and **diff_eff** as the relative power into the output beam. Electrical band-shaping effects can be modeled as a pre-processing step through convolution or the FFT-based filtering code described earlier (in 1-D).

BIBLIOGRAPHY

A. Biran and M. Breiner, *MATLab for Engineers*, Addison-Wesley, Reading, MA (1995).

M. A. Heald, "Computation of Fresnel Diffraction," *Am. J. Phys.* **54**(1), 980–983 (1986).

MATLAB Reference Guide: High-Performance Numeric Computation and Visualization Software, The MathWorks, Inc., South Natick, MA (1992–1993).

The Student Edition of MATLAB User's Guide: High-Performance Numeric Computation and Visualization Software, The Math Works, Inc., South Natick, MA (1995).

O. B. Shukla, "Development of an Algorithm to Simulate Optical Processing Arcitectures," Part I, The Johns Hopkins University G. W. C. Whiting School of Engineering Part-Time Programs in Applied Science and Engineering, Special Projects Course Report, August 2, 1988.

O. B. Shukla, "Development of an Algorithm to Simulate Optical Processing Arcitectures," Part II, The Johns Hopkins University G. W. C. Whiting School of Engineering Part-Time Programs in Applied Science and Engineering, Special Projects Course Report, September 29, 1988.

O. B. Shukla, "Feasibility of Using a Liquid Crystal Television (LCTV) for Optical Applications," JHU/APL Technical Memorandum # F1F(1)89-U-236, December 1, 1989.

O. B. Shukla, "Modeling and Simulation of the Nonlinear Response of Spatial Light Modulators," JHU/APL Technical Memorandum # F1F(1)90-U-172, June 3, 1990.

K. Sigmon, *MatLab Primer*, CRC Press, Boca Raton, FL (1994).

R. G. Wilson, S. M. McCreary, and J. L. Thompson, "Optical Transformations in Three-Space: Simulations with a PC," *Am. J. Phys.* **60**(1), 49–56 (1992).

R. G. Wilson, *Fourier Series and Optical Transform Techniques in Contemporary Optics, An Introduction*, John Wiley & Sons, New York (1995).

Appendix D
HINTS AND SOLUTIONS TO SELECTED PROBLEMS

In almost any technical field problem solving is an essential part of instruction and self-learning. There are many types of problems, including:

1. Purely numerical substitution and calculation problems
2. Analytical derivation problems in which the desired result is known
3. Analytical derivation problems in which the desired result is unknown
4. Problems whose results are best determined through computer modeling
5. Problems requiring some literature research

Each type of problem has its utility and relevance, and an attempt has been made to provide some of each type in this book. Generally, for calculation problems, solutions are not frequently provided because they are straightforward and are usually intended to familiarize the student with typical values and physical dimensions of parameters and results. Analytical derivation problems for which the result is stated are also not frequently worked-out in detail, because that is often the *whole* purpose of the problem: to discover the best and most appropriate methodology for solving it. This is also true when the desired result is not indicated, but because such problems give the student even fewer clues to the approach and end result, some assistance is provided here. For computer modeling problems the necessary tools have been provided in Appendix C. Finally, some problems require that the student engage in a literature search to find a helpful approach or a discussion of similar problems. Obviously, the bibliography sections at the end of each chapter are provided for this purpose. Part of the rationale for the this text's approach to *selected* hints and solutions is to convey to the student the experience of research, to wit, "going down alleys to find they are blind."

CHAPTER 1

1. (a) Recall that

$$\mathrm{comb}(x) = 2\pi \lim_{N \to \infty} \sum_{n=N}^{N} \delta(n - x)$$

and that
$$\sum_{n=1}^{N} a^n = \frac{1-a^N}{1-a}$$

(b) Use
$$\delta(x) = \int_{-\infty}^{\infty} \exp(-j2\pi kx)\, dk$$

Then use
$$\int_{-\infty}^{\infty} \delta[f(x)]\, dx = \sum_{k} \frac{1}{|f'(x_k)|}$$

(c) Look at the right-hand side, which is
$$2\pi \lim_{x \to \infty} \int_{0}^{x} J_o(2\pi \rho r)\, dr$$

Express $\delta(u, v)$ defined as the model:
$$\lim_{N \to \infty} \frac{N J_1[2\pi N (u^2 + v^2)^{1/2}]}{(u^2 + v^2)^{1/2}}$$

in polar form. Express
$$\int_{-\infty}^{\infty} \delta(u, v)\, du\, dv = 1$$

in polar form. $\delta(u, v)$ must $\to \delta(\rho)/\pi\rho$ to preserve the property of the Dirac delta function, such that
$$\int_{-\infty}^{\infty} \delta(u, v)\, du\, dv \neq 0$$

due to the Jacobian of transformation.

5. Use the convolution theorem to get
$$G(u, v) = F(u, v) \sum_{n} \sum_{m} \delta\left(u - \frac{n}{x_0}, u - \frac{m}{y_0}\right)$$

7. (a) The Gaussian is separable in x and y. Complete the square in the argument of the exponential. Self-imaging.

10. First note that:
$$\exp[-j2\pi r\rho \cos(\theta - \phi)] = \exp\{-j2\pi r\rho \sin[[\theta - \phi] + \pi/2]\}$$

Then expand this as a Bessel series and substitute in the Fourier–Bessel integrand and use the definition of the Dirac delta function to simplify it to
$$G(\rho, \phi) = 2\pi \int_{0}^{1} r\, dr\, J_1(2\pi \rho r) \cos \phi$$

Check the limiting form of the series representations for the result for $\rho \to 0$. Look at the result only along the x and y axes and then along radial lines to visualize your results.

13.
$$\mathbf{A} = \begin{bmatrix} \cos\theta & \sin\theta \\ -\sin\theta & \cos\theta \end{bmatrix} \Rightarrow \mathbf{A}_T^{-1} = \begin{bmatrix} \cos\theta & \sin\theta \\ -\sin\theta & \cos\theta \end{bmatrix}$$

Calculate: $\mathbf{x}_T \cdot \mathbf{A}_T^{-1} \mathbf{u}'$ and substitute into the exponential argument of the Fourier transform integral.

18. Use the equivalent of the convolution theorem for Hankel transforms to show

$$H[\delta(r-a)] = 2\pi a J_o(2\pi a\rho)$$

21. Use the polar form of $t = j\exp(jx)$ and express $\exp[(t - 1/t)a/2]$ as a Bessel series. Then take the Fourier transform of both sides.

22. For this problem, the unique part of the solution can be addressed by just looking at the convolution and correlation in the x axis, since in the y axis the convolution and correlation are the same (yielding a triangle function whose base is $2y_o$ for equal rect functions), and $f(x, y)$ is separable. Then the convolution in x is

$$z(x) = \int_{-\infty}^{\infty} f(x') \operatorname{rect}\left(\frac{x - x'}{x_o}\right) dx'$$

$$z(x) = \begin{cases} 0 & x < 0 \\ x^2/2 & 0 \leq x < x_o \\ x_o x - x^2/2 & x_o \leq x < 2x_o \\ x_o x - x^2/2 + 3x_o^2/2 & 2x_o \leq x < 3x_o \\ 0 & x \geq 3x_o \end{cases}$$

The correlation is similar but without the coordinate inversion; that is,

$$R(x) = \int_{-\infty}^{\infty} f(x') \operatorname{rect}\left(\frac{x' - x}{x_o}\right) dx'$$

CHAPTER 2

3. First note that the arcsine function is odd. Therefore,

$$\sin^{-1}[R_{xx}(\tau)] = \sum_{n=1}^{\infty} b_n \sin[n\pi R_{xx}(\tau)]$$

Integrate by parts to explicitly derive b_n, substitute, and then take the Fourier transform to get the power spectrum.

5. Using Equation (2.36), we obtain

$$\overline{y^2} = \frac{1}{2\pi} \int_{-\infty}^{\infty} |H(\omega)|^2 \frac{N}{2} d\omega$$

the peak value of $H(\omega)$ is $H(0)$, which is the height of the rectangular region equivalent in area to that of $H(\omega)$. The width is the noise equivalent bandwidth (NEBW), where

$$\text{NEBW} = \int_0^{\infty} \frac{|H(\omega)|^2}{|H(0)|^2} d\omega$$

Therefore,

$$\overline{y^2} = |H(0)|^2 \frac{N}{2} \text{NEBW}$$

Given $h(t)$, then
$$H(\omega) = \frac{\alpha H(0)}{j\omega + \alpha}$$
Therefore, NEBW $= \pi\alpha/2$, whereas the 3-dB BW $= \alpha$, where $|H(\alpha)|^2 = |H(\omega)|^2/2$. Therefore,
$$\frac{\text{NEBW}}{\text{3-db BW}} = \frac{\pi}{2}$$

8. $s(x, y)$ and $h(x, y)$ are uncorrelated. Therefore,
$$R_{rr}(x', y') = +R_{ss}(x', y') + R_{nn}(x', y') + \text{cross-terms}$$
Show that cross-terms are zero. Note that
$$R_{nn}(x', y') = \sigma^2 \delta(x', y')$$
for white noise.

9. The correlation $R_{ss}(x', y')$ can be defined as a filtering process represented formally by the expression
$$R_{ss}(x', y') = f(x', y') \otimes h(x', y')$$
which is an integral, where
$$f(x', y') = s(x', y') + n(x', y')$$
and
$$h(x', y') = s^*(-x', -y')$$
You must complete the square in the exponential argument of the integrand.

CHAPTER 3

1. Since $f(x, y)$ is separable, only a 1-D Fresnel transform needs to be done. Expand $\sin(2\pi ux)$ in terms of exponentials. You must complete the square on the complete integrand of the inverse Fourier transform.

2. Expand the denominator of the integrand as a partial fraction and change variables.

3. Recall that $H(f) = G(f) + jB(f)$, where $G(f)$ is even and $B(f)$ is odd. Use the original definition of the Hilbert transform pair [Equations (3.55) and (3.56)] and the above-mentioned symmetry properties.

4. Transform to polar coordinates.

10. Recall the Mellin transform:
$$F_M(s) = \int_{-\infty}^{\infty} f(x) x^{s-1} \, dx \quad \Rightarrow$$

$$F_M(1) = \sqrt{\pi} \text{ (area)}, \quad F_M(2) = 0 \text{ (mean)}, \quad F_M(3) = \frac{\sqrt{\pi}}{2} \text{ (variance)}$$

for $f(x) = \exp(-x^2)$, and so on

11. The Mellin transform is

$$F_M\{f_1(ax)\} \equiv M_2(\omega) = \int_0^\infty f(x) x^{j\omega-1} \, dx$$

Then changing variables yields

$$|M_a(\omega)| = \left| \frac{2}{\omega} \sin\left[\frac{\omega}{4} \ln\left(\frac{x_o}{2}\right)\right] \right|$$

$$R_{M_aM}(x') = \text{FT}^1\{M_a^*(\omega)\} = R_{MM}(x' - \ln a)$$

Therefore,

$$R_{M_aM}(x') = \text{tri}\left(\frac{x}{x_o}\right) \otimes \delta(x - \ln a)$$

for

$$f_1(x) = \text{rect}\left(\frac{x}{x_o}\right)$$

13. Use the projection-slice theorem, the convolution theorem, and the uniqueness of the Fourier transform pair.

16. The Abel transform is

$$f_A(x) = 2 \int_x^\infty \frac{\frac{\rho}{\pi r} J_1(2\pi \rho r)}{\sqrt{r^2 - x^2}} r \, dr$$

Change variables and use a table of integrals. Reexpress the Bessel functions in trigonometric form, or use equivalent of projection-slice theorem.

17. Start with

$$f(u, v) = \int_{-\infty}^\infty dx \int_{-\infty}^\infty dy \int_{-\infty}^\infty dt \, f(x, y) \exp(-j2\pi t) \delta(t - ux - vy)$$

19. Use a convergence factor to make the integrands integrable.

20. Use Equation (3.61) where the expression for a single line is

$$\delta(p_o - x \cos \phi_o - y \sin \phi_o)$$

provided that $p \neq p_o$ and $\phi \neq \phi_o$, and a cutoff is included in the form of a circ function, although it can be left in the general form of a function $A(x, y)$. This functional form will emerge from the integral with arguments determined by the two lines (the Radon kernel and the line to be transformed). To evaluate and interpret your results, consider the integral for the case where $\phi = \phi_o$ but $p \neq p_o$ and $A = 1$ along the single line $y = (p_o - x \cos \phi_o)/ \sin \phi_o$.

24. Use Equation (3.85):

$$W(a_x, a_y, b_x, b_y) = (a_x a_y)^{1/2} \int_{-\infty}^{\infty} \int_{-\infty}^{\infty} S(u,v) H(a_x u, a_y v)$$
$$\cdot \exp[-j2\pi(b_x u + B_y v)]\, du dv$$

For 1-D the $H(a_x u, a_y v)$ are given by Equations (3.89), (3.91), and (3.93). You may wish to consult the references on wavelets, as needed.

CHAPTER 4

1. The magnification of the telescope is determined by $M = \beta/\alpha$. β corresponds to the eye ($\sim 4 \times 10^{-4}$ rad), and α corresponds to the objective. $\alpha = 1.22\lambda/D_o$ by diffraction.

3. Use the Fresnel equations at normal incidence. Also note Equation (4.31) in the limit as $\alpha \to 0$, substituting for R and T the appropriate expressions.

5. In general, $v_g = d\omega/dk$ and $\omega = ck$, where v_p = phase velocity $= c/n$. Therefore, $\omega = ck/n$ embodies the dispersion.

9. The E fields are
$$E_x = E_{ox} \cos(kz - \omega t)$$
and
$$E_y = E_{oy} \cos(kz - \omega t + \phi)$$

Expand E_y and isolate ϕ, eliminate any explicit dependence on $kz - \omega t$, and thus arrive at an equation for the ellipse at an orientation α with respect to the x axis from which α can be derived.

10. From geometry $\phi + (\pi/2 + \theta_r) + (\pi/2 - \theta_r) = \pi$, where $\theta_r = \phi/2$ and $\delta_m = 2\beta$. Then use Snell's law.

12. Use the Lensmaker's formula. There are two possible solutions.

14. Use the Gaussian lens formula and the constraint of common object and image planes for both lenses.

16. Consider an annular differential source of area ΔA and radiance B in the (x, y) plane seen from a point P at R as shown below.

The radius of the ring is
$$z = r_o \tan \theta$$
from which dz can be determined and hence the source differential area $dA_s = 2\pi z dz$. Thus the irradiance at the receiver is
$$dH = B_o \frac{\cos^4 \theta}{r_o^2} dA_s$$

382 HINTS AND SOLUTIONS TO SELECTED PROBLEMS

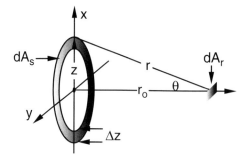

So

$$H = \int_0^{\theta_{max}} dH$$

where

$$\theta_{max} = \arcsin\left(\frac{D^2/4}{F^2 + D^2/4}\right)$$

For small D, compare to

$$\frac{\pi B_o D^2/4}{f^2}$$

18. This is known as the *Brewster anamorph pair*. Both prisms are tilted away from the mid-position between them by equal but opposite amounts.

21. Use the paraxial ray approximation and look at the argument of the plane wave expression for lines that make the phasor zero.

CHAPTER 5

1. Use the paraxial ray approximation. In the expression for the plane wave—that is, $\exp(j\mathbf{k} \cdot \mathbf{R})$—approximate the scalar product as a sum of linear terms.

3. Follow the analysis for the two-slit case (pp. 118–119) but define three phase differences and reduce expression to a simplified trigonometric form.

5. Transition to the far-field occurs at $z \gg D^2/2\lambda$, but the transition between Fresnel and Fraunhofer occurs at $z \sim \pi D^2/\lambda$.

9. The aperture at $z = 0$ is

$$\text{circ}\left(\frac{\sqrt{x_1^2 + y_1^2}}{D}\right)$$

For N locations,

$$t(x_1, y_1) = \left[\text{circ}\left(\frac{\sqrt{x_1^2 + y_1^2}}{D}\right) * \delta(y_1) \sum_n \delta(x_1 - nd) \right] \text{rect}\left(\frac{x_1}{Nd}\right)$$

To get the far-field pattern you must calculate

$$I(x_o, y_o) = |U(x_o, y_o)|^2$$

where $U(x_o, y_o)$ is given by Equation (5.34). Look at the array and element factors. Use the convolution theorem and the property of replication.

11. To get the total diffraction pattern intensity, you must use the linear superposition of $I(P)$ for radius a aperture and radius b stop.

13. (a) Express $t(x, y)$ in phasor form. Use Equation (5.29) and the convolution theorem. Then to get the Fresenel pattern apply inverse Fourier transform to the previous result.

    (b) Simplify the expression for the field for terms quadratic in f_o using de Moivre's formula. Then approximate for $m \ll 1/2$ and interpret your results.

15. Define $\mathbf{A} = \mathbf{r}_{o1}$ and $\mathbf{B} = \hat{\mathbf{n}}$ for the diagram shown below.

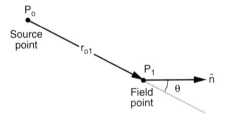

CHAPTER 6

2. The object plane is bounded by

$$t_o(x_o, y_o) = \text{circ}\left(\frac{\sqrt{x_o^2 + y_o^2}}{d}\right)$$

and the pupil function, $P(x, y)$, is $t_o(x_o + x_f, y_o + y_f)$. It is not necessary to solve for the expression of Equation (6.20). Just consider the geometry of the single lens Fourier transform configuration and interpret Equation (6.20).

3. Specify the expressions for the field at planes P_1, P_2, and P_3. Look at the expression for P_2 and assume its pupil can be ignored. Complete the square in the exponential argument and reduce the result to get the power spectrum for $t_1(x', y')$.

4. Apply the two-lens combination formula along the y axis with focal lengths f_1 and f_{2y}. Apply the constraint that the sum of object and image distances (s_o and s_i) for a single lens acting in the x axis

must equal $f_1 + f_{2y} + d$ with the proviso that s_o is consistent with imaging in the y axis also, even though the Fourier transform magnitude is determined in the y axis.

5. Equivalent transmittance to "any variations of intensity" is the first harmonic; that is, $t_1(x) = \cos(2\pi f x)$. Then look at the Fourier transform of $t_1(x)$ in each case.

8. (a) Express $t(r)$ in terms of exponentials. (b) Recall that for a lens, α must equal $k/2f$.

9. Consult E. L. O'Neill, "Transfer Function for an Annular Aperture," *JOSA* **46**, pp. 285–288 (1956); and errata, *JOSA* **46**, p. 1096 (1956).

10. A monochromatic plane wave given by $\exp(j\mathbf{k} \cdot \mathbf{r}_o)$ at oblique illumination has $\mathbf{k}$ given by

$$k\cos\theta\hat{\mathbf{i}} + k\sin\theta\hat{\mathbf{j}}$$

Use this in the expression for $T(f_x, f_y)$. You should get three terms. Use the coherent cutoff spatial frequency $\rho_o = 1/2\lambda d_i$ and the transfer function. Calculate the intensity at maximum half-angle set by the lens diameter and $2f$ (distance between object and "image" planes).

18. Look at the modulus squared of the field with and without spectral averaging when

$$u_1(x_1, y_1) = \cos(2\pi u_o x_1) \cdot 1 \cdot \text{rect}\left(\frac{x_1}{x_o}\right)$$

that is, it is sufficient to consider one dimension.

CHAPTER 7

1. Use the relations

$$l_c = \frac{\lambda^2}{\Delta\lambda}, \quad \lambda = c/f, \quad \text{and} \quad |\Delta\lambda| = c\frac{\Delta f}{f^2}$$

2. Calculate the Fourier transform of $E(t)$, viz., $E(\omega)$ to get $|E(\omega)|^2/|E(0)|^2$.

3. Model the Fabry–Perot cavity with reflectances r_1 and r_2 and transmittances t_1 and t_2 to yield $E_{\text{out}}/E_{\text{in}}$. Equate E_{out} and E_{in} in the condition of oscillation.

12. Referring to the charge transfer model, we seek the transfer function $H(\omega)$ relating input q_{in} and output q_{out} for n transfers. Write down the corresponding difference equation for the kth cell at the nth point in time:

$$q_{k+1}(nT) = n\varepsilon q_{k+1}(nT - T) + (1 - n\varepsilon)q_k(nT - T)$$

This can be solved for $H(z)$ such that for small enough ε, $n\varepsilon$ is also small enough to allow the approximation that

$$z_n = \frac{1 - n\varepsilon z^{-1}}{1 - n\varepsilon} \cong [1 + n\varepsilon(1 - z^{-1})]z = \exp[-n\varepsilon(1 - z)]$$

In the z plane we have $q_{\text{in}}(z) = q_{\text{out}}(z)z^{-N}$ for N cells (or delays). Therefore,

$$H(z) = z_n^{-N}$$

so that $H(\omega) = H(z)$ on the unit circle; that is, $z = \exp(-j\omega T)$

CHAPTER 8

1.
$$I(f_x) = \left| \int_{-\infty}^{\infty} U_{in}(x) \exp(-j2\pi f_x x) \, dx \right|^2$$

where
$$U_{in}(x) = \sqrt{\eta_o}\,\text{rect}\left(\frac{x - D/2}{D}\right) \frac{1}{\sqrt{\pi}r_o} \exp\left[-\frac{(x-D/2)^2}{r_o^2}\right]$$

2.
$$U_a(x) = 10^{-\alpha f^2 x/20V} \quad \text{or} \quad \exp(-\rho x)$$

where
$$\rho = \frac{\alpha f^2}{20V \log_{10} e}$$

Then evaluate
$$U_{out} = \sqrt{\eta_o}\, U_{in} U_a$$

6. Refer to D. L. Hecht, "Spectrum Analysis Using Acousto-optic Devices," *Opt. Eng.* **16**, 461 (1977) and use
$$\eta = \eta_1 \left(1 - \frac{1}{3}\eta_1 - \frac{2}{3}\eta_2\right)$$

where η_1 is the diffraction efficiency for frequency 1, and η_2 is the diffraction efficiency for frequency 2.

14. The output wave is similar in form to the spectrum of an FM system. See the discussion on pp. 57–58, Chapter 2.

CHAPTER 9

1. **(a)** Start with getting the result
$$a(t) \cong a_o \exp(j\omega t)[1 + \alpha s_1(x + v_s t)]$$

   (b) Calculate
$$\text{FT}\left\{a(t)\text{rect}\left(\frac{t - T/2}{T}\right)\right\}$$

(c) Then
$$i(x) \cong ja_o \exp(j\omega t)\alpha s_1(x + v_s t)$$
(e) Finally calculate $FT\{i'(x)\}$
(f) First calculate $FT\{i(x)\}|_{u=0}$, then look at the cross-term.

3. Use Equation (9.19).

4. Simply follow through the analysis on pp. 243–244 using $s(t)$ from Problem 3.

6. Final lens Fourier transforms the function $E_{\text{trans}}(x, t)$ (apart from a quadratic phase factor). The first term in $E_{\text{trans}}(x, t)$ has the spatial carrier $\exp(-j2\pi f_o x)$ and the third term contains $\exp(-j2\pi 2 f_o x)$. Only the second term has spatial frequency content around zero, which apart from the rect window function, is just $r_{12}(t) \exp(-j2\pi v_o x)$. Therefore the detector eliminates the phase factor.

CHAPTER 10

2. Look at the nulls in one axis of the Mellin transform.

3. (a) The grating function is
$$t(x, y) = \left[\text{rect}\left(\frac{x}{X}\right) * \sum_m \delta(x - mx_o) \cdot \text{rect}\left(\frac{x}{X}\right) * \sum_n \delta(y - ny_o)\right]$$
(b) Assume $f_1 = f_2 = f$. Calculate $U(x, y)$ for discrete spatial frequencies $(u, v) = (k_o x/f, k_o y/f)$.
(c) The spatial filter required is a rectangular slit whose orientation, width, and location should be specified.

5. The brute force approach to this problem is to write out explicitly $U_2(s_2, y_2)$ in terms of $t(x, y)$ and $U_3(s_3, y_3)$ in terms of $U_2(s_2, y_2)$ according to the methodology described in Chapter 6, pp. 139–143, and then to combine the results, rearranging the integrals and collecting terms. A far simpler method is to recognize that at P_3, to within a phase factor:
$$FT\{t(x_1, y_1)\} = T\left(\frac{u_3}{\lambda f}, \frac{y_3}{\lambda f}\right)$$
such that at P_4 we see
$$FT\left\{\sqrt{a_o}(x_3^2 + y_3^2) T\left(\frac{x_3}{\lambda f}, \frac{y_3}{\lambda f}\right)\right\}$$
Then when U_4 is calculated to within a constant, it is the Fourier transform of $T(u, v)(u^2 + v^2)$, which is equivalent to the Laplacian of $t(x_1, y_1)$.

7. First note that
$$g(x, y) = [f(x, y) + n(x, y)] \otimes h(x, y)$$
where $h(x, y)$ is of the form
$$c_1 \delta(x, y) + c_1 f(x + x_o, y) + c_2 f^*(-x + x_o, -y)$$

Therefore, $g(x_1, y_1)$ will have three terms.

8. The Fourier plane amplitude signal (at plane P_2) is
$$F(u, v) = F_1(u, v)\exp(-jx_o u) + F_2(u, v)\exp(jx_o u)$$

After square-law detection, the irradiance at P_2 to be read-out at P_3 is
$$F(u, v) = |F_1(u, v)|^2 + |F_2(u, v)|^2 + 2|F_1(u, v)||F_1(u, v)|\cos[2x_o u - \phi_1(u, v) - \phi_2(u, v)]$$

9. To obtain $I_i(x_i, y_i)$ first change variables to simplify the integral I to
$$I = \int_{-1/2}^{1/2} \frac{dz}{u - z}$$

which has a pole at $z = u$. Therefore it must be treated using contour integration. There are three cases that must be considered: when the pole is to the left, inside, and to the right of the region bounded by $(-1/2, 1/2)$. In each case the integrated result
$$I = \ln(u - z)\Big|_{-1/2}^{1/2}$$

must be evaluated for the sign of the numerator and denominator in the argument of the logarithm. The case of interest is when the pole is inside $(-1/2, 1/2)$, in which the contour is

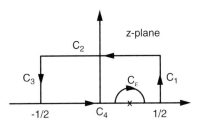

It is useful to change variables for the different contours to ultimately arrive at the desired result.

CHAPTER 11

3. Look at expressions for $F(0)$ and $f(0)$ using Fourier and inverse Fourier transforms, respectively. Note that $F(0)$ and $f(0)$ have upper bounds set by values of these integrals for $x = 0$ and $f = 0$, respectively.

4. Recognize that at plane P_3 to within a phase factor we have
$$FT\{t(x_1, y_1)\} = T\left(\frac{x_3}{\lambda f}, \frac{y_3}{\lambda f}\right)$$

8. The signal given by Equation (11.67) is spatially integrated in range by the first Bragg cell, namely,
$$s_1(x, t') = \int_{-\infty}^{\infty} s(x'/v, t')\exp[-jb(x' + x)^2/V_a]\,dx'$$

The azimuth modulation in the second Bragg cell is detected interferometrically, hence

$$I_1(x, t') = |S_1(x, t') + \exp(j\omega t')|^2$$

which contains three terms.

The SAR image is then focused in azimuth (t') by temporal correlation with the reference function given by Equation (11.66), namely,

$$I_3(x, y) = \int_{-\infty}^{\infty} I_1(x, t') I_2(y + v't') \, dt'$$

This will then yield Equation (11.67).

9. In general we know

$$s(x, y) = \int_{-\infty}^{\infty} \int_{-\infty}^{\infty} S(u, v) \exp[j2\pi(ux + vy)] \, du dv$$

so that we can directly substitute into the expression for $l(x)$.

12. (a) An ideal monochromatic wave is $u(x) = \exp(j2\pi kx)$.
    (b) A rectangular pulse is $u(x) = \text{rec}(x/x_o)$.
    (c) An infinitely long linear FM signal is $u(x) = \exp(j2\pi ax^2)$.

15. The output light distribution is simply the real part of the product of $x_1(t)$ (in complex form) with the incident plane wave expression. $V(z)$ will consist of three terms.

17. Note the following computations:

$$|\Delta t| = \frac{1}{10\Delta f} = \frac{1}{2 \times 10^8 \text{ Hz}} = 5 \text{ nsec}$$

$$|\Delta \tau| = \left|\frac{-d \cos\theta}{c} \Delta\theta\right| = \frac{d|\Delta\theta|}{c} \quad \text{for small } \theta$$

$$|\Delta\theta| = \frac{c}{d}|\Delta\tau| = \frac{3 \times 10^8 \text{ m/sec}}{10 \text{ m}} (5 \times 10^{-9} \text{ sec}) = 0.15 \text{ rad } = 8.6°$$

$$\text{Accuracy} = \frac{1}{50}|\Delta\theta| = 3 \times 10^{-3} \text{ rad } = 0.17°$$

$$\text{Array size} = \# \text{pixels} \times \text{pixel size} = 512 \times 25\mu\text{m} = 12.8 \text{ mm}$$

Using LiNb$_2$O$_3$ with $v_s \approx 7 \times 10^3$ m/sec $\Rightarrow$ that the maximum possible extent of the correlation pattern is

$$v_s \tau / 2 = (7 \times 10^3 \text{ m/sec})(3 \times 10^3 \text{ m/sec})/2 = 0.1 \text{ mm}$$

where the factor of 1/2 accounts for counterpropagating waves. Thus the required lens magnification is

$$\frac{12.8 \text{ mm}}{0.1 \text{ mm}} = 128$$

18. The Wigner–Ville distribution is defined as

$$W(t, \omega) = \int_{-\infty}^{\infty} f_1^*(t - \tau'/2) f_2^*(t + \tau'/2) \exp(-j\omega\tau') \, d\tau'$$

By direct substitution, you will see the cross-term.

CHAPTER 12

3. Calculate the area of overlap of two identical rectangles centered on the origin for one not rotated and the other rotated by ϕ.

7. Consult p. 105 to recall the properties of the Dove prism. To estimate the linear array readout rate, use the available frame time and the number of pixels that must be processed in the appropriate angular sector (π rad). Since the Radon transform is an integration, the required dynamic range is the ratio of the maximum possible signal at $\theta = \pi/4$ to the minimum possible signal at $\theta = 0$.

8. Use Equations (12.16) and (12.17) or use Equation (12.19).

10. The general expression for the moments of a distribution $p(x)$ is

$$E[x^n] = \overline{x^n} \int_{-\infty}^{\infty} x^n p(x) \, dx$$

In the case of the Poisson distribution, this will be a discrete sum.

13. The processing steps go as follows. LED1 is pulsed when the diffraction grating corresponding to x_1 moves in front of it. LED1 is pulsed-on in proportion to a_{11}. The diffraction efficiency is proportional to x_1 in the AO cell. Then detector 1 is illuminated with light proportional to $a_{11}x_1$. After this the x_1 grating segment moves in front of LED2 while the x_2 segment moves in front of LED1. Now LED1 is pulsed in proportion to a_{12} and LED1 in proportion to a_{21}. Now detector 1 gets light proportional to $a_{11}x_1 + a_{12}x_2$, which equals y_1, while detector 2 gets light proportional to $a_{12}x_1$. After segment x_2 has moved in front of LED2, the pulse from LED2 (in proportion to a_{22}) yields an output at detector 2 proportional to $a_{21}x_1 + a_{22}x_2 = y_2$. The student can sketch this process in the architecture physically for each step in the process and extend the results to the 3×3 matrix case.

16. Use Equation (12.32) where
$$x_i = \text{bipolar version of } A_1$$
$$x_j = \text{bipolar version of } A_2$$

The output is the complement of one of the inputs.

17. Refer to Example 12.7.

INDEX

Acousto-optic
  angular spectrum, 209
  Bragg cells, 200
  deflection, 206
  figure of merit, 209
  materials, 211
  modulation, 204
  spectrum analyzer, 235, 239
Airy
  pattern, 130
  function, 103
Ambiguity
  diagram, 281
  of Gaussian, 284
  processor, 290, 291
  of rectangular pulse train, 285
Anamorphic
  magnification, 114
  prism pair, 105
Angle
  Brewster's, 99
  critical, 99
  of deflection, 105
  of incidence, 98
  of reflection, 98
  of refraction, 98
Angular correlation, 273
Aperture time, 293
Apodization, 234
Architecture
  coherent, 231, 248
  correlator, 244
  Fourier Transform, 236, 237
  incoherent, 244
  space integrating, 230, 239
  space-time integrating, 250
  spectrum analyzer, 232
  time integrating, 230, 242
Associative memory, 330
  bidirectional, 341
  holographic, 340
  linear, 331
Azimuth

chirp, 300
  reference function, 301

Backprojection, 78, 90
Biaxial crystal 97
Birefringence, 96
Bragg
  acoustic attenuation, 211
  angle, 203
  as beam deflector, 206
  cell, 200
  commercial grade cells, 212
  condition, 201
  diffraction, 202
  diffraction efficiency, 208
  Doppler shift, 204
  dynamic range, 210
  as modulator, 206
  signal analysis, 205
  time-bandwidth product, 207
Brewster's angle, 99
BSO, 220

Cauchy's formula, 114
Charge Coupled Device (CCD), 186
  2-D array, 185
  linear self-scanned array, 184, 186
  noise contributiond, 190
  shift-and-add (TDI) mode, 187
Charge Injection Device (CID), 185, 189–190
Charge transfer inefficiency, 185
Classifier, 312
Coherence, 120
  complete, 120
  length, 121
  mutual function, 120
  partial, 120
  spatial (lateral), 121
  tempoal (longitudinal), 120
Coherent line source, 126
Collimator, 110
Conductance, 74
Contrast reversal, 261

INDEX

Convolution, 3
  fast, 8
  integral, 3
Coordinates
  polar, 9, 11
  rectangular, 9
Correlation
  angular, 273
  Binary, 267
  of circular pillbox, 6
  coefficient, 41
  peak-to-sidelobe ratio, 53
  phase only, 266
  properties, 3, 48, 349
  of rectangular pillbox, 3
Correlator
  coherent, 248
  incoherent architecture, 244
  joint transform, 246
  multi-channel, 246
  shadow-casting, 247
  space integrating, 248
  temporal scanning, 245
  time integrating, 249
  time-space integrating, 251
Cosine-fourth law, 113
Covariance, 41

Debye-Sears (Raman-Nath) regime, 200
Deformable mirror device, 219
Density function, 37
  binomial, 42
  bivariate Gaussian, 42
  Gaussian, 39, 42
  Poisson, 43
  Rayleigh, 44
  uniform, 45
Detection
  heterodyne, 181
  phase sensitive, 62
  square-law, 181
  TDI, 187
  theory, 47
Detector
  cutoff wavelength, 179
  detectivity, 183
  equivalent noise bandwidth, 182
  linear array, 184
  matrix array, 184
  NEP, 182
  noise voltage, 183
  photoconductive, 180
  photovoltaic, 180
  responsivity, 181
  single, 179
Diffracting aperture, 123
Dichroic, 97
Diffraction, 122
  Airy pattern, 130
  array factor, 129
  from circular aperture, 129
  double-slit, 119
  efficiency, 207
  element factor, 129
  far-field, 130
  first-order, 201
  Fraunhofer, 125
  Fresnel, 131
  multiple slit, 128
  near-field, 130
  from rectangular aperture, 129
  scalar theory, 122
  single-slit, 127
  zeroth-order, 206
Dirac delta function, 2
  1-D, 20
  properties, 348
Discrete Fourier transform, 325
Dispersion, 95
Distortion
  geometric, 268
  invariance, 270
  invariant transformation, 270
Doppler effect, 95
Dynamic range, 255

f-number, 108
Fabry-Perot
  cavity, 164
  resonator, 103
Faraday effect, 217
Feature extraction, 313
  angular correlation, 322
  distortion invariant, 313
  Fourier transform, 313
  moments, 314
  Radon transform, 317
  Walsh-Hadamard transform, 320
  wedge-ring detector, 313
Filter
  dichroic, 97
  Gabor, 83
  high-pass, 29, 260, 266
  low-pass, 74, 260
  matched, 45
  spatial, 166, 260
  Vanderlugt, 258
Fourier
  analysis, 8
  coefficients, 262
  magnitude, 32
  phase, 32
  synthesis, 8
  transform
    1-D, 350
    2-D, 7
    of delta function arrays, 26
    low bit-level, 239
    of simple synthetic images, 27
    polar form, 10
    properties, 7, 349
    rectangular form, 9
    separable, 349
Fraunhofer
  approximation, 123
  diffraction, 125
  region, 133
  zone, 132
Frequency
  of co-occurrence, 38
  electrical, 151
  spatial, 7, 151
  temporal, 151
Fresnel
  approximation, 123
  diffraction, 131

equation, 100
integral, 68
region, 133
reflection coefficient, 100
transform, 66
zone, 133
zone plate, 132
Fresnel-Fraunhofer transition, 69
Function
  Airy, 103
  Bessel, 11, 351
  circular pillbox, 6, 11
  comb, 21, 350
  Dirac delta, 2
  Gaussian, 30, 39, 42
  impulse, 17
  jinc, 12, 351
  joint distribution, 39
  marginal distribution, 40
  rect, 4, 9, 350
  rectangular pillbox, 4
  self-imaging, 30
  sgn, 337
  sinc, 9, 350
  sombrero, 12
  triangle, 79

Generic signal processing architectures, 290
Geometrical optics, 97, 106
Geometric series, 126

Half-wave plate, 97
Hamming distance, 331
Hankel transform, 11
Heterodyne, 56
Hilbert transform, 69, 352
Histogram
  1-D, 40
  2-D, 40
Holographic
  matched filter, 258
  SLM, 198
Homodyne, 56
Hopfield net, 334
Huygen's wavelets, 125

Imaging, 142
  coherent, 143
  incoherent, 143
  Schlieren, 263
Impulse
  2-D, 2
  arrays, 23, 77
  function, 17
  lines, 19, 22
  pair, 24
  response, 2
Interference, 118
Intermodulation products, 210
Invariance, 268
  distortion, 270
  linear shift, 2
  rotation, 272
  scale, 80, 272
  translation, 272
Irradiance, 112
Isoplanatic regions, 1, 146

Jacobian, 44
Joint power spectrum, 53

Joint transform correlator, 53

Keplerin beam expander, 110
Kramers-Krönig relations, 65

Laser
  diode, 168
    arrays, 176, 178
    array angular distribution, 177
    beam divergence, 172
    beam forming optics, 174
    coherence length, 176
    duty cycle, 175
    energy levels, 169
    heterojunction, 170
    linewidth, 166
    modulation efficiency, 173
    output power, 172, 174
    parameters, 175
    single, 173
    spectral bandwidth, 166, 170
    structure, 173
    surface emitting, 177
    threshold current, 172
  Gaussian beamshape, 165
  helium-neon, 166
  multi-mode, 171
  semiconductor, 170
  single-mode, 171
  spatial filtering of, 166
Lenslet arrays, 111
Lens, 107
  arrays, 112
  combination, 109
  condenser pair, 110
  configurations, 109
  cylindrical, 298
  f-number, 108
  focal length, 108
  Fourier transform property, 139
  Gaussian formula, 108
  imaging properties, 142
  lateral magnification, 108
  multiple, 111
  negative, 108
  numerical aperture, 108
  positive, 108
  pupil function, 140
  quadratic phase-shift, 137
  simple, 107
  telephoto, 110
  thick, 107
  thin, 108
  vignetting effect, 141
Lensmaker's formula, 139
Light emitting diode (LED), 168
Likelihood ratio, 47
Line spread function, 79
Linear
  algebra, 324
  chirp, 58
  coordinate transformations, 13, 326
  superposition, 2
  system, 1
  transformation theorem, 14
Liquid crystal
  and birefringence, 212
  construction, 214
  electrically addressed, 216
  and hybrid field effect, 213

operation, 213
parameters, 215
spatial light modulator, 211
television, 216
and twisted nematic effect, 212
Log polar transform, 272
Logarithmic scaling, 81
Lorentzian line shape, 164
Low bit-level image, 240

Magneto-optic
binary state operation, 218
crystal, 216
spatial light modulator, 216
transmission model, 217
Magnification, 14
anamorphic, 14
angular, 111
lateral, 108
Matched filter, 45, 259
binary phase-only, 264, 267
correlator, 265
derivation, 45
phase only, 266
Matrix-vector multiplier, 324
Mellin transform, 65, 271, 352
Microchannel spatial light modulator, 218
Mixing, 55
Modulation, 55
amplitude, 56
chirp, 59
double sideband, 56
frequency, 57
Modulation transfer function, 148
CCD pixel geometric, 195
CCD transfer inefficiency, 195
Moments
central, 80
invariant, 314, 341
statistical, 37
Multiple quantum well device, 221

Neural network
bidirectional ssociative memory (BAM), 341
holographic, 339
Hopfield, 334
multi-layer, 334
optical, 329
paradigm, 330
Noise
equivalent power (NEP), 182
Johnson (thermal), 190
photon (shot), 190
Non-linearity
binary, 51, 55
kth-law, 55
logarithmic scaling, 49, 81
sigmoidal, 52
square-law, 50
transfer characteristic, 49, 52
transformations, 49

Optical density, 102
Optical transfer function, 146

Paraxial approximation, 108
Pattern recognition, 312
Perceptron, 332
Phase-only correlator, 266
Phase transfer function, 150

Photon-phonon interaction, 203
Physical optics, 118
Piezo-electric transducer, 204
Pockels read-out modulator (PROM), 220
Point spread function, 144
Polarization, 96
circular, 97
left-hand circular, 96
p-wave, 100
partial, 101
right-hand circular, 96
s-wave, 100
transverse electric, 99
transverse magnetic, 99
Poynting vector, 96
Prisms, 105
multiple Dove, 106
multi-faceted arrays, 107
Process
non-linear, 49
stationary, 39
strict stationarity, 40
wide-sense stationarity, 40
Probability
binomial, 42
bivariate Gaussian density, 42
cumulative, 37
density function, 37
of detection, 48
of false alarm, 48
Gaussian density, 39
hypothesis, 47
joint distribution, 38
log likelihood, 47
marginal, 40
Poisson density, 43
Rayleigh density, 44
uniform density, 45
Projection, 75
Pulse
compression, 60, 154
train, 285

Quarter-wave plate, 97
Q-value of Bragg cell, 261

Radar
ambiguity function, 281
pulse, 280
resolution, 283
synthetic aperture, 292
Radiometric quantity
irradiance, 112
power, 112
Radon, Johann, 65
Radon Transform, 65, 75
Raster scanning, 21
Reconstruction, 78
Rectilinear glass structures, 97
flat plate, 101
prism, 104
array, 106
Dove, 106
right-angle, 105
Rotation, 16, 76

Sampling, 23
Schlieren imaging, 263
Schwarz inequality, 46
SEED devices, 222

Sidelobe, 47
Signal
    analytic, 72
    causal, 72
    complex, 71
    positive frequency, 71
    real, 71
Signal-to-bias ratio, 255
Signal-to-noise ratio, 46
Sinogram, 77
Skew-symmetric, 48
Slow-shear wave, 207
Snell's law, 98
Software for modeling and visualization, 356
Space integrating, 230
Spatial
    filtering, 260
    multiplexing, 269
Spatial light modulator
    acousto-optic Bragg cell, 200
    characteristics, 200
    deformable mirror device, 219
    electrically addressed, 198
    generic, 199
    liquid crystal, 211
    magneto-optic (MOSLM), 216
    microchannel (MSLM), 218
    multiple quantum well (MQW), 221
    optically addressed, 198
    Pockels readout (PROM), 220
    Preobrasovatel Izobrazheniy (PRIZ), 220
Spatial multiplexing, 268
Spectrum analysis
    acousto-optic based, 232
    autoregressive, 328
    error vs number of bits, 239
    folded, 289
    liquid crystal based, 237
    narrowband acousto-optic, 235
    time integrating, 242
    space integrating, 239
Stationarity
    strict, 39
    wide-sense, 39
Statistics
    2-D histogram, 40
    ensemble, 39
    ergodic, 39
    first-order and second-order, 38
Stochastic processes, 36
Susceptance, 74
Synthetic aperture radar, 292
    azimuth reference function, 303
    fast axis, 296
    film-based optical processor, 298
    focused, 294
    imaging geometry, 293
    and interferometric detection, 303, 306, 308
    phase history, 297
    point spread function, 297
    Programmable AO processor, 304
    slow axis, 296
    transmitted waveform, 299

Telescope, 111
    angular magnification, 110
    astronomical, 110
    keplerian, 109
    resolving power, 11

Theorem
    projection-slice, 76
    Wiener-Khintchin, 61
Thumbtack ambiguity, 283
Time integrating, 230
Time-bandwidth product, 191, 207, 254
Transfer function
    binary, 51, 55
    coherent, 146
    incoherent, 147
    modulation, 147
    non-linear, 49
    optical, 146
    phase, 147
    sigmoidal, 52
Transform, 64
    Abel, 79
    Fourier, 7
    Fourier-Bessel, 10
    Fresnel, 66
    Hankel, 11
    Hartley, 90
    Hilbert, 69
    inverse Abel, 87
    inverse Radon, 78
    Mellin, 79
    Radon, 75
    rotation-invariant, 272
    scale-invariant, 81, 272
    shift-invariant, 272
    wavelet, 83
Transformation
    distortion invariant, 270
    linear, 324
Triple-product processor, 289

Uniaxial crystal, 97

Vector-matrix, 326
    and autoregressive spectrum analysis, 328
    Fourier transform mask, 327
    linear processing operations, 326
    multiplier, 336
    Stanford design, 324
Velocity
    acoustic, 204
    group, 95
    phase, 94
Verdet constant, 217
Video feedback, 274
Wave
    electromagnetic, 93
    equation, 93
    extraordinary, 97
    ordinary, 97
    plane, 94
    spherical, 94, 122, 143
Wavelet
    Gabor, 83, 86
    Haar, 84
    Huygen's, 125
    Mexican hat, 85
    Morlet, 85
    transform, 83
Widrow-Hoff technique, 333
Wigner distribution, 288

Young's experiment, 119